EGYPTIAN OEDIPUS

EGYPTIAN OEDIPUS

Athanasius Kircher and
the Secrets of Antiquity

DANIEL STOLZENBERG

THE UNIVERSITY OF CHICAGO PRESS
CHICAGO AND LONDON

The University of Chicago Press, Chicago 60637
The University of Chicago Press, Ltd., London
© 2013 by The University of Chicago
All rights reserved. Published 2013.
Paperback edition 2015
Printed and bound by CPI Group (UK) Ltd, Croydon, CR0 4YY

22 21 20 19 18 17 16 15 14 13   3 4 5

ISBN-13: 978-0-226-92414-4 (cloth)
ISBN-13: 978-0-226-27327-3 (paper)
ISBN-13: 978-0-226-92415-1 (e-book)
10.7208/chicago/9780226092415 1.001.0001

Library of Congress Cataloging-in-Publication Data

Stolzenberg, Daniel.
  Egyptian Oedipus : Athanasius Kircher and the secrets of antiquity / Daniel Stolzenberg.
    pages cm.
  Includes bibliographical references and index.
  ISBN 978-0-226-92414-4 (cloth : alk. paper) — ISBN 0-226-92414-9 (cloth : alk. paper) — ISBN 978-0-226-92415-1 (e-book) — ISBN 0-226-92415-7 (e-book)
  1. Kircher, Athanasius, 1602–1680. Oedipus aegyptiacus. 2. Egyptian language—Writing, Hieroglyphic. 3. Occultism. I. Title.
PJ1093.K643S76 2013
4939.1092—dc23
                                                                    2012022898

♾ This paper meets the requirements of ANSI/NISO Z39.48-1992 (Permanence of Paper).

HIEROGLYPHICS Language of the ancient Egyptians, invented by the priests to conceal their shameful secrets. To think that there are people who understand them! But perhaps the whole thing is just a hoax?
—Gustave Flaubert, *Dictionary of Received Ideas*

CONTENTS

A Note on Quotations and Translations ... ix
List of Abbreviations ... xi

Introduction: Oedipus in Exile ... 1
1. Esoteric Antiquarianism ... 36
2. How to Get Ahead in the Republic of Letters ... 71
3. Oedipus in Rome ... 104
4. Ancient Theology and the Antiquarian ... 129
5. The Discovery of Oriental Antiquity ... 151
6. Erudition and Censorship ... 180
7. Symbolic Wisdom in an Age of Criticism ... 198
8. Oedipus at Large ... 226
Epilogue: The Twilight of Tradition and the Clear Light of History ... 254

Acknowledgments ... 261
Bibliography ... 265
Index ... 297

A NOTE ON QUOTATIONS AND TRANSLATIONS

Since virtually all the early modern books that I quote, including some of the rarest and most obscure, have recently become easily available online, especially through open access sites like Gallica, the Hathi Trust, the Munich Digitization Center, the Internet Archive, and Google Books, in the footnotes I have provided original text only for significant quotations from unpublished manuscripts. Unless otherwise indicated, all translations are my own. For ease of reading, I have translated the titles of books into English; the original titles are given in the footnotes.

ABBREVIATIONS

APUG	Archivio della Pontificia Università Gregoriana
ARSI	Archivum Romanum Societatis Iesu
ASV	Archivio Segreto Vaticano
BAV	Biblioteca Apostolica Vaticana
BNF	Bibliothèque Nationale de France
BNCR	Biblioteca Nazionale Centrale di Roma
BISM	Bibliothèque interuniversitaire (Montpellier), Section Médecine
LAR	Athanasius Kircher, *Lingua Aegyptiaca restituta* (Rome, 1643)
PC	Athanasius Kircher, *Prodromus Coptus* (Rome, 1636)
OA	Athanasius Kircher, *Oedipus Aegyptiacus* (Rome, 1652–54)
OP	Athanasius Kircher, *Obeliscus Pamphilius* (Rome, 1650)
Vita	Athanasius Kircher, *Vita Admodum Reverendi P. Athanasii Kircheri* in *Fasciculus Epistolarum*, ed. Hieronymus Langemantel (Augsburg, 1684)

INTRODUCTION

Oedipus in Exile

The wisdom of the Egyptian philosophers truly seems to be something much too divine to be understood by any little man; for, in my opinion, the Egyptian system of the world, which was based on the laws of attraction and repulsion, seems to be the closest of all to the truth. This opinion of mine now has the consent of all Europe, which approved it not so long ago, but attributed it to Newton, in his calculus. But Kircher came before Newton; and lest someone thinks that I am daydreaming, I would have him read carefully and with an unprejudiced mind those things that Kircher wrote in the last chapter of *Coptic Forerunner* and *Egyptian Oedipus*.
—Adam František Kollár (1790)[1]

MALTA, 1637

The summer of 1637 found Athanasius Kircher (1601/2–1680) marooned on a small Mediterranean island. After three felicitous years in Rome, the Jesuit priest had been transferred to Malta, a European outpost between Sicily and North Africa governed by the Catholic Knights of St. John of Jerusalem. He arrived at the end of May in the retinue of a young German prince, Landgrave Frederick of Hesse-Darmstadt. Following high-level ne-

---

1. Kollár, *Ad Petri Lambecii commentariorum libros VIII. Supplementorum liber* (1790), 357. Earlier, Kollár was less enthusiastic. In his revised edition of Peter Lambeck's bibliography of the Imperial Library at Vienna, he wrote: "The incredible temerity of the polymath Kircher in these studies is too well known to everyone to need to be confirmed by this example. It is irksome to examine and refute each thing that he imagines here, since I do not have time to consider difficult trifles and rightly consider it to be a foolish work of absurdities." Lambeck, *Commentariorum de Augustissima Bibliotheca liber* (1766–82), vol. 1, 192–94.

gotiations with Rome, Frederick had converted from Lutheranism to Catholicism. In reward, Pope Urban VIII promised him the Grand Cross of the Order of St. John and the lucrative office of coadjutor of the Grand Priory of Germany. This required Frederick to take up temporary residence on the island, and Kircher, a fellow German, was sent along to serve as the landgrave's confessor and teach mathematics at Malta's Jesuit college. It was a poor match. Kircher's talents and ambitions were scholarly rather than pastoral, while Frederick was a rowdy youth and Malta a cultural backwater.[2]

While Kircher made desultory efforts to instill Frederick with Catholic piety, back in the mainstream of European intellectual life, René Descartes (1596–1650) published his first book. *A Discourse on the Method for Conducting One's Reason Rightly and for Searching for Truth in the Sciences*, issued anonymously in Leiden in June 1637, announced a radical program to overhaul the whole of human knowledge. Surveying the state of learning, and looking back on his own education at the renowned Jesuit college of La Flèche, Descartes detected nothing but error and uncertainty. "There has been no body of knowledge in the world," he lamented, "which was of the sort that I had previously hoped to find."[3] To escape this impasse, he argued, it would be necessary to start from scratch, treating all received wisdom as so many prejudices and constructing a secure body of knowledge on foundations in no way dependent on tradition. In place of books, schools, and the accumulated learning of millennia, Descartes substituted a method based on the principle of accepting nothing as true that could not be demonstrated by a sequence of clear and distinct ideas. Beginning with only the indubitable *cogito ergo sum* (I am thinking, therefore I exist), he proved the reality of God, the human soul, and inert matter that could be studied through a mathematical science of nature. For the rest of the seventeenth century Cartesian philosophy was a lightning rod—scourge of traditionalists and rallying cry of *moderni*.

Descartes famously attributed his breakthrough to a period of forced isolation. Serving in the army of Prince Maurice of Nassau, he was detained one winter in Germany, where he found himself with abundant free time and few external distractions, such as books or conversation partners. One momentous day, alone with his thoughts in a stove-heated room, he experienced an intellectual epiphany that led to the insights on which he would base his life's work. In Descartes' vision, the most profound and certain

---

2. On Kircher in Malta, see Bartòla, "Alessandro VII" (1989); Borg, *Fabio Chigi* (1967); Zammit Ciantar, "Athanasius Kircher in Malta" (1991).

3. Descartes, *Discourse on Method* (1980), 5.

knowledge of the universe could be achieved by a solitary individual, unburdened of prior knowledge and properly exercising his reason.[4]

Kircher did not find isolation so stimulating. In Rome, he had thrived at the bustling Collegio Romano, the flagship school of the Society of Jesus and crossroads of its international missionary network, and moved in the city's leading cultural circles. Supported by powerful patrons, he was at work on an ambitious research project whose goal was nothing less than to decipher the Egyptian hieroglyphs. He had just published the first fruit of this research—a treatise on Coptic, which brought him a taste of literary fame—when Malta intervened.[5] Taking stock of his new surroundings, Kircher must have felt that a hitherto promising career had taken a sudden wrong turn. He watched with envy as Lucas Holstenius, another German émigré intellectual in the landgrave's entourage, hightailed it back to Rome at the first opportunity. Kircher passed the time devising a "physico-mathematical" instrument, hunting for manuscripts, exploring mysterious subterranean chambers, and befriending the learned papal nuncio and future pope, Fabio Chigi.[6] But, cut off from the libraries, collections, and community that had sustained him, his research came to a standstill. Feeling his talents waste away with the days, he sent letters to Rome, entreating his Jesuit superiors and influential protectors to release him from his exile. Eventually, his pleas were heard: the general of the Society dispatched a substitute German priest-mathematician, and in 1638 Kircher returned to the Eternal City to resume his studies.[7]

THE HIEROGLYPHIC SPHINX

Like Descartes, Kircher went on to achieve considerable, if less enduring, fame in the international community of scholars that called itself the Republic of Letters. But his idea of the scholarly enterprise was different. To a certain extent, he represented the bookish learning, rooted in ancient tradition, which intellectual reformers like Descartes fought against in the name

---

4. Descartes acknowledged the need for collaboration in building on the foundations obtainable by solitary ratiocination. See ibid., 33–34.
5. Kircher, *Prodromus Coptus* (1636).
6. The instrument is described in Kircher, *Specula Melitensis* (1638); republished in Schott, *Technica curiosa* (1664), 427–77. A facsimile of the rare first edition is appended to Zammit Ciantar, "Athanasius Kircher in Malta" (1991).
7. See Muzio Vitelleschi to Kircher, Rome, 30 July 1637, and 7 January 1638, APUG 561, fols. 18^r and 19^r; the latter is printed in Bartòla, "Alessandro VII" (1989), 83. See also Holstenius to Barberini, Naples, 7 September 1637, in ibid., 80; Barberini to Holstenius, Rome, 24 September 1637, in ibid., 82; Francesco Barberini to Fabio Chigi, Rome, 10 October and 15 October 1637, in Borg, *Fabio Chigi* (1967), 312, 318.

of a new philosophy. (In a 1643 letter to Constantijn Huygens, Descartes described Kircher as "more charlatan than scholar," and refused to read his books.)[8] While Kircher, too, was devoted to natural sciences, these disciplines were only part of his encyclopedic vision. A quintessential polymath, of that soon-to-be-extinct academic species that eschewed specialization and aspired to master the entire panorama of human knowledge, he was renowned for the vast range of his scholarly output. Magnetism, music, optics, archeology, chemistry, geology, linguistics, cryptography, Lullism, and China were only some of the subjects to which he devoted substantial studies. Like other Jesuits, Kircher sought an accommodation between tradition and innovation, striving to reconcile the Aristotelian philosophy officially espoused by the Society of Jesus with new intellectual trends.[9] This balancing act could support a considerable load of novelty. In his major astronomical work, *Ecstatic Journey* (Rome, 1656), for example, Kircher paid heed to the condemnation of heliocentrism by endorsing Tycho Brahe's compromise cosmology. (Mathematically identical to the Copernican model, it placed the sun and moon revolving around an immobile, central earth while the other planets orbited the sun.) But Kircher audaciously placed this geocentric planetary system within a quasi-infinite universe reminiscent of the views of Giordano Bruno, the Neapolitan heretic burned in Rome in 1600.[10] Even when challenging orthodoxy, however, Kircher remained faithful to the veneration of antiquity that was the common legacy of humanism and scholasticism.

This book explores one part of Kircher's encyclopedic corpus: his study of Egypt and the hieroglyphs. After two decades of toil, Kircher brought this project to completion in his largest and most challenging work, *Egyptian Oedipus*, issued in four volumes in Rome in 1655.[11] With the title of his magnum opus, Kircher characteristically paid honor to himself. Like Oedipus answering the riddle of the Sphinx, Kircher believed he had solved the enigma of the hieroglyphs (fig. 1). Together with its companion volume,

---

8. Findlen, "Janus Faces of Science" (2000), 223.
9. On Jesuit science see: Hellyer, *Catholic Physics* (2005); Feingold, ed., *Jesuit Science and the Republic of Letters* (2003); Feingold, ed., *New Science and Jesuit Science* (2003); Romano, *La contre-réforme mathématique* (1999); Fabre and Romano, eds., *Les jésuites dans le monde moderne* (1999); Gorman, "Scientific Counter-Revolution" (1999); Giard, ed., *Les jésuites à la Renaissance* (1995); Baldini, *Legem impone subactis* (1992).
10. Kircher, *Itinerarium exstaticum* (1656). See Ziller Camenietzki, "L'extase interplanetaire d'Athanasius Kircher" (1995); Rowland, "Athanasius Kircher, Giordano Bruno" (2004); Siebert, *Die grosse kosmologische Kontroverse* (2006).
11. Kircher, *Oedipus Aegyptiacus* (1652–54). The work was printed over several years. Its three "tomes" bear the dates 1652, 1653, and 1654, but it was completed and made available to the public all at once in 1655.

Fig. 1. Kircher as the Egyptian Oedipus before the hieroglyphic sphinx. See chapter 4 for an explanation of the symbolism. Athanasius Kircher, *Oedipus Aegyptiacus* (Rome, 1652–54), vol. 1, frontispiece. Courtesy of Stanford University Libraries.

Pamphilian Obelisk,[12] *Egyptian Oedipus* presented Kircher's Latin translations of hieroglyphic inscriptions—utterly mistaken, as post–Rosetta-Stone Egyptology would reveal—preceded by treatises on ancient Egyptian history, the origins of idolatry, allegorical and symbolic wisdom, and numerous non-Egyptian textual traditions that supposedly preserved elements of the "hieroglyphic doctrine." In addition to ancient Greek and Latin authors, Kircher's vast array of sources included texts in Oriental languages, including Hebrew, Arabic, Aramaic, Coptic, Samaritan, and Ethiopian, as well as archeological evidence such as inscriptions, statues, amulets, idols, vases, sarcophagi, and monuments (figs. 2–5). The resulting amalgam is, without doubt, impressive. But it can also bewilder. Pondering its "elephantine" volumes, historian Frank Manuel memorably called *Egyptian Oedipus* "one of the most learned monstrosities of all times."[13]

This book presents a new interpretation of Kircher's project in terms of an encounter between two early modern intellectual traditions: erudition (antiquarian research and philology) and occult philosophy (the Renaissance Neoplatonic tradition, based on a lineage of esoteric wisdom attributed to extremely ancient pagan wise men). Kircher's spectacular shortcomings have made it difficult to appreciate how much he participated in the important scholarly developments of his time. Once his proper measure is taken, he proves a useful figure for reassessing important aspects of seventeenth-century scholarship. By reading his hieroglyphic studies as a work of erudite historical research, instead of philosophy, I show that Kircher differed fundamentally from earlier writers in the so-called Hermetic tradition, whose work he has been seen as continuing, and that he shared more with his contemporaries than has usually been acknowledged. *Egyptian Oedipus* was not quite so monstrous as Manuel imagined. In particular, I argue that Kircher's use of occult philosophy in the service of antiquarian research was not anomalous, and that the prevailing chronology of the fate of occult philosophy must be revised. Behind Kircher's two greatest failures—his incredible translations of hieroglyphic inscriptions and his reliance on spurious documents—lay widely accepted principles about symbolic communication and the transmission of ancient knowledge. As a case study of seventeenth-century scholarship, this book illuminates a complex moment when empiricism and esotericism coexisted, and shows how the discipline of Oriental studies was born from an early modern Mediterranean world in which texts, artifacts, and scholars circulated between Christian and Islamic civilizations.

---

12. Kircher, *Obeliscus Pamphilius* (1650).
13. Manuel, *Eighteenth Century Confronts the Gods* (1959), 190–91.

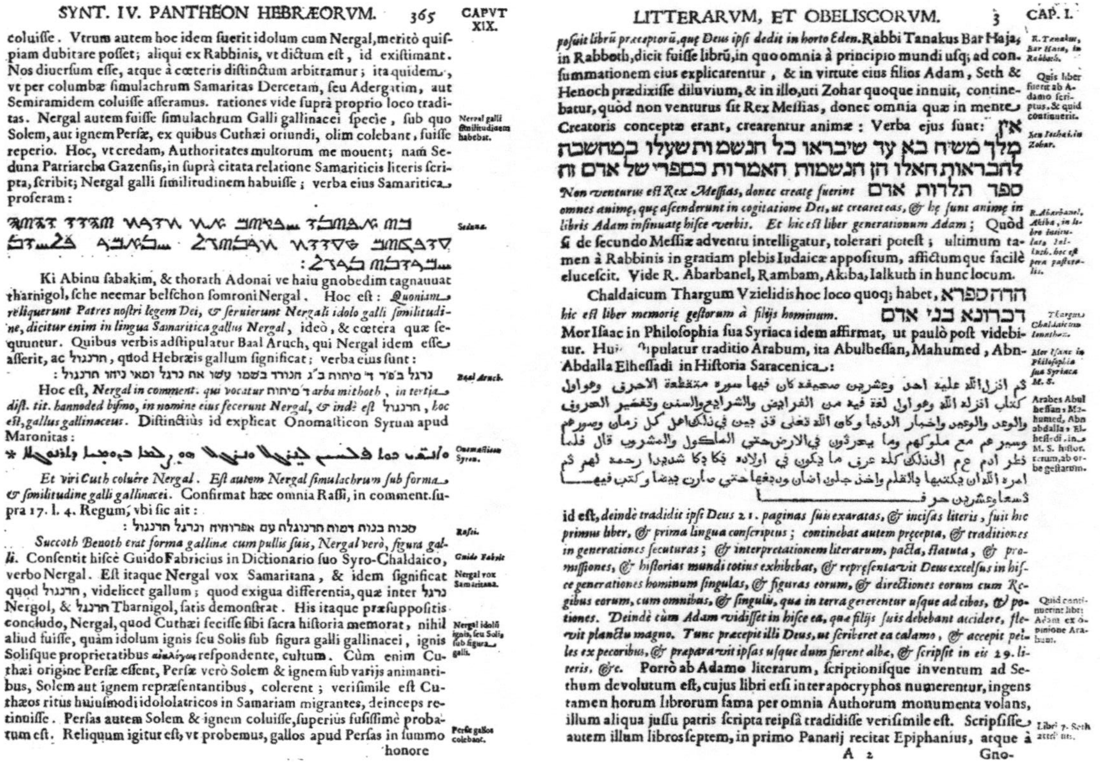

Fig. 2. Quotations in Samaritan, Syriac, Hebrew, and Arabic embedded in Kircher's Latin exposition. Such typography was expensive and technologically challenging in the seventeenth century. Athanasius Kircher, *Oedipus Aegyptiacus* (Rome: 1652–54), vol. 1, 365; *Obeliscus Pamphilius* (Rome: 1650). 3. Courtesy of Stanford University Libraries.

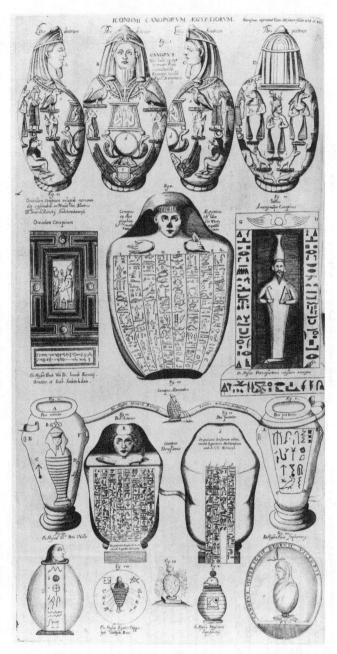

Fig. 3. Canopic jars from various collections. Athanasius Kircher, *Oedipus Aegyptiacus* (Rome: 1652–54), vol. 3, fp. 434–35. Courtesy of Stanford University Libraries.

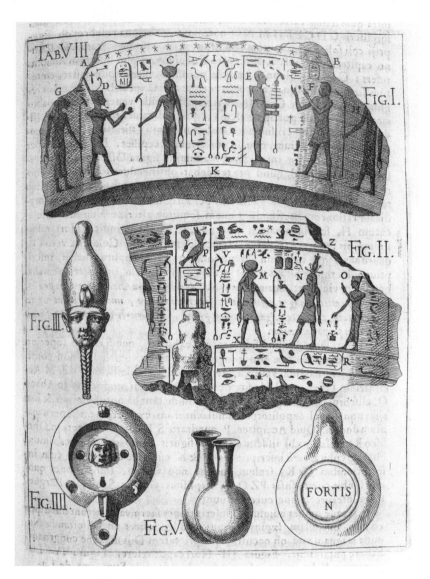

Fig. 4. Fragments of hieroglyphic inscriptions and other Egyptian antiquities. Athanasius Kircher, *Oedipus Aegyptiacus* (Rome: 1652–54), vol. 3, 385. Courtesy of Stanford University Libraries.

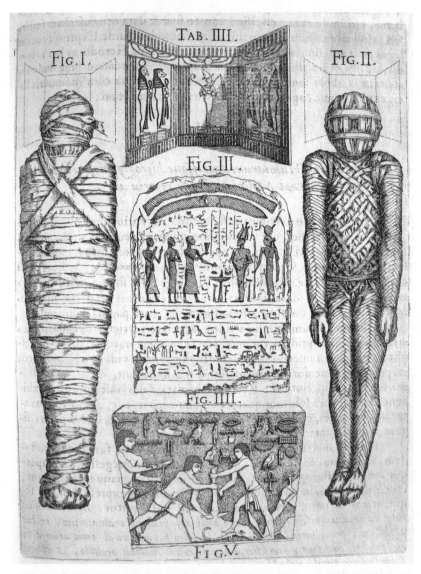

Fig. 5. A mummy and inscriptions documenting Egyptian funerary practices. Athanasius Kircher, *Oedipus Aegyptiacus* (Rome: 1652–54), vol. 3, 417. Courtesy of Stanford University Libraries.

A SEVENTEENTH-CENTURY LIFE[14]

Athanasius Kircher was born in 1601 or 1602 in Geisa, a small town in Thuringia in central Germany.[15] From the age of ten, when his parents sent him to study in nearby Fulda, he spent his life, first as student then as teacher, at Jesuit schools in Germany, France, Malta, and Rome.[16] In 1618 he entered the Society of Jesus as a novice, beginning the intensive curriculum of philosophy and theology prescribed for future priests.[17] Over the next decade his training took him to Paderborn, Cologne, Koblenz, Heiligenstadt, Aschaffenburg, and Mainz. After taking holy orders in 1629, Kircher spent his tertianship (the year after ordination designated for withdrawal and spiritual contemplation) in Speyer, where he happened upon a book of engravings of Roman obelisks in the college library, which sparked his desire to decipher the hieroglyphs.[18] During university Kircher cultivated the two fields that would anchor his long, polymathic career: mathematical sciences and Oriental languages. His efforts were recognized in 1630 with an appointment as professor of mathematics, Greek, and Hebrew at the Jesuit college in Würzburg.

Kircher's admission into the Society in 1618 coincided with the Bohemian revolt, which soon escalated into the pan-European conflict known as the Thirty Years' War, devastating much of Germany. When Paderborn fell to Protestant forces in 1620, Kircher narrowly escaped with his life. By 1629 he concluded that his prospects were brighter elsewhere. Bitterly commenting that Catholicism had enjoyed more success in India in seven

---

14. On Kircher's life, the principal source is his Latin autobiography, Kircher, *Vita Admodum Reverendi P. Athanasii Kircheri* (1684), published posthumously by Hieronymus Langenmantel. Totaro, *L'autobiographie d'Athanasius Kircher* (2009) provides a critical edition with French and Italian translations, but the introductory essay should be read skeptically. An annotated English translation is now available in Fletcher, *Study of the Life and Works of Athanasius Kircher* (2011), 459–551.

15. Kircher was unsure of his birth year. He gave it as 1601 in Kircher, *Magnes* (1641), after index, but as 1602 in his *Vita*. Early archival records have 1601 (e.g., ARSI Rom. 57, fol. 153^v; Rom. 58, fol. 15^v; Rom. 59, fols. 12^r, 198^r). In 1664 Kircher wrote the canon of Geisa hoping to learn the truth, but by then his hometown's baptismal records had been lost. See Richter, "Athanasius Kircher und seine Vaterstadt" (1927). On Kircher's youth, see Jäger, "Athanasius Kircher, Geisa, und Fulda" (2002).

16. The literature on the early modern Jesuits has grown vast. An excellent entry point is a pair of conference proceedings: O'Malley et al., *The Jesuits* (1999); and O'Malley et al., *The Jesuits II* (2006).

17. ARSI Rhen Inf. 38, fols. 212^v, 242^v, 259^r; Rhen Inf. 37, fol. 197; O'Neill and Domínguez, eds., *Diccionario Histórico* (2001), vol. 3, 2196.

18. ARSI Rhen. Sup. 25, fol. 49^r; *Vita*, 36. The book was likely Herwart von Hohenburg, *Thesaurus Hieroglyphicorum* ([1610]).

years than in his homeland in seventy, he begged to be transferred abroad, lest "my mind waste away, cramped inside Germany's barren and dusty wasteland."[19] His superiors were unmoved. But in 1631, after the advancing armies of Gustavus Adolphus closed Jesuit colleges in Würzburg and elsewhere, Kircher fled to France, never to see Germany again. Assigned a new teaching post in Avignon,[20] Kircher remained in France less than two years, but they were decisive ones, due to his encounter with the brilliant Provençal impresario of learning, Nicolas-Claude Fabri de Peiresc, who became his mentor and orchestrated his transfer to Rome in 1633. There, Kircher's life as a wandering scholar came to an end; after the trip to Malta in 1637–38, he rarely traveled beyond the Roman Campagna.

Kircher arrived in Rome at a delicate moment. Only months before, Galileo had been convicted by the Holy Office on charges of "vehement suspicion of heresy," ushering in a period of intellectual conservatism. The "marvelous conjuncture," during which the papacy of Galileo's supporter, Maffeo Barberini, fleetingly aroused hopes for an official Catholic embrace of Copernicanism, had passed.[21] But the new alignment proved auspicious for a rising star with the right intellectual skills and political dexterity. The Society of Jesus had been ambivalent about the Galileo affair. While Jesuit scholars played a leading role in the prosecution, prominent Jesuit mathematicians had been among Galileo's early supporters, and, as Kircher confided to Peiresc in 1633, some of them privately sympathized with the new cosmology.[22] Kircher, who inherited the mathematician's studio in the Collegio Romano, previously occupied by the great Christoph Clavius (architect of the Gregorian calendar) and the brilliant but discrete Christoph Grienberger, was deft at pursuing scientific investigations that were timely while avoiding controversy.[23] Arriving in Rome in autumn of 1633, Kircher might have appeared to Urban VIII as an ideal replacement for his recently

---

19. "Quare obnixè oro atque obsecro, per amorem Dei, ac sanctissime Virginis Matris, perquam illud ingens consequendae Apostolicae persecutionis desiderium, ne preces meas, ac obtestationes irritas esse patiatur, ne quaeso permittat, ut animus meus inter Germaniae huius sterilis ac aronosae angustias constrictus contabescat, exera<m> animum meum hactenus vinctum cum totum in obsequium divinae maiestatis effundendo, ne ego in offerendo me deo promptior liberaliorque, quam deus in beneficiis mihi refundendis esse videtur." Kircher to Father General [Vitelleschi], Mainz, 12 January 1629, ARSI Rhen. Sup. 42 ("Indiam petentes"), fols. 20^r–21^v. See below, n. 55, for more of this letter.

20. ARSI Rhen. Sup. 25, fol. 68^v; Lugd. 14, fol. 239^v, 19^r; fol. 14^r.

21. Redondi, *Galileo Heretic* (1987); Biagioli, *Galileo, Courtier* (1993).

22. Peiresc to Gassendi, 6 September 1633, in Tamizey de Larroque, ed., *Lettres* (1888–98), vol. 4, 353–54. On Galileo and the Jesuits, see Feldhay, "Use and Abuse of Mathematical Entities" (1998); Feldhay, *Galileo and the Church* (1995).

23. Gorman, "Mathematics and Modesty in the Society of Jesus" (2003), 13.

disgraced favorite, a practitioner of the up-to-date fusion of mathematics and experimental physics known as "physico-mathematics," but with solid orthodox credentials.[24] Perhaps more attractively, his research agenda included investigations of an altogether different sort. During his first years in Rome, under Barberini patronage, Kircher mostly put aside scientific pursuits to study Oriental manuscripts and Egyptian antiquities.

Kircher had an uncommon gift for ingratiating himself among the rich and powerful, a useful talent encouraged by the Society of Jesus. Before completing his studies in Mainz, he had won the favor of the local archbishop, who took Kircher into his service after witnessing an impressive theatrical display that he designed.[25] In the years that followed, Kircher ascended from the favor of provincial notables to the highest levels of courtly patronage, as the Barberini were succeeded by Innocent X and other popes, the Habsburg royal family, Queen Christina of Sweden, and Duke August of Braunschweig-Lüneburg, among others.[26] Backed by powerful supporters, Kircher lived a life of which most scholars could only dream, enjoying a special status within the Collegio Romano, somewhat akin to a post at an institute of advanced studies. With only occasional teaching duties, and aided by a succession of younger Jesuit assistants, he could devote his ample energies to studying, experimenting, collecting, corresponding, and publishing, as well as strengthening his ties to the aristocratic world.

Under Kircher's care, the mathematician's studio gradually transmogrified into the *Musaeum Kircherianum*, one of Europe's most famous collections of natural history, antiquities, scientific instruments, machines, and other wonders (fig. 6).[27] Kircher's quarters in the Collegio Romano became the command center of the collaborative enterprise that Paula Findlen has aptly called the "Kircherian machine."[28] From his privileged node at the center of the Catholic world, Kircher functioned as a conduit of information

---

24. Only after he had achieved a certain stature and level of protection from outside the Society did Kircher broach controversial astronomical matters in Kircher, *Itinerarium exstaticum* (1656). The fact that the controversy was kept under wraps and resolved in Kircher's favor testifies to his dexterity at exploring the boundaries of acceptable Jesuit science. See references, n. 10, above. On "physico-mathematics," see Dear, *Discipline and Experience* (1995).

25. *Vita*, 32–33.

26. On Kircher's patronage strategies, see Baldwin, "Pious Ambition" (2003); Baldwin, "Reverie in a Time of Plague" (2004).

27. Mayer-Deutsch, *Musaeum Kircherianum* (2010); Lo Sardo, ed., *Athanasius Kircher* (2001); Findlen, "Science, History, and Erudition" (2001); Findlen, "Scientific Spectacle in Baroque Rome" (1995); Findlen, *Possessing Nature* (1994); Casciato, Ianniello, and Vitale, eds., *Enciclopedismo in Roma barocca* (1986).

28. Findlen, "Last Man Who Knew Everything" (2004), 5.

Fig. 6. Kircher greeting visitors in his famous museum at the Collegio Romano. The wood models of obelisks were recently rediscovered. Without pedestals, they measure about one meter tall. Giorgio de Sepibus, *Romani Collegii Societatus Jesu Musaeum celeberrimum* (Amsterdam: 1678), frontispiece. Courtesy of Stanford University Libraries.

from the Society of Jesus's incomparable global communication network to the international and multiconfessional Republic of Letters.[29] Intellectual leadership was a key component of the Jesuits' apostolic mission, and both Kircher and his superiors recognized his cosmopolitan role as a valuable asset for the Catholic cause. *Egyptian Oedipus* concluded with Kircher's personal elaboration of the Jesuit motto: "All for the greater glory of God, and the improvement of the Republic Letters."

Kircher exchanged letters with correspondents throughout Europe and the world, and his museum became a meeting place for scholars of all faiths who visited Rome.[30] But it was to books above all that he owed his fame. He published more than thirty, beginning with *Art of Magnesia* (Würzburg, 1631), a minor academic dissertation on magnetism, and ending with *Tower of Babel* (Amsterdam, 1679), an exploration of counterfactual architecture and historical linguistics based on the biblical story of the confusion of tongues. Beginning with *The Great Art of Light and Shadow* (Rome, 1646), an influential study of optics and solar timekeeping, Kircher's signature publications were large-format, heavily illustrated encyclopedic studies that balanced learning (they were all composed in Latin, though a few saw vernacular translations) and spectacle (a born showman, Kircher knew how to entertain as well as edify).[31] They proved so popular that a leading Dutch printer, Jan Jansson van Waesberghe, offered Kircher a lifetime contract.[32] Starting with *Underground World* (1666), a best-selling treatise on baroque earth science, most of Kircher's books were printed by Protestants in Amsterdam, the center of the European book trade.[33]

In the autobiography that Kircher wrote near the end of his life, the great self-promoter was surprisingly taciturn about most of his scholarly career. Major works like *China Illustrated* (Amsterdam, 1667) and *The Great Art of Knowing* (Amsterdam, 1669) were passed over in silence, and *Underground World* was mentioned only in passing to refer readers to its account

---

29. Gorman, "Angel and the Compass" (2004); Harris, "Confession-Building" (1996); Osorio Romero, ed., *Luz imaginaria* (1993).

30. Much of Kircher's surviving correspondence has been digitized by the Athanasius Kircher Correspondence Project, initiated by Michael John Gorman and Nick Wilding at the Institute and Museum of the History of Science in Florence, and continuing under the direction of Paula Findlen at Stanford University.

31. Kircher, *Ars magnesia* (1631); Kircher, *Turris Babel* (1679); Kircher, *Ars magna lucis* (1646).

32. "Estratto dalle lettre di Sig. Jansonio ed Eliseo Weyerstraed Mercanti de libri in Amsterdam intorno la vendita de libri del P. Atha. Kircher," 29 July 1662, APUG 563, fol. 244^r.

33. Kircher, *Mundus subterraneus* (1665).

of God snatching him from the flames of Vesuvius.[34] Instead, Kircher narrated his numerous youthful scrapes with death: nearly crushed by a mill wheel, stampeded by horses, miraculously cured of gangrene, trapped on an ice floe in the frozen Rhine while fleeing an invasion, and almost lynched by Protestant soldiers. Above all, he dwelt on his discovery and renovation of a ruined shrine in Mentorella, in the Roman countryside, where Saint Eustace had been converted by a vision of a cross between the horns of a stag. Kircher saw his life as guided by special providence and believed that God had chosen him to achieve great things. In the scholarly realm, he discerned the divine plan most clearly in his hieroglyphic studies. The project that he affectionately called "my Oedipus" was the only aspect of his scholarship discussed at length in the autobiography. Together, the restoration of the hieroglyphic doctrine and the restoration of the shrine to Saint Eustace encapsulated Kircher's self-image: a paragon of the Jesuit scholar, harmoniously balancing piety and learning (fig. 7).

THE AGE OF ERUDITION

While the eighteenth century looked back upon Descartes as a harbinger of enlightenment who boldly cast off sterile tradition to clear the way for a modern science based on reason, Kircher was remembered as the butt of jokes satirizing the excesses of old-fashioned scholarship. The *Yverdon Encyclopedia*, published in Switzerland in the 1770s, included an entry on Kircher, which repeated what were by then well-worn anecdotes:

> Everything that bore the mark of antiquity was divine in his eyes. His extreme passion for everything ancient exposed him to many pranks. They say that some young fellows, aiming for a laugh at his expense, had many imaginary characters engraved on an unshaped stone and buried this stone in a spot where they knew there was going to be construction. Sometime after, excavation took place, and the stone was found and brought to Father Kircher as something unusual. The father, overjoyed, set to work interpreting its characters with abandon, and finally succeeded in giving them the most beautiful meaning you can imagine.

After reporting another hoax in the same vein (like the first, it was borrowed from J. B. Mencken's *Charlatanry of the Learned*), the article concluded: "Despite all this, Kircher's *Oedipus Aegyptiacus, Mundus Subterraneus,*

---

34. Kircher, *China illustrata* (1667); Kircher, *Ars magna sciendi* (1669).

Fig. 7. Kircher painted from life around the time he published *Pamphilian Obelisk*. Anonymous portrait of Athanasius Kircher, oil on canvas, c. 1650. Rome, Galleria Nazionale di Arte Antica.

etc., are expensive and sought-after, and one cannot deny this father's massive erudition."[35]

Even the grudging acknowledgment of Kircher's "massive erudition" may have been damning with faint praise. In the eighteenth century "erudition" had come to signify a style of learning that had fallen from grace in enlightened precincts of the Republic of Letters. Jean d'Alembert's manifesto, *Preliminary Discourse to the Encyclopedia* (1751), set forth the classic statement of this attitude, dividing the learned world into three realms, each associated with a corresponding mental faculty: erudition with memory, philosophy (including mathematics and the natural sciences) with reason, and *belles lettres* (literature, especially poetry) with imagination. Memory and erudition occupied the lowest rank in d'Alembert's taxonomy. Recounting the emergence of European minds from medieval barbarism, he described how the revival of learning began, appropriately, with erudition—the study of languages and history—since this was based on the simplest of mental activities: fact collecting and memory. Ignoring nature, d'Alembert wrote, erudite scholars

> thought they needed only to read to be learned; and it is far easier to read than to understand. And so they devoured indiscriminately everything that the ancients left us in each genre . . . These circumstances gave rise to that multitude of erudite men, immersed in the learned languages to the point of disdaining their own, who knew everything in the ancients except their grace and finesse, as a celebrated author has said, and whose vain show of erudition made them so arrogant because the cheapest advantages are rather often those whose vulgar display gives most satisfaction.[36]

These harsh words were partially balanced by d'Alembert's acknowledgment that the popular scorn heaped upon erudition was excessive, and in a subsequent *Encyclopedia* article he took a softer tack, affirming erudition's potential to contribute further to the advancement of knowledge.[37] Justified or not, such contempt was fashionable, and d'Alembert observed that young scholars had turned away from erudite research, flocking instead to

---

35. , "Kircher, Athanase" (1770–75). Cf. Mencken, *Charlatanry of the Learned* (1937), 85–86. These defamatory stories followed an ancient trope: see Grafton, *Forgers and Critics* (1990), 3–4.

36. d'Alembert, *Preliminary Discourse* (1995), 63.

37. d'Alembert, "Erudition" (1751–65). On the *Encyclopédistes*' relationship to erudition, see Edelstein, "Humanism" (2009), and Edelstein, *The Enlightenment* (2010), 44–51, which present d'Alembert's attitude as more sympathetic.

fields of learning that appeared fresher and more fruitful: mathematics and the natural sciences.

"Erudition" as a distinctive type of scholarship, as opposed to its original meaning of learning in general, seems to have been coined in the eighteenth century as a pejorative. Such, at any rate, was the opinion of Edward Gibbon, who penned a spirited defense of erudition against d'Alembert's calumnies.[38] Stripped of its negative connotation, erudition provides a useful label for the cluster of historically minded, early modern disciplines based on knowledge of languages, literature, and antiquities. If d'Alembert's grim diagnosis described the mid-eighteenth century scene, especially in France, a century earlier the situation could not have been more different. Kircher lived at the climax of a golden age of erudition, when new methods and materials were transforming the study of the past. Even as the natural sciences began their steady ascent to intellectual supremacy, antiquity still beckoned. Erudition was not yet dry as dust. On the contrary, it glittered with the promise of new discoveries and profound intellectual rewards.

Erudition's roots lay in Renaissance humanism, which revived the study of Latin and Greek literature, developed critical methods for emending and interpreting texts, and valorized classical antiquity as a model of virtue, wisdom and style. Ironically, the humanists' quest to bring the literary culture of antiquity back to life ultimately led them to recognize the futility of that endeavor. Poring over classical literature, Renaissance scholars confronted the irreducible difference between the ancient and modern worlds. To fully grasp ancient texts, they discovered, it was necessary to understand the unfamiliar culture that had produced them. Not least, this entailed an appreciation of the historicity of language, which became the basis of the art of textual criticism. Between the fifteenth and seventeenth centuries, practitioners of critical philology developed increasingly sophisticated techniques for authenticating claims made in and about old texts, Lorenzo Valla's iconic debunking of the Donation of Constantine being an early example. By the time Kircher began his studies, many scholars were less invested in imitating ancient authors and more concerned with interpreting their texts as evidence of the human past. The result was a new type of historical research based on reconstructing the social and cultural contexts of former times.[39]

---

38. Pocock, *Barbarism and Religion* (1999), 141.
39. Grafton, *What Was History?* (2007); Grafton, *Forgers and Critics* (1990); Kelley, *Foundations of Modern Historical Scholarship* (1970); Burke, *Renaissance Sense of the Past* (1969); Pocock, *Ancient Constitution* (1957), 1–29.

In forging this new kind of history, textual criticism was abetted by another discipline that evolved out of Renaissance humanism: antiquarianism.[40] Antiquaries were experts in the tangible remains of the past: "antiquities."[41] Their quarry included old manuscripts, but above all, they pioneered the study of material artifacts such as coins, cameos, inscriptions, and ruins—any "primary source," physical or textual, that might provide empirical evidence of past ages. Following ancient models like Varro, antiquarian scholarship often focused on the institutions, customs, rites, and topography of former times. In contrast to the classical tradition of humanist historical writing, which aimed to produce modern political and military histories in the style of ancient historians like Polybius and Tacitus, who had recounted events recent to their own time, antiquarianism was characterized by a fundamental curiosity about epochs distinct from "modern" times: "antiquity" and, also, the "Middle Ages."[42]

Today, antiquarianism, often prefaced by the adjective "mere," is used derisively to refer to a kind of historical research mired in trivia and devoid of larger significance—an echo of d'Alembert's dismissal of the "vain show of erudition." But in the sixteenth and seventeenth centuries, antiquarianism and its partner in erudition, philology, were on the cutting edge of scholarship, tackling matters of vital cultural importance. In a society as oriented toward tradition as early modern Europe, the past—the time of origins—was a crucial source of legitimacy for all manner of modern claims. As states tried to centralize power, for example, royal scholars combed the medieval archives for evidence of monarchical rights while the nobility employed similar tactics to bolster opposing feudal claims.[43] Erudition also played a central role in the theological controversies that permeated European culture in the centuries after the Reformation. As Protestant and

---

40. The classic studies are Momigliano, "Ancient History" (1950); Momigliano, "Rise of Antiquarian Research" (1990). See also: Stenhouse, *Reading Inscriptions* (2005); Herklotz, *Cassiano dal Pozzo* (1999); Schnapp, *Discovery of the Past* (1997); Haskell, *History and Its Images* (1993); Jacks, *The Antiquarian* (1993); Weiss, *Renaissance Discovery* (1969), and works cited in the following notes.

41. In early modern Italian, "antiquario" usually referred to collectors, artists, and architects who handled antiquities, rather than to scholars. I follow the now common usage in referring to scholars of antiquities as antiquaries. See Claridge, "Archaeologies" (2004).

42. Momigliano's stark dichotomy between antiquarianism and history has been challenged by recent studies. In particular, Grafton, *What Was History?* (2007) shows how, during the sixteenth century, Renaissance theorists of the *ars historica* shifted the emphasis from writing new histories to reading ancient ones, eroding the boundary between antiquarianism (or erudition) and history. See also Franklin, *Jean Bodin* (1963), 83–88.

43. Kelley, *Foundations of Modern Historical Scholarship* (1970); Baker, *Inventing the French Revolution* (1990), ch. 2; Soll, "Antiquary and the Information State" (2008).

Catholic scholars sought the upper hand in debates over scriptural interpretation and church history, "sacred philologists," working at the confluence of antiquarianism, Oriental philology, and biblical hermeneutics, generated some of the most innovative scholarship of the period.[44]

In sum, erudition was an array of scholarly practices aimed at knowledge of the past, an "archeology of past states of society and culture,"[45] defined in terms of their distinctive institutions, beliefs, and customs. Critical philology and antiquarianism were its two pillars.[46] Building on Arnaldo Momigliano's pioneering studies, recent research has revealed early modern erudition as a laboratory of modern approaches to historical evidence and a progenitor of disciplines such as archeology, anthropology, sociology, religious studies, and cultural history.[47] The idea of a fundamental clash between modern science and humanist scholarship, iconically embodied by Descartes, does not hold up as a generalization about early modern learning. On the contrary, with their shared concern for testimony, facticity, and discovery, erudite historical research and experimental natural science had much in common, which explains why they were frequently practiced by the same individuals, notably Leibniz and Newton.[48] Fundamentally concerned with evidence, erudition was empiricism applied to the study of the past.[49]

EASTERN PROMISES

In Kircher's day, no branch of erudition appeared more promising than the emerging discipline of Oriental studies.[50] An outgrowth of classical philol-

---

44. Miller, "London Polyglot Bible" (2001); Stolzenberg, "John Spencer" (2012).

45. Pocock, *Barbarism and Religion* (1999), 5.

46. I thus use "erudition" to cover the territory that scholars often refer to as "antiquarianism." In proposing erudition as a preferable umbrella term, I mean to reinforce Ingo Herklotz's argument about the crucial relationship between antiquarianism and philology (Herklotz, "Arnaldo Momigliano's 'Ancient History'" [2007]), as well as the recent qualifications of Momigliano by Miller, "Introduction" (2007) and Grafton, *What Was History?* (2007). My thinking has also been influenced by Pocock's study of Gibbon.

47. Miller, ed., *Momigliano and Antiquarianism* (2007).

48. Spitz, "Significance of Leibniz" (1952); Manuel, *Issac Newton, Historian* (1963). On seventeenth-century empiricism, see Daston, "Moral Economy" (1995), 12–18.

49. On the links between empiricism in the study of the past and the study of nature, see Pomata and Siraisi, eds., *Historia* (2005); Cerruti and Pomata, eds., *Fatti: Storie dell'evidenza empirica. Theme Issue. Quaderni Storici* (2001); Siraisi, *History, Medicine, and the Traditions* (2007); Grafton, "Where Was Salomon's House" (2009); Freedberg, *Eye of the Lynx* (2002).

50. I use terms like "Orient" and "Oriental languages" neutrally, following actors' categories. The postcolonial framework for studying Orientalism, inspired by Edward Said, is not terribly helpful for making sense of Renaissance and seventeenth-century philology. I plan to address the relationship between early modern and modern Orientalism in a future work. I am

ogy, the modern study of Near Eastern languages was born at the turn of the sixteenth century, when Renaissance scholars began to study Hebrew.[51] In the wake of the Reformation, as the correct reading of scripture became a matter of increasingly high stakes, Hebrew, as well as Aramaic, Samaritan, Ethiopian, Armenian, and other languages that preserved versions of scripture and documents of the early church, became essential weapons of theological warfare. The desire of scholars to understand the Bible in historical context fed into a broader interest in the history of the Near East, inspiring the study of other Oriental languages, most importantly Arabic.[52] Oriental philology was driven largely by an inward-looking impulse: Europeans' desire to understand their own heritage.[53] But its realization was only possible because of increasing commerce between Europe and the Islamic world, which facilitated the circulation of information, materials, and people around the Mediterranean world.[54]

Oriental studies attracted Kircher from an early age. According to his autobiography, he took Hebrew lessons from a rabbi as a schoolboy in Fulda. As a university student he devoted himself to Oriental languages, beginning with Hebrew and branching out to Syriac and Arabic. In 1629, while preparing for ordination, he sent a petition to the general of the Society of Jesus requesting assignment as a missionary. While offering himself "indifferently" for service wherever there was the opportunity to promote the greater glory of God, he expressed his preference for North Africa and the Near East.[55]

---

much in agreement with Suzanne Marchand's views in the introduction and first chapter of Marchand, *German Orientalism* (2009).

51. Grafton and Weinberg, *I Have Always Loved the Holy Tongue* (2011); Coudert and Shoulson, eds., *Hebraica Veritas* (2004); Sutcliffe, *Judaism and Enlightenment* (2003), part 1; Cavarra, ed., *Hebraica* (2000); Burnett, *Johannes Buxtorf* (1996); Manuel, *Broken Staff* (1993); Zinguer, *L'hébreu* (1992); Rooden, *Theology, Biblical Scholarship, and Rabbinical Studies* (1989); Katchen, *Christian Hebraists* (1984).

52. Miller, *Peiresc's Orient* (2012); García-Arenal and Rodríguez Mediano, *Un Oriente español* (2010); Dew, *Orientalism* (2009); Hamilton, *Copts and the West* (2006); Hamilton, *William Bedwell* (1985); Toomer, *Eastern Wisedome* (1996); Contini, "Gli inizi della linguistica siriaca" (1996); Piemontese, "Grammatica e lessicografia araba" (1996); Tamani, "Gli studi di aramaico" (1996). Russell, ed., *"Arabick" Interest* (1994); Dannenfeldt, "Renaissance Humanists" (1955); Levi della Vida, *Ricerche* (1939).

53. I refer specifically to the discipline of Oriental philology and not European literature on the Near East in general. Beyond academic Orientalism, there was great interest in Islam and the Ottoman Empire. See, e.g., Meserve, *Empires of Islam* (2008).

54. Hamilton, Van Den Boogert, and Westerweel, eds., *Republic of Letters and the Levant* (2005).

55. Kircher to Father General [Vitelleschi], Mainz, 12 January 1629, ARSI Rhen. Sup. 42 ("Indiam petentes"), fols. 20ʳ–21ᵛ. The secondary literature commonly states that Kircher wished to be a missionary in China, but the archival record documents his persistent prefer-

Despite his claim that he had studied Oriental languages to prepare for missionary work, one rather suspects that he sought to become a missionary in order to increase his knowledge of Oriental languages and literature. The petition was denied. Kircher's superiors, noting his melancholic temperament, intellectual talent, and lack of practical experience, had marked him out for a career of teaching and scholarship, not missionary service.[56]

Languishing on Malta, it was not for Rome or any European center of learning that Kircher pined. Perched in the middle of the Mediterranean, he fixed his gaze on the other shore. Again he appealed to his Jesuit superiors, declaring himself willing to undertake any assignment that would bring him "to Egypt or the Holy Land, in order to see those countries and improve his knowledge of Oriental languages and studies."[57] Making his case to the officials in Rome who controlled his fate, he emphasized the relatively easy journey from Malta to Alexandria and described the scholarly riches that he was sure to encounter. Were his wish granted, he would examine "the countless antiquities surviving in Egypt" and explore its "very ancient libraries, most abundantly furnished with the rarest books," among them, manuscripts in Coptic, Arabic, Greek, Hebrew, and other languages—even papyri covered in hieroglyphs.[58]

---

ence for the Near East. At the beginning of this letter, Kircher specified his desire to go the "the province of the Orientals or Abyssinians"; at its end he spoke more broadly, but the order is telling: "Iterum me offero indifferenter ad quasvis provincias, ibi maior honori Dei promonendi occasio, ad Abassiam [upper Ethiopia], Arabiam, Palestinam, Constantinopolim, Persiam, Indiam, Chinam, Iaponiam, Americam, maximè orientaliores quarum vernaculas, me brevisimo tempore dei gratiâ adiutum . . . ."

56. In the 1633 triennial catalog of the College of Avignon, Kircher's superiors assessed him as follows: "Talent: good; judgment: good; discretion: some; experience of things: not great; accomplishment in letters: great; natural complexion: melancholic; for which ministry of the Society he has talent: teaching." ARSI Lugd. 19, fol. 17^r. See the similar reports for 1639, 1642, 1649, and 1652 in ARSI Rom. 57, fol. 210^v, 58, fol. 121^v, 59, fols. 95^r, 278^r.

57. Lucas Holstenius (on behalf of Kircher) to Francesco Barberini, Naples, 7 September 1637, in Bartòla, "Alessandro VII" (1989), 80.

58. "Cum ex variis non Mahumetanis tantum; sed et Christianis, quia et ab < . . . > Patre nostro qui Aegyptum paenè totam ad nili usque Cataractes lustravit, mira perceperim cùm de antiquitatibus innumeris Aegypto superstitibus; tùm de Bibliothecis antiquissimis, librorum reconditissimorum copia instructissimis, animus eius videndae desiderio aliàs aestuans dictarum rerum relatione iam paene in incendium erupit manifestissimum. Nihil igitur gratius in hoc mundo contingere posset quam si ego Illmae d.mae vrae opera ab Emmo Cardinale vel Sacra Congregatione aut etiam Adm. R.de P. Generale, obtinere possem, tam laudabilem expeditionem ut videlicet αυτόπτιις ea omnia intueri liceret, quae absens tanta admiratione contemplor, librosque Coptos, Arabicos, Graecos, hebraeos, aliosque, quorum supra memoratus Pater magnam se copiam in Matria totius Aegypti celeberrima Bibliotheca vidisse retulit, ab interitu vindicare possem." Kircher to dal Pozzo, Malta, 15 August 1637, BISM ms. H 268, 9^{rv}. Kircher's mention of hieroglyphic papyri reportedly housed in a Cairo library is from *LAR*, 512. See also Barberini to Holstenius, Rome, 24 September 1637, and Muzio Vitelleschi to Kircher, 7 January 1638, both in Bartòla, "Allesandro VII," 81–83.

The ascendancy of Neoclassicism in the eighteenth century was attended by a waning of enthusiasm for Oriental studies, which was only to be reversed in the nineteenth century. By contrast, Kircher and other seventeenth-century scholars were convinced that invaluable literary treasures stood on the threshold of European discovery. This was the same moment when Europeans like Francis Bacon and Descartes envisaged a "great instauration" of natural science and philosophy. Although less well remembered, a similar sense of anticipation suffused the study of ancient history. In fact, Kircher and other Oriental philologists explicitly compared the new insights awaiting detection in Near Eastern literature to the epoch-making astronomical discoveries of Galileo and the discovery of the Americas.[59] What stood between them and the new historical horizon was access to sources. The library and the cabinet of antiquities would be their observatory, knowledge of Oriental languages their telescope.

OCCULT PHILOSOPHY

Although Kircher claimed that Oriental texts were the key to his breakthrough, the hieroglyphic doctrine that he described in *Egyptian Oedipus* was, in its essentials, an elaboration of the occult philosophy that previous scholars had reconstructed from Greek and Latin texts. As is well known, the rediscovery of Plato in the Italian Renaissance was deeply influenced by late antique Neoplatonism. Marsilio Ficino, the most influential and profound of the early modern Platonists, viewed the Platonic tradition through the writings of Plotinus, Porphyry, Proclus, Synesius, and Iamblichus. Reading the Platonic dialogues as esoteric allegories, these late antique men had revered Plato as a theologian as much as a philosopher. They considered him not the originator but perhaps the greatest expositor of a much older tradition, a *prisca theologia* or ancient theology originating with legendary, often Oriental sages such as Orpheus, Hermes Trismegistus, and Zoroaster. Following Ficino, most sixteenth- and seventeenth-century Europeans made no distinction between Plato's philosophy and what we now call Neoplatonism. The term *Platonici* evoked followers not only of Plato and his self-identified successors in the Academy, like Plotinus, but also of his imaginary predecessors, like Hermes Trismegistus.[60]

---

59. *OA* I, a4^v–b1^r. John Selden made the same analogy in a 1646 letter quoted in Toomer, *Eastern Wisedome* (1996), 69.

60. Copenhaver and Schmitt, *Renaissance Philosophy* (1992), 14–18, 134–63; Allen, *Synoptic Art* (1998), 1–49; Hanegraaff, "Tradition" (2005); Stuckrad, *Locations of Knowledge* (2010),

Unlike Plotinus and his disciples, however, the Renaissance Platonists were Christians (Ficino was a priest) who sought a pious philosophy that would harmonize pagan wisdom with Christianity. This synthesis was facilitated, somewhat ironically, by recourse to early Christian writers, who had endorsed the genealogy of the *prisca theologia*, but with a very different aim. In battling the claims of pagan philosophers—especially the charge that Christianity had stolen teachings from Plato and other pagan traditions—church fathers such as Eusebius, Lactantius, and Augustine co-opted the Neoplatonists' doxography to argue that it was Plato and the Greeks who were unoriginal, and that resemblances between Christian and Greek doctrines were due to the influence on the latter of Moses or earlier Hebrew patriarchs via the ancient theologians (*prisci theologi*).[61] To these church fathers, the *prisca theologia* was an apologetic strategy to defend the originality and profundity of Christianity against a still powerful pagan tradition. For Renaissance Platonists like Ficino, Giovanni Pico della Mirandola, Cornelius Agrippa, and Agostino Steuco, on the other hand, the idea of the *prisca theologia* validated pagan wisdom against claims of Judeo-Christian exclusivity. They tended to vacillate, however, between asserting an independent tradition of pagan illumination and affirming the claim that the *prisca theologia* originated with Moses or earlier biblical personalities.[62]

I refer to this early modern tradition as "occult philosophy" in preference to the term "Hermeticism," which overly privileges the role of Hermes Trismegistus.[63] (I use "Hermetism" more narrowly to refer to beliefs about Hermes Trismegistus and literature attributed to him.) Unlike its synonym, "Renaissance Neoplatonism," "occult philosophy" emphasizes the essential claim of a genealogy of esoteric knowledge predating Plato, supposedly hidden beneath symbols and allegories. Since "occult philosophy," which is a key analytical term in this book, may hold various and vague associations in readers' minds, I wish to emphasize that I use it consistently and

---

25–42. See also Mulsow, "Ambiguities of the Prisca Sapientia" (2004); Celenza, "Search for Ancient Wisdom" (2001); Hankins, *Plato in the Italian Renaissance* (1990), vol. 2, 460–64; and the classic study, Yates, *Giordano Bruno* (1964).
    61. Walker, *Ancient Theology* (1972), 1–21.
    62. Stuckrad, *Locations of Knowledge* (2010), 30–33.
    63. Copenhaver, "Iamblichus, Synesius and the 'Chaldean Oracles'" (1987); Copenhaver, "Hermes Trismegistus" (1988); Copenhaver, "Natural Magic, Hermetism, and Occultism" (1990). "Hermeticism" also tends to conflate the philosophical tradition that stems from Ficino (which I am calling occult philosophy) and the alchemical tradition associated with Paracelsus and Rosicrucianism, which were based on different Hermetic texts and had distinct histories. See Leinkauf, "Interpretation und Analogie" (2001); Mulsow, "Das schnelle und langsame Ende" (2002); Ebeling, *Secret History* (2007), 89–90.

narrowly to refer to the early modern Neoplatonist tradition as defined in this section, and not as a woolly catch-all or synonym for broader categories such as "occult sciences," "occultism," or "magic."[64] Occult philosophy is part of the history of Western esotericism, but it is far from coextensive with that larger field.[65]

The *prisca theologia* provided occult philosophy's legitimating pedigree. Its canon consisted of ancient Platonist literature as well as mystical works attributed to ancient theologians. In addition to texts such as the Hermetic Corpus, the Chaldean Oracles (ascribed to Zoroaster), and the Orphic hymns, which were first taken up by late antique Neoplatonists, early modern thinkers expanded the canon of esoteric wisdom to include other traditions, most significantly the Jewish Kabbalah. Early Christian Neoplatonic texts, like the writings of pseudo-Dionysius, were also treated as sources of occult philosophy. The content of occult philosophy varied significantly among interpreters, as did the details of its lineage and canon, but its foundation was a Neoplatonic metaphysics based on a hierarchy of levels of reality, emanating from a transcendent God. In its most schematic form, these were defined as the terrestrial, celestial, and angelic (or intellectual, or archetypal) worlds (fig. 8). This metaphysics supported a distinctive Neoplatonic semantics and hermeneutics, in which symbols and allegories were privileged as "nondiscursive" forms of communication, capable of mediating the ultimate truths that existed in the ineffable realm of ideas. It also sustained belief in certain kinds of magic—both "natural," based on manipulating the links of sympathy and antipathy that connected the terrestrial and celestial realms, and theurgy, which sought the aid of angelic intelligences.

The relationship of occult philosophy to orthodox Christianity and to Aristotelianism was ambiguous. Important elements of ancient Neoplatonism contradicted Christian doctrine, but these could be explained as failings of pagan thinkers, who, despite their excellence, could only see

---

64. See Hanegraaff, "Occult / Occultism" (2005), 887. Thus, I do not consider alchemy and astrology as part of occult philosophy, although they could be incorporated into its historical and theoretical framework, as Kircher did in making them subfields of the "hieroglyphic doctrine." The various "occult sciences," too often conflated, led distinct if at times overlapping careers in early modern Europe. See Newman and Grafton, "Introduction" (2001).

65. "Western esotericism" refers to a series of interrelated currents in European culture since approximately the Renaissance, including Hermeticism, occult philosophy, natural magic, and alchemy. See Hanegraaff, "Esotericism" (2005); Hanegraaff, "Forbidden Knowledge" (2005); and Stuckrad, *Locations of Knowledge* (2010), who both describe "esotericism" as the retrospective product of polemical discourses. Cf. Faivre, *Access to Western Esotericism* (1994).

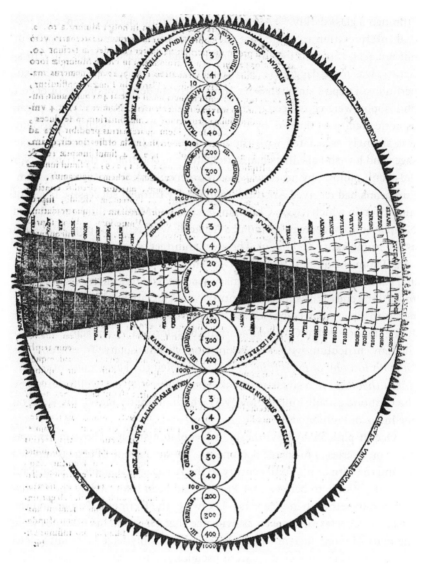

Fig. 8. Woodcut diagram illustrating Kircher's discussion of the numerological mysteries of "hieroglyphic arithmetic." From top to bottom and right to left, it depicts the hierarchical relationship of the angelic, celestial, and elemental worlds. Athanasius Kircher, *Oedipus Aegyptiacus* (Rome: 1652–54), vol. 2, part 2, 15. Courtesy of Stanford University Libraries.

"through a glass darkly."[66] Alternatively, one could engage in generous textual interpretation to argue that these contradictions were only apparent but not real; this solution, by so closely equating pagan wisdom to Christianity, was more dangerous, as it tended to make scriptural revelation superfluous. Despite such tensions, most early modern proponents of occult philosophy were convinced of the compatibility of Christianity and pagan esoteric traditions. Indeed, by the seventeenth century, occult philosophy was strongly associated with the controversial claim that ancient pagan sages and Jewish kabbalists had anticipated the doctrine of the Trinity. Nor did occult philosophy necessarily seem irreconcilable with Aristotelianism. Europeans had received Aristotle's philosophy through successive layers of mediation in which pagan, Christian, Jewish, and Muslim thinkers had used Neoplatonic concepts to harmonize Aristotle with religious principles such as the immortality of the soul. The great architect of Christian Aristotelianism, Thomas Aquinas, relied on two Christian Neoplatonists, Augustine and pseudo-Dionysius, as his primary theological authorities.[67] It was only toward the end of the sixteenth century, when dissatisfaction with Aristotle began to ripen, that Francesco Patrizi, professor of Platonic philosophy at Rome, proposed that Christian learning be reformed by replacing the Aristotelian curriculum with Neoplatonist occult philosophy. To this end, his *New Philosophy of Everything* (1591) assembled the key documents of the *prisca theologia* in a single textbook. His efforts were not well received by the Catholic establishment, however, which promptly placed his work on the Index of Prohibited Books.[68]

Occult philosophy's synthesis of Neoplatonism and Christianity was based on a series of misdated documents. The attribution of texts composed around the turn of the sixth century to the Areopagite disciple of Saint Paul made it possible to believe that Proclus had been informed by the Dionysian corpus rather than vice versa. Likewise, the Hermetic Corpus and the Chaldean Oracles were taken as evidence that Plato and his disciples were the heirs of more ancient sages who could plausibly have been influenced by Moses, when in fact they were pseudonymous texts composed under the influence of Platonism in the first centuries of the Christian era. The authority of the Kabbalah similarly derived from assigning late antique and

---

66. Allen, "At Variance" (2008), 32.
67. Dauphinais, David, and Levering, eds., *Aquinas the Augustinian* (2007), xii.
68. Patrizi, *Nova de universis philosophia* (1591); Leijenhorst, "Francesco Patrizi's Hermetic Philosophy" (1998), 125–47; Vasoli, "Francesco Patrizi" (1980). For a contrasting view of Patrizi, see Kristeller, *Eight Philosophers* (1964), 11–26. See also the acute essay, Dietz, "Space, Light, and Soul" (1999).

medieval texts to authors in distant antiquity. During the early modern period, these chronological errors began to be corrected. Lorenzo Valla first expressed doubts about the Dionysian corpus in the fifteenth century, but it took several centuries for its authority to erode, especially among Catholics. More swift was the demise of the credibility of the Hermetic Corpus after Isaac Casaubon's demonstration in 1614 that these supposed testaments of remote antiquity must have been composed after the time of Christ.

EMPIRICISM AND ESOTERICISM

This book is about the conjuncture of empiricism and esotericism in seventeenth-century scholarship. In contrast to prior studies, which have explained Kircher's investigation of the hieroglyphic doctrine as a vehicle to promote Hermetic philosophy, I argue that *Egyptian Oedipus* is better understood as an antiquarian treatise, which made use of occult philosophy as a historical framework to explain ancient objects and inscriptions. The idea of a nexus between erudite historical research and occult philosophy may seem paradoxical. After all, erudition revolutionized historical scholarship with empirical methods for evaluating textual and material evidence, whereas occult philosophy was based on erroneous historical claims, supported by misdated, pseudonymous documents. Casaubon's redating of the Hermetic Corpus has become iconic precisely because it symbolizes modern, critical scholarship demolishing the authority of a premodern tradition of esoteric wisdom. Kircher notoriously gave credence to the Hermetic Corpus a generation after Casaubon. As a result he has been seen as an anachronism himself, clinging to an idea of sublime primordial wisdom that had been upended by the modern spirit of criticism. But this view misjudges both Kircher and his age.

Kircher was selective in applying the methods of erudite research, in some respects practicing state-of-the-art scholarship, while in others ignoring critical standards of evidence. But he was not unique in thinking that occult philosophy offered valuable aid to the investigator of ancient paganism and the earliest ages of history. In the seventeenth century, contrary to a widespread assumption, erudite scholarship contributed to occult philosophy's ongoing vitality, even as it undermined the authority of key esoteric texts. Proving that Hermes Trismegistus had not written the *Corpus Hermeticum* was not the same as proving that the Egyptian sage and his esoteric wisdom had not existed. Thus, long after Casaubon, European scholars continued to equate the history of philosophy with the history of divine wisdom, lending credence to "perennial philosophy," the idea that the most perfect knowledge existed at the beginning of time.[69] The enduring belief that Adam and his progeny possessed profound knowledge not only of God, but also of natural philosophy, dialectics, and the mechanical arts, and that the history of its transmission could be described by appealing to biblical narratives, such as the story of Noah and his sons, created an environment hospitable to the notion of a gentile *prisca theologia*.[70]

In the 1650s, despite cracks in the edifice, occult philosophy did not seem obviously irreconcilable with critical and empirical historical research. Ultimately, Kircher's attempt to constructively fuse erudition and esotericism may have been doomed to failure, but it was not out of step with his times. The truly consequential turning point came later, as the result of a more gradual process that reached fruition in the eighteenth century. It consisted in the marginalization not only of occult philosophy but of all versions of perennial philosophy in favor of theories of progress, which made the passage of time the precondition for the development of complex bodies of knowledge. In the end, empiricism was at odds with esotericism—and much else. But what distinguished the seventeenth century was precisely the coexistence of beliefs and modes of thought that the Enlightenment would sunder.

A comparison with the history of natural science is helpful. Recent scholarship has shown how, contrary to received wisdom, the emergence of new empirical methods for studying the natural world did not lead straightforwardly to the rejection of traditional beliefs in magic and marvels. On the contrary, the early phase of experimental science in the seventeenth century was characterized by intense, perhaps unprecedented, interest in phenomena such as witchcraft and the curing of wounds by sympathetic medicine. Effective protocols for assessing knowledge claims based on observation and experiment were far from self-evident. They took time to develop, and in the interval, empiricism encouraged rather than discouraged the study of phenomena that would soon be redefined as illusory or pseudoscientific.[71] This book makes the case for an analogous phenomenon in the realm of historical scholarship. Ultimately the critical and empirical methods of erudition were an important factor in the marginalization of occult philosophy

---

69. Schmitt, "Perennial Philosophy" (1966); Malusa, "Renaissance Antecedents" (1993); Schmidt-Biggemann, *Philosophia Perennis* (2004).

70. Bizzocchi, *Genalogie incredibili* (1995), shows how aristocratic demand for "incredible" genealogies played a role in sustaining this historical framework.

71. Daston and Park, *Wonders and the Order of Nature* (1998); Clark, *Thinking with Demons* (1997); Daston, "Marvelous Facts" (1991). See also Eamon, *Science and the Secrets of Nature* (1994); Harkness, *John Dee's Conversations with Angels* (1999).

and other traditional beliefs about the past. But, for an important period in the seventeenth century, erudition and occult philosophy collaborated. Descartes was typical in treating marvelous facts, such as a corpse bleeding in the presence of its murderer, as parts of the natural order that his new science must explain.[72] But he was an outlier, even among the intellectual avant-garde, in his categorical rejection of the authority of tradition. More representative was Isaac Newton, who believed in a version of the *prisca theologia* and saw his theory of universal gravitation, like his alchemical investigations, in terms of the recovery of ancient wisdom.[73]

HISTORY OF THE BOOK

Long relegated to the historiographic margins, in recent years Kircher has experienced a remarkable reversal of fortune.[74] Rising academic interest in Kircher has coincided with major trends in the history of early modern science and scholarship: the desire to move beyond the great men identified with the origins of modernity, increased attention to the vitality of older forms of knowledge and the proliferation of hybrids of tradition and innovation, and appreciation of the pervasive role of religion and apologetics in early modern thought. At the same time, the study of material culture, sociability, institutions, and scholarly practices has inspired research on topics such as collecting, correspondence, scientific societies, and the history of the book. Kircher, due to the character of his life and work as well as the abundance of surviving documentation, has proven an ideal subject for these new historiographic approaches.[75]

---

72. Daston, "Language of Strange Facts" (1998), 21.
73. McGuire and Rattansi, "Newton" (1966).
74. See, for example, Sarah Boxer, "A Postmodernist of the 1600s is Back in Fashion," *New York Times*, 25 May 2002.
75. These connections are especially evident in the contributions to Findlen, ed., *Athanasius Kircher* (2004) and in Findlen, *Possessing Nature* (1994). For Kircher's life and work see the splendid recent overview, Godwin, *Athanasius Kircher's Theatre of the World* (2009); Rowland, *Ecstatic Journey* (2000); Leinkauf, *Mundus Combinatus* (1993); Pastine, *Nascita dell'idolatria* (1978); Reilly, *Athanasius Kircher* (1974). The following monographs focus on specific aspects of Kircher's studies: Mayer-Deutsch, *Musaeum Kircherianum* (2010); Siebert, *Die grosse kosmologische Kontroverse* (2006); Englmann, *Sphärenharmonie und Mikrokosmos* (2006); Marrone, *I geroglifici* (2002); Rivosecchi, *Esotismo in Roma barocca* (1982); Scharlau, *Athanasius Kircher als Musikschriftsteller* (1969). Edited volumes devoted to Kircher also include Vercellone and Bertinetto, eds., *Athanasius Kircher* (2007); Beinlich et al., eds., *Magie des Wissens* (2002); Beinlich, Vollrath, and Wittstadt, eds., *Spurensuche* (2002); Lo Sardo, ed., *Athanasius Kircher* (2001); Stolzenberg, ed., *Great Art of Knowing* (2001); Fletcher, ed., *Athanasius Kircher* (1988); Casciato, Ianniello, and Vitale, eds., *Enciclopedismo in Roma barocca* (1986). The late John

This book is a microhistory of the making and meaning of Kircher's hieroglyphic studies. Intensively investigated by appropriate methods, a book like *Egyptian Oedipus* is a powerful lens, capable of bringing into focus both fine-grained details and larger patterns of the culture to which it belonged. These methods include the traditional tactics of the intellectual historian—close reading of texts, identifying sources and influences—as well as approaches associated with cultural history, such as the study of practices, social networks, and material culture. In order to make sense of Kircher's publications, I situate his thought within intellectual genealogies going back to the Renaissance and late antiquity. At the same time, I anchor his work to a specific time and place—Rome in the middle of the seventeenth century—and trace the filaments that linked his chamber in the Collegio Romano to the worldwide web of the Republic of Letters. To understand *Egyptian Oedipus* we need to understand the broader European culture of erudition, but also the distinctive strain that flourished in Rome.

Above all, this book is a study of knowledge in the making. It treats Kircher's scholarship not as a static system of ideas, but as a process, to a large extent collaborative, that unfolded over time. I have sought insight into his hieroglyphic studies in the communities and institutions that made them possible and also set their limits—libraries, museums, the Society of Jesus, the Holy Office, patrons, collaborators, and critics—and I have tried to reconstruct his scholarly practices, "right down to the lowly and delicate technical details" (to borrow a phrase of Marc Bloch).[76] This approach depends on supplementing published sources with archival evidence. The investigator of Kircheriana is fortunate: as a member of a religious order still in existence, who spent most of his life in a city dominated by the world's oldest continuously functioning institution—and one of the most bureaucratic—Kircher left behind a hefty paper trail. The chapters that follow exploit the testimony of Kircher's surviving correspondence, extant portions of the manuscript of *Egyptian Oedipus*, Jesuit administrative records, and other unpublished materials. The result, I hope, is not only a more convincing interpretation of Kircher's studies, but also a depiction of the man and his world with sufficient color, texture, and shadow to give a sense of life to the thoughts and actions of a distant age.

This book has definite chronological and thematic limits. It focuses on

---

Fletcher paved the way for modern Kircher studies. Happily, his 1966 thesis has at last been published: Fletcher, *Study of the Life and Works of Athanasius Kircher* (2011). Unhappily, it did not appear in time to be consulted in the preparation of this book.

76. Bloch, *Historian's Craft* (1953), 12.

Kircher's career from 1632, when we encounter the first traces of his study of Egyptian hieroglyphs, until 1655, when he published *Egyptian Oedipus*. During these years we can see Kircher become Kircher—the famous scholar, the "Roman Oracle," curator of curiosities, and one-man publishing machine—gradually assuming the larger-than-life persona he would wear for the rest of his life. In speaking of Kircher's hieroglyphic studies, I refer to the full range of subject matter that informed his investigation of the "hieroglyphic doctrine," including his study of Coptic and unfinished Arabic translation projects. I make no claim to offer a comprehensive study of Kircher's scholarship: only passing mention is made to his work in the mathematical sciences during these years, and his career after 1655 is omitted.[77] Even within this restricted purview, my treatment of Kircher's hieroglyphic studies is necessarily selective; but I have based my choice of topics on judgments about the significance of his project as a whole. My approach is thus very different not only from studies that have considered his hieroglyphic studies more partially, but also from those that have treated his massive oeuvre in its entirety. Inevitably, something is lost by examining only one side of a polymath, but much is gained as well. Only by limiting my scope in this way have I been able to carry out a detailed investigation of Kircher's scholarship in the making.

Chapter 1, "Esoteric Antiquarianism," situates *Egyptian Oedipus* in its most important literary contexts: Renaissance Egyptology, including philosophical and archeological traditions, and early modern scholarship on paganism and mythology. It argues that Kircher's hieroglyphic studies are better understood as an antiquarian rather than philosophical enterprise, and it shows how much he shared with other seventeenth-century scholars who used symbolism and allegory to explain ancient imagery. The next two chapters chronicle the evolution of Kircher's hieroglyphic studies, including his pioneering publications on Coptic. Chapter 2, "How to Get Ahead in the Republic of Letters," treats the period from 1632 until 1637 and tells the story of young Kircher's decisive encounter with the arch-antiquary Peiresc, which revolved around the study of Arabic and Coptic manuscripts. Chapter 3, "Oedipus in Rome," continues the narrative until 1655, emphasizing the networks and institutions, especially in Rome, that were essential to Kircher's enterprise. Using correspondence and archival documents, this pair of chapters reconstructs the social world in which Kircher's studies

---

77. Kircher published two other Egyptological works after *Oedipus Aegyptiacus*—Kircher, *Ad Alexandrum obelisci interpretatio* (1666), and Kircher, *Sphinx mystagoga* (1676)—but they added little that was new.

were conceived, executed, and consumed, showing how he forged his career by establishing a reputation as an Oriental philologist.

The next four chapters examine *Egyptian Oedipus* and *Pamphilian Obelisk* through a series of thematic case studies. Chapter 4, "Ancient Theology and the Antiquarian," shows in detail how Kircher turned Renaissance occult philosophy, especially the doctrine of the *prisca theologia*, into a historical framework for explaining antiquities. Chapter 5, "The Discovery of Oriental Antiquity," looks at his use of Oriental sources, focusing on Arabic texts related to Egypt and Hebrew kabbalistic literature. It provides an in-depth look at the modus operandi behind Kircher's imposing edifice of erudition, which combined bogus and genuine learning. Chapter 6, "Erudition and Censorship," draws on archival evidence to document how the pressures of ecclesiastical censorship shaped Kircher's hieroglyphic studies. Readers curious about how Kircher actually produced his astonishing translations of hieroglyphic inscriptions will find a detailed discussion in chapter 7, "Symbolic Wisdom in an Age of Criticism," which also examines his desperate effort to defend their reliability. This chapter brings into sharp focus the central irony of Kircher's project: his unyielding antiquarian passion to explain hieroglyphic inscriptions and discover new historical sources led him to disregard the critical standards that defined erudite scholarship at its best. The book's final chapter, "Oedipus at Large," examines the reception of Kircher's hieroglyphic studies through the eighteenth century in relation to changing ideas about the history of civilization.

THE SPACE BETWEEN

Suspended between East and West, Kircher longed to push onward from Malta to Egypt and the Holy Land. But the island in the middle of the Mediterranean was the closest he ever came physically to the lands of his learned dreams. Instead, back in Rome, immersing himself in manuscripts and antiquities preserved in the city's collections, and ruminating on the Egyptian obelisks scattered among its piazzas, he embarked on a virtual tour through Oriental antiquity, recording his discoveries in the erudite travelogue called *Egyptian Oedipus*. Even had he been able to visit the Near East, it is doubtful that the encounter would significantly have changed his ideas, shaped as they were by a mindset suspended between two ways of thinking about the past: the traditional reverence for antiquity, of which occult philosophy, with its ideal of esoteric wisdom passed on from the first age of the world, was a particular variety; and a more skeptical, empirical approach to history that developed out of critical philology and antiquarianism. Much of the tension in his work, and its historical interest, can be traced to the tug of these opposing forces.

Kircher, whose career spanned the half century before 1680, died at the cusp of a period of unusually rapid cultural mutation. The age before, precisely because of the seismic shift that came after, has challenged historical interpretation. Half a century of intensive research on the Scientific Revolution has done much to clarify the relationship between the pre-Newtonian and post-Newtonian natural sciences, but parallel developments in the human sciences remain more opaque. Posterity has not esteemed Kircher as one of the seminal figures in the genealogy of modernity—a Galileo, a Descartes, a Hobbes—but in his day he was one of Europe's most successful scholars. He embodied the contradictions of a moment when recognizably modern ways of thinking about the past had become available, yet older and conflicting models remained appealing and, to many, persuasive. As such, he allows us to explore a side of intellectual history too often lost to view. Without this view we cannot fully grasp the work of the canonical thinkers, much less understand the age on its own terms.[78]

78. See Feingold, "Grounds for Conflict" (2003), esp. 122–23.

CHAPTER ONE

Esoteric Antiquarianism

This science has its visionaries as well as all others. There are several, for example, that will find a mystery in every tooth of *Neptune's* trident, and are amazed at the wisdom of the ancients that represented a thunder-bolt with three forks, since, they will tell you, nothing could have better explained its triple quality of piercing, burning and melting. I have seen a long discourse on the figure and nature of horn, to shew it was impossible to have found out a fitter emblem for plenty than the *Cornu-copia*. These are a sort of authors who scorn to take up with appearances, and fancy an interpretation vulgar when it is natural.
—Joseph Addison[1]

KIRCHER'S HIEROGLYPHS

Egyptian Oedipus is a big book. Its dense assemblage of Latin scholarly exposition, excerpts from primary sources, images of archeological artifacts, and diagrams fills more than two thousand pages, spread over four in-folio volumes. A typographical tour de force, employing dozens of exotic fonts to display Oriental text and lavishly furnished with woodcuts and engravings, it took three years for the Roman printer, Vitale Mascardi, to see it through the press (figs. 2–5). In an age accustomed to scholarly grandiosity, its scale impressed. To the modern reader, however, it is the book's substance that most astonishes. *Egyptian Oedipus* promises a complete "restoration of the hieroglyphic doctrine," all the lost secrets of religion and science that ancient Egyptians supposedly encoded on their monuments. The massive final volume gathers almost every hieroglyphic inscription

1. Addison, "Usefulness of Medals" (1721), 447.

known to Europeans at that time, as well as other ancient artifacts, including mummies, sarcophagi, Canopic jars, sphinxes, idols, lamps, and amulets, found in Rome, other Christian cities, Istanbul, and Egypt. Kircher glosses each object with a learned explanation of its ancient significance. Without a Rosetta Stone, he translates the hieroglyphic inscriptions, character by character, into Latin prose.

But *Egyptian Oedipus* hardly confines itself to matters Egyptian. Kircher interprets the hieroglyphs by comparing Egyptian inscriptions with evidence from other traditions that supposedly preserve elements of the "hieroglyphic doctrine." The book contains extensive discussions of topics such as pagan religion from Mexico to Japan, ancient Greek esoteric texts like the Orphic hymns and Pythagorean verses, Jewish Kabbalah, Arabic magic, ancient alchemy, astrology, and astral medicine. Kircher calls his scholarly procedure the "combinatory method," because it juxtaposes diverse texts and artifacts meant to illuminate one another. Such a method is potentially sound; indeed, the comparative study of texts, inscriptions, and artifacts was one of early modern erudition's significant contributions to historical scholarship.[2] But one can't help but be struck by the inappropriate heterogeneity of the fragments out of which Kircher would reconstruct the ancient Egyptian world. To harmonize the "sacred history" of the Bible with the "profane history" of pagan civilizations, Kircher relies on a hermeneutics of symbolism and allegory. Properly interpreted, the seemingly "absurd" myths of the Greeks, Egyptians, and other heathens express a monotheistic theology that prefigures many of the tenets of Christianity. Among the many levels of meaning contained in the story of Isis and Osiris, for example, Kircher detects the doctrine of the Holy Trinity (fig. 9).[3]

To the modern reader, the historical narrative at the heart of Kircher's study—a saga of secret knowledge lost, found, and lost again, enacted by heroic wise men, devious priests, and powerful magicians—has a patently fabulous character. At the dawn of time, Kircher, explains, Adam, instructed by God and the angels, and guided by experience acquired during his centuries-spanning life, possessed perfect wisdom, which he passed on to his children. Noah and his sons preserved antediluvian knowledge from destruction by the Flood, which Kircher placed 1,656 years after the creation of the world and 2,394 years before the birth of Christ. But Noah's son Ham polluted the Adamic wisdom with magic and superstition. Eventually the great Egyptian sage Hermes Trismegistus purified the antediluvian

---

2. Miller, "Antiquary's Art" (2001). See also Burke, "Images as Evidence" (2003).
3. See Schmidt-Biggemann, "Hermes Trismegistos" (2001).

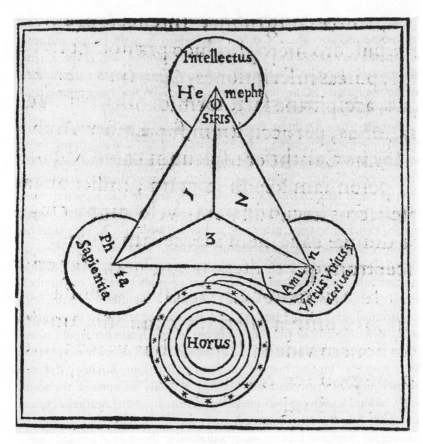

Fig. 9. Diagram of Kircher's Trinitarian interpretation of the myth of Isis and Osiris. Athanasius Kircher, *Obeliscus Pamphilius* (Rome: 1650), 213. Courtesy of Stanford University Libraries.

wisdom from its corrupt admixture, and invented hieroglyphic writing to preserve it for posterity. But later Egyptian priests mixed the Hermetic wisdom with magic and superstition, creating, yet again, an ambiguous legacy, which was passed on to other nations.[4]

Kircher undertook his investigation almost two centuries before Jean-François Champollion (1790–1832) famously solved the enigma of hieroglyphic writing. Using the bilingual Rosetta Stone as his key, Champollion demonstrated that hieroglyphs, despite their figurative form, represented the sounds of spoken language, albeit in a more complex manner than a

4. See below, ch. 4, "Kircher's Hermetic History."

purely alphabetic writing system.⁵ Modern Egyptology reveals that ancient obelisks were memorials whose inscriptions record the identity of the kings who built them and the gods to whom they were dedicated. For example, the obelisk that now stands in the private garden of the Villa Celimontana in Rome was originally erected in Heliopolis in the thirteenth century BCE by Ramesses II and dedicated to the god Horus. We know this because its inscription reads, "Horus, strong bull, beloved of Maat, *Usr-Maat-Ra Setep-en-Ra* (Ra-strong-in-truth-chosen-of-Ra), king of Upper and Lower Egypt, son of the Sun, Ramesses II" (fig. 10).⁶ Kircher, who scoffed at the idea that obelisks recorded such mundane details, interpreted the same hieroglyphs as symbols of esoteric wisdom. He arrived at a longer translation:

> Supramundane Osiris, concealed in the center of eternity, flows down into the world of the Genies, which is the most near, similar, and immediately subject to him. He flows down into the divinity Osiris of the sensible World, and its soul, which is the Sun. He flows down into the Osiris of the elemental World, Apis, beneficent Agathodemon, who distributes the power imparted by Osiris to all the members of the lower world. His minister and faithful attendant, the polymorphous Spirit, shows the abundance and wealth of all necessary things by the variety that he brings about and oversees. But since the beneficent power of the polymorphous Spirit may be impeded in various ways by adverse powers, the sacred Mophto-Mendesian table, which acquires the moist strength and fertility of the Nile, in order for the flow of good things to be performed without impediment, must be worn for protection. But since the polymorphous Spirit is not capable of thoroughly completing this task, the cooperation of Isis, whose dryness tempers the moisture of Mendes, is needed. To obtain it, the following sacred table of Osiris is composed, on which are taught the things to be done in sacrifices and the way to perform the Komasian rites. Through this table and the sight of it supramundane Osiris shows the desired abundance of necessary things.⁷

Faced with Kircher's relentless mobilization of erudition in the name of such nugatory results, one may be tempted to dismiss him as a learned

---

5. Hieroglyphic writing uses a combination of three kinds of signs: ideograms, phonetic signs, and determinatives. See Davies, *Egyptian Hieroglyphs* (1987); Beinlich, "Athanasius Kircher" (2002), 85–86.

6. Marucchi, *Gli obelischi egizi di Roma* (1898), 101–3.

7. *OA* III, 325. In rendering Kircher's obelisk translations into English, I have strived to preserve their turgidity.

Fig. 10. The obelisk of Villa Celimontana. Athanasius Kircher, *Oedipus Aegyptiacus* (Rome: 1652–54), vol. 3, 322, detail. Courtesy of Stanford University Libraries.

charlatan or a crank. More than a few readers have done so, from the seventeenth century to the present, usually emphasizing his unwavering faith in texts attributed to Hermes Trismegistus despite Isaac Casaubon's demolition of their authenticity decades earlier. And yet, despite widespread recognition of his flaws, Kircher was one of the most famous scholars of his time and his hieroglyphic studies were at the center of his reputation, both ill and good. *Egyptian Oedipus* was taken seriously by seventeenth-century scholars. Indeed, as chapter 8 documents, it informed scholarship well into the eighteenth century. If we want to understand how early modern Europeans thought about the past, we should take it seriously too.

RENAISSANCE EGYPTOLOGY

Egyptian Oedipus was the culmination of two centuries of European fascination with ancient Egypt and hieroglyphic writing.[8] The kingdom of the Pharaohs had enthralled the Greeks since Herodotus, in the fifth century BCE, described an ancient land of marvels and sacred mysteries which had passed its science and religion on to younger civilizations such as Greece.[9] As Renaissance scholars immersed themselves in Greek and Roman literature, they reactivated the classical image of Egypt, which merged with and at times submerged the biblical view of Egypt as the land of exile and idolatry. In the early fifteenth century, Italian humanists rediscovered important works about Egypt by Pliny, Herodotus, Strabo, Diodorus, and Plutarch, among others. Especially decisive in shaping the Renaissance image of Egypt was the revival of Neoplatonism, spearheaded by Marsilio Ficino. The Florentine philosopher translated and disseminated works by Plotinus and other Neoplatonists of late antiquity, who described the hieroglyphs as a symbolic language encoding the profound mysteries of ancient Egyptian sages. Plutarch's *On Isis and Osiris* and Macrobius's *Saturnalia* provided detailed allegorical interpretations of Egyptian religion that harmonized with Platonic theology and metaphysics, while Iamblichus, Porphyry, and Proclus venerated Egypt and the Orient as the homeland of true philosophy and pious magic.

Most influential were two Greek texts that found their way to Italy from the disintegrating Byzantine Empire and were purported to be translations from ancient Egyptian. In the 1420s, a manuscript of the *Hieroglyphica* of

---

8. The essential study of developments through the mid-sixteenth century is Curran, *Egyptian Renaissance* (2007), to which I am greatly indebted. See also Iversen, *Myth of Egypt* (1993); Dannenfeldt, "Egypt and Egyptian Antiquities" (1959).

9. Burstein, "Images of Egypt" (1996).

Horapollo was discovered on the island of Andros and brought to Florence. Attributed to an Egyptian priest—but most likely composed in the fourth century CE by a Greek author with little knowledge of the Egyptian language— the *Hieroglyphica* was a guidebook to interpreting hieroglyphic symbolism. Translated into Latin and various vernaculars, Horapollo saw many editions, exerting a profound influence on the European understanding of the hieroglyphs.[10] In 1460, at the command of Cosimo de' Medici, Ficino interrupted his translation of Plato to undertake a Latin version of another recently discovered manuscript, which promised to reveal the ultimate source of Platonic wisdom. The *Corpus Hermeticum* was a collection of treatises on philosophical and theological themes, believed to have been written in remote antiquity by the legendary Egyptian sage Hermes Trismegistus. In reality, these texts were composed in Greek in the first centuries of the modern era, which is why their views so uncannily "anticipate" Platonist and Christian teachings from the same milieu.[11] The Hermetic Corpus became the locus classicus for the Renaissance conception of Egyptian wisdom, establishing Hermes Trismegistus as a key figure in the pious pagan tradition anticipating Christianity, known as the *prisca theologia*, or ancient theology.[12]

Simultaneously with the literary encounter, humanist scholars, especially in Rome, began systematically to investigate the physical remains of the ancient world, which included a significant quantity of Egyptian material. Most spectacular were the obelisks that Augustus and subsequent emperors had carried across the Mediterranean to display as symbols of imperial majesty, eventually adorning the ancient city with close to fifty monuments. By the beginning of the sixteenth century, only the Vatican obelisk still stood, but the remains of others lay scattered about the town. Statues of sphinxes, lions, and smaller artifacts were also to be found both above and below ground, legacies of imperial Rome's infatuation with Egyp-

---

10. Boas, *Horapollo* (1993); Giehlow, "Hieroglyphenkunde" (1915); Sider, "Horapollo" (1960–92).

11. Copenhaver, ed., *Hermetica* (1992); Gentile and Gilly, eds., *Marsilio Ficino* (1999); Yates, *Giordano Bruno* (1964).

12. *Corpus Hermeticum* refers specifically to sixteen Greek treatises (Ficino translated fourteen, and two more were added to sixteenth-century editions). In early modern Europe they were often referred to collectively as *Pimander*, the title Ficino gave to the first dialogue in his collection. Their reception and publication history was tied to another Hermetic treatise, the *Asclepius*, which only survived in its ancient Latin translation. I use the term "Hermetic Corpus" to refer to the *Corpus Hermeticum* and *Asclepius* collectively. I use "Hermetica" to refer to all ancient and medieval writings attributed to Hermes, which greatly exceed the Hermetic Corpus. For the reception of Hermetic literature from antiquity to the twentieth century, see Ebeling, *Secret History* (2007).

tian culture which gave rise to the Temple of Isis in the Campus Martius. Over time, the provenance of these objects was forgotten. Most famously, the Vatican obelisk was long mistaken for a funerary monument bearing the ashes of Julius Caesar. It was not until the 1420s that Italian antiquaries identified these monuments as Egyptian and their inscriptions as the famous "sacred letters" described by Pliny and Ammianus Marcellinus, and explained by Horapollo.[13]

These literary and material encounters inspired a Renaissance tradition of Egyptology focused on hieroglyphs, with two chief dimensions: one philosophical, centered on Hermetic wisdom and hieroglyphic symbolism, and one archeological, involving the study of Egyptian antiquities.[14] The influence of Horapollo and the Neoplatonists was profound but paradoxical. Their accounts of hieroglyphic symbolism were widely taken as authoritative, yet they failed miserably as a key to decipher actual inscriptions. None of the ancient works explained how individual symbols worked together to convey complex ideas, and hardly any of Horapollo's hieroglyphs corresponded to those on Rome's monuments. Instead, inspired by the compelling but erroneous notion of hieroglyphs as a philosophical language, universally intelligible to the learned elite, humanists like Leon Battista Alberti devised new, modern "hieroglyphs." The result was a tension between the symbolic and archeological approaches: the philosophical theory that hieroglyphs were universally intelligible was irresistible, but it was awkwardly belied by experience of actual Egyptian inscriptions, which remained inscrutable. Cognitive dissonance was contained not by revising the theory but by avoiding confrontation. For the most part, writers on "hieroglyphic" symbolism, although often avid investigators of other antiquities, declined to study genuine Egyptian inscriptions, while students of Egyptian epigraphy refrained from interpretation, contenting themselves with collection and description.[15]

The philosophical approach generated the extraordinarily popular early modern emblem tradition, encompassing several codified genres of symbolic devices including emblems, *imprese*, heraldry, and "hieroglyphs." Among the most influential works in this field was Pierio Valeriano's *Hieroglyphica* (1556), a work of sixty chapters, each explaining a different "hieroglyphic"

13. Curran, *Egyptian Renaissance* (2007), chs. 2–3.
14. Since Karl Giehlow's groundbreaking work a century ago, scholars of Renaissance Egyptology have emphasized the philosophical and emblematic approach to hieroglyphic studies over the more empirical tradition. A major achievement of Curran's study is to reconstruct the history of Egyptological epigraphy and archeology from the late fifteenth to mid-sixteenth centuries, and to integrate this material with the history of the symbolic tradition.
15. Curran, *Egyptian Renaissance* (2007), esp. 76.

symbol.¹⁶ Valeriano thought of hieroglyphs not as specifically Egyptian, but rather as a "divine language of symbols" common to many ancient cultures. Despite being a serious student of ancient material culture, when it came to authentic Egyptian hieroglyphs, Valeriano adopted what Brian Curran calls a "hands-off" attitude.¹⁷ Frequently reprinted and translated, Valeriano's work soon rivaled Horapollo's treatise, with which it was often bound, as the standard guide to hieroglyphic symbolism. Enthusiasm for hieroglyphs and emblems—the two terms were practically synonyms—spread quickly from its Italian epicenter to the rest of Europe, producing an astounding number of publications during the early modern period.¹⁸ Although in practice, modern hieroglyphs and emblems often communicated pedestrian didactic messages, their authority was rooted in the idea that the ancients had communicated profound wisdom by means of symbols.¹⁹ There was considerable overlap between emblem literature and the mythographic manuals that appeared around the same time, which explained the symbolic meaning of ancient pagan gods, paying considerable attention to Egyptian deities.²⁰

Typical of the sixteenth century's expansive conception of the hieroglyph was Jan van Gorp's *Hieroglyphica*, posthumously published in Antwerp in 1580. Gorp's work (which argued that of all surviving languages, Flemish was closest to the primordial language of Adam) treated symbolism in general and focused more on Hebrew than on Egyptian material. The most successful seventeenth-century addition to the genre of the hieroglyphic dictionary was *On the Symbolic Wisdom of the Egyptians* by the Jesuit Nicolas Caussin. True to the genre, and despite its title, Caussin's work was not about specifically Egyptian material. Like Valeriano, Kircher respectfully observed, it treated "not so much the hieroglyphic doctrine of the ancients, as the emblematic doctrine assembled from the histories of all ages."²¹

Parallel to the proliferating literature on hieroglyphic symbolism, the study of Egyptian antiquities carried on, primarily in Rome, as part of the humanists' ongoing investigation of the ancient world's material remains.

---

16. Valeriano, *Hieroglyphica* (1556).
17. Curran, *Egyptian Renaissance* (2007), 227–34. See also Curran, "Reticence and Hubris" (1998/1999).
18. Manning, *The Emblem* (2002); Praz, *Studies in Seventeenth-Century Imagery* (1964).
19. Seznec, *Survival of the Pagan Gods* (1995), 102–3.
20. The most important of these, all by Italians, appeared at midcentury: Giraldi, *De deis gentium* (1548); Conti, *Mythologiae* (1551); and Cartari, *Imagini delli Dei de gl'Antichi* (1556). See Seznec, *Survival of the Pagan Gods* (1995); Duits and Quiviger, eds., *Images of the Pagan Gods* (2009).
21. Gorp, *Hieroglyphica* (1580); Caussin, *De symbolica Aegyptiorum sapientia* (1631); *OP*, 117.

Already in the first half of the fifteenth century the pioneering Roman antiquary Flavio Biondo had called attention to the city's ruined obelisks and their inscriptions. Later in the century, the historian and forger Annius of Viterbo pursued the archeological study of supposedly hieroglyphic inscriptions, although the chief objects of his investigation were not of genuine Egyptian provenance. By the 1480s, however, other scholars, less famous but more rigorous, took to the field and made copies of authentic hieroglyphic inscriptions. By the middle of the sixteenth century, the sketchbooks of antiquaries like Pirro Ligorio were filled with images of Egyptian objects, including accurate renderings of hieroglyphs. These labors, however, seem to have had more influence on visual artists, notably Raphael, than scholars.[22]

During the papacy of Sixtus V (1585–90), Renaissance Egyptology gave rise to a remarkable program of archaeology and urban renewal. Inspired by Counter-Reformation zeal, Sixtus set out to purge Rome's public spaces of residual paganism, in part through demolition, but mostly through reappropriation. Superstitious monuments, cleansed by exorcism, crowned with crosses or statues of saints and ornamented with pious inscriptions, would be transformed into symbols of Catholicism's inexorable triumph over its enemies. Thus Sixtus revived the tradition of his imperial forebears, adorning the Eternal City with renovated Egyptian obelisks. First, in 1586, he moved the one standing obelisk from its location at the site of the ancient Vatican circus to a more prominent location in St. Peter's Square. In the remaining years of his pontificate, Sixtus erected three more obelisks, in front of the churches of Santa Maria Maggiore, San Giovanni Laterano, and Santa Maria del Popolo. Over the following centuries, other popes and private citizens followed suit, and today thirteen ancient obelisks adorn Roman piazzas.[23] The continuity with the past that these projects sought to evoke was captured in the timeline at the beginning of Kircher's *Pamphilian Obelisk*, which began immediately after the biblical Flood and united Egyptian pharaohs, Roman emperors, and Catholic popes in a single tradition.[24]

Michele Mercati's *On the Obelisks of Rome*, published in 1589 on the occasion of the renovation of the Vatican obelisk, was the first systematic treatise devoted specifically to Egyptian antiquities. In his subject matter, Mercati was an important precursor to Kircher, but methodologically he

22. Curran, "Reticence and Hubris" (1998-99), 172–77; Curran, *Egyptian Renaissance* (2007), 99–105, 237–43. On the influence of Ligorio's manuscripts in the Barberini circle, see Russell, "Pirro Ligorio" (2007).

23. Curran et al., *Obelisk* (2009); Cipriani, *Gli obelischi* (1993) ch. 1; Iversen, *Obelisks in Exile* (1968).

24. *OP*, e3^r–f3^v.

came closer to the reticent attitude of the eighteenth-century Egyptologists, who will be discussed in chapter 8. Mercati was most comfortable describing what he could observe directly or report from ancient authorities. Pessimistic that hieroglyphs could ever be deciphered, he thought that as much as could be known had to be found in ancient texts. On the testimony of Hermes Trismegistus, Clement of Alexandria, Iamblichus, and others, he affirmed that ancient Egyptian priests had invented hieroglyphs as a symbolic language to record their sacred sciences, especially astrology and magic, at which they excelled. Although he offered some information about individual symbols, he did not attempt to interpret surviving inscriptions.[25] In other words, Mercati did not disagree significantly with Kircher about the nature of hieroglyphic wisdom, but his closeness to the physical evidence seems to have discouraged him from speculative interpretations.[26]

By the beginning of the seventeenth century, there were signs of a rapprochement between Egyptology's archeological and philosophical traditions. In 1605, the Paduan antiquary, Lorenzo Pignoria, published a treatise on the famous *Mensa Isiaca* or Bembine Table, discovered in Rome in 1525, which depicted Egyptian cult scenes (figs. 11, 26). Unlike Mercati, Pignoria was willing to hazard an interpretation of the table's symbols, but his identifications of individual figures were explicitly tentative, and he did not attempt to explain how they related to one another semantically. As with Mercati, Pignoria's style of empiricism set boundaries to his hermeneutic horizon. Another noteworthy effort to explain the Bembine Table was not characterized by such restraint. Around 1610, Johann Georg Herwart von Hohenburg published a pioneering atlas of hieroglyphic inscriptions, including the *Mensa Isiaca*. In 1626, Herwart's son published his father's interpretation of the table, which was based on the unique and much-derided thesis that the central mystery of ancient symbolic wisdom was the science of magnetism and navigation by compass.[27] It is worth noting that these early efforts at interpreting hieroglyphs focused on an inscription that we now know not to be hieroglyphic at all, consisting of purely iconic images.

The contradiction between the prevalent theory of how to read hieroglyphs symbolically and the recalcitrance of actual Egyptian inscriptions, which had previously kept a distance between discussions of "hieroglyphic" wisdom and

---

25. Mercati, *De gli obelischi* (1589); Curran et al., *Obelisk* (2009), 154–55.
26. A parallel case was Kircher's English contemporary, John Greaves, whose in situ study of Egyptian monuments, *Pyramdiographia* (1646), eschewed interpretation of hieroglyphic inscriptions. See Shalev, "Measurer of All Things" (2002).
27. Pignoria, *Vetustissimae tabulae explicatio* (1605); Herwart von Hohenburg, *Thesaurus Hieroglyphicorum* ([1610]); Herwart von Hohenburg, *Admiranda ethnicae theologiae* (1626).

Fig. 11. Lorenzo Pignoria was an important predecessor of Kircher, who studied Egyptian hieroglyphs in an antiquarian context. This engraving is from a seventeenth-century reprint of his interpretation of the Bembine Table, which included an appendix with commentaries on ancient amulets by Kircher and Jean Chifflet. Lorenzo Pignoria, *Mensa Isiaca* (Amsterdam: 1669), frontispiece. Courtesy of Stanford University Libraries.

studies of actual Egyptian antiquities, had not gone away. In the case of Pignoria's attempt to bridge the gap, the tension was not overly exacerbated due to the cautious and provisional nature of his explanations. As for Herwart, his interpretative strategy was so idiosyncratic—reading the *Mensa Isiaca* as a global nautical map—that no one was likely to take it as a test of the standard theory of hieroglyphic symbolism. It was Kircher who would fully unite Egyptology's archeological and philosophical traditions, if only fleetingly, by conducting an empirical investigation, unprecedented in scope, of ancient Egyptian remains, which he explained in terms of Hermetic wisdom and Neoplatonic symbolism—not tentatively, but (so he claimed) definitively. To achieve this union, Kircher had no choice but to resort to brute force; that is to say, scholarly rigor had to give way before the hieroglyphs would yield up their secrets. In *Egyptian Oedipus* the old tension would not be resolved but be taken to the breaking point. Prior studies have rightly placed Kircher's hieroglyphic studies in the context of Renaissance Egyptology, but they have overemphasized the philosophical dimension at the expense of the archeological, even though it is precisely the conjunction of the two that defines their character. The result has been a lopsided portrayal of Kircher and his project, which obscures his relationship to seventeenth-century scholarship.

THE HERMETIC PHILOSOPHER

In her classic study, *Giordano Bruno and the Hermetic Tradition*, Frances Yates described Athanasius Kircher as a "reactionary Hermetist." She meant that Kircher tenaciously held to the Renaissance Neoplatonists' belief in the legendary Egyptian sage Hermes Trismegistus and the tradition of pious pagan wisdom that he symbolized even after Isaac Casaubon's 1614 critique demonstrating the *Hermetica* to be products of the early Christian era rather than testaments from the time of Moses or earlier.[28] This image has dominated modern interpretations of Kircher, which have explained his work—especially his studies of the hieroglyphs and ancient wisdom—by appealing to the Renaissance "Hermetic tradition" and by placing Kircher in the context of a Counter-Reformation culture, which supposedly clung to Hermeticism after it had lost currency elsewhere.[29]

---

28. Yates, *Giordano Bruno* (1964), 416–23. On Casaubon's critique, see Grafton, *Defenders of the Text* (1994), chs. 5 and 6; Mulsow, ed., *Ende des Hermetismus* (2002). I discuss Kircher's response to Casaubon in chapter 7.

29. Kircher is described as a "reactionary" Hermeticist by Pastine, *Nascita dell'idolatria* (1978), 14; Rivosecchi, *Esotismo in Roma barocca* (1982), 54; Rossi, *Dark Abyss of Time* (1984), 123.

Giovanni Cipriani, for example, builds on Yates in his study of obelisk renovation in baroque Rome. Devoting the book's second half to "the age of Kircher," Cipriani describes *Egyptian Oedipus* as the culmination of a tradition of "Ficinian Hermeticism" within early modern Catholicism, and explains Kircher's investigation of the hieroglyphs as an ideological enterprise to support the Counter-Reformation Church's claim to universal authority. He likens Kircher to Francesco Patrizi, who in the late sixteenth century hoped to reform Catholic learning by replacing Aristotelianism with Hermetic philosophy. Dino Pastine likewise places Kircher at the forefront of a seventeenth-century revival of religious Hermeticism, describing his scholarship as "a vast apologetic production intended to give theoretical support to the Society's effort to expand the boundaries of the Catholic world."[30] R. J. W. Evans paints another portrait of Kircher as a Catholic Hermetic philosopher, concluding his study of early modern Habsburg culture with a summary of *Egyptian Oedipus*, which he describes as the embodiment of the intellectual ideals of the Counter-Reformation.[31] Variations on this theme can found in many other studies.[32] It has come to constitute the received view.[33]

Like Ficino, Patrizi, and other Neoplatonists of the fifteenth and sixteenth centuries, Kircher believed in a pre-Christian tradition of esoteric wisdom shared by the wise men of different pagan cultures, and he placed the Egyptian sage Hermes Trismegistus at its head. Furthermore, Kircher's reconstruction of the *prisca theologia*—despite its frequent unorthodoxy—resonated with the global ambitions of the Catholic Church by demonstrating mankind's common spiritual past and the vestiges of truth preserved among gentiles. By interpreting *Egyptian Oedipus* as an ideologically motivated defense of Hermetic philosophy, these studies make its appeal comprehensible while marginalizing its most outlandish component, the translations of hieroglyphic inscriptions, which turn out to have been a kind of pretense: they were not the goal of Kircher's project, but, as Valeriano Rivosecchi puts it, "the most spectacular means he could imagine to re-launch a philosophical tradition." Even Erik Iversen, whose treatment of Kircher has the merit of focusing on his archeology, insists on the predominantly

---

30. Cipriani, *Gli obelischi* (1993); Pastine, *Nascita dell'idolatria* (1978), 22–27; quotation at 22.

31. Evans, *Making of the Habsburg Monarchy* (1979), esp. 433–42.

32. E.g., Iversen, *Myth of Egypt* (1993), 92–102; Rivosecchi, *Esotismo in Roma barocca* (1982), esp. ch. 2; Frigo, "Il ruolo della sapienza egizia" (2007).

33. Stausberg, *Faszination Zarathushtra* (1998), 481–84, for example, relies on Cipriani, Rivosecchi, and Evans and describes Kircher's Egyptian studies as motivated by the goal of legitimating the religious and political claims of the papacy.

ideological character of Kircher's project, declaring that, compared to his primary philosophical and theological goals, "the actual interpretation of the hieroglyphs as such became of secondary importance."[34]

Despite its prima facie plausibility, this view of Kircher's hieroglyphic studies has serious weaknesses. To begin with, it exaggerates the role of Neoplatonist occult philosophy in early modern Catholicism, while underestimating its enduring role in European culture more generally.[35] The official endorsement of Thomism by the Tridentine Church, and by the Society of Jesus in particular, rendered occult philosophy problematic, especially after the provocations of Francesco Patrizi and Giordano Bruno associated Hermeticism with anti-Aristotelianism and heresy.[36] Furthermore, if Kircher's chief aim was to provide ideological sustenance to Jesuit missionary endeavors, it is odd that nowhere in his hieroglyphic studies does he make that explicit. Although the claim that pagan traditions contained hidden truths compatible with Christianity had general proselytizing value, Kircher focused on a region, Egypt, that was of small relevance to Jesuit missionaries. As David Mungello observes, the relatively marginal role that Kircher assigned China in his universal history was at odds with the more generous assessment of his colleagues in the China mission.[37] Nothing prevented Kircher from producing a history of esoteric wisdom oriented toward China and other major sites of evangelism, had that been his priority. In fact, later Jesuits rewrote Kircher's history in just this way, substituting China for Egypt in the genealogy of pagan wisdom (see figs. 40, 41).[38]

But the chief limitation of these studies is the assumption that Kircher's hieroglyphic studies were essentially a philosophical project. This is true also of some authors who do not describe his hieroglyphic studies as an effort to promote Hermetic philosophy. Thomas Leinkauf, for example,

---

34. Rivosecchi, *Esotismo in Roma barocca* (1982), 52; Iversen, *Myth of Egypt* (1993), 96.

35. The traditional Christian view of the Hermetic Corpus and the Sibylline Oracles, which held them to be ancient testimonies in which pagans, despite themselves, predicted the coming of Christianity, led Catholic ecclesiastical historians to defend their antiquity, but not necessarily to esteem their philosophical and theological content. In certain times and places, interest in Neoplatonist philosophy and magic could mesh with the agenda of the Counter-Reformation Church, as R. J. Evans argues was the case among Catholic elites in the Holy Roman Empire (Evans, *Making of the Habsburg Monarchy* [1979]). But this affinity was in tension with other tendencies in the Tridentine Church. It is surely too much to describe the *Corpus Hermeticum* as "a cornerstone of Counter-Reformation culture" (uno dei cardini della cultura controriformista); Marrone, *I geroglifici* (2002), 48.

36. Leijenhorst, "Francesco Patrizi's Hermetic Philosophy" (1998); Yates, *Giordano Bruno* (1964); Rowland, *Giordano Bruno* (2008).

37. Mungello, *Curious Land* (1985), 172–73.

38. See below, chapter 7, "The Persistence of Occult Philosophy."

without a doubt one of Kircher's most profound readers, is less interested in authorial intentions than in the underlying structure of Kircher's thought. He finds its key not in Hermeticism or occult philosophy per se, but in the nexus of a Lullist combinatory method and a Cusanian monist metaphysics. Leinkauf's study, which treats the totality of Kircher's publications as an intellectual system, is rich with insights.[39] But I would question whether a work like *Egyptian Oedipus* is best understood in philosophical terms. What if the predominant view of Kircher's hieroglyphic studies has it backward? Might occult philosophy have been the *means*, and deciphering the hieroglyphs Kircher's *end*?

THE ESOTERIC ANTIQUARY

One day in 1654 a noble stranger arrived at the door of the Collegio Romano bearing a mysterious gem, which he asked to be delivered to Athanasius Kircher. The stone had been found during the construction of a church in Assisi and was carved with Greek letters and secret symbols. Although the stranger had been to every city in Italy showing it to scholars of Greek, no one could make sense of its inscription. Kircher gave the stone one glance and deciphered its meaning. It was a Gnostic amulet, he pronounced, representing the solar genie (fig. 12). The Greek letters had to be read isopsephically—that is, each letter stood for a number—hence the inability of ordinary Greek scholars to comprehend them. Kircher took out his pen, resolved the letters into their corresponding numeric values, and dashed off an authoritative interpretation.[40]

This anecdote appears in the preface that Kircher's assistant Kaspar Schott contributed to *Egyptian Oedipus*, closely supervised, no doubt, by the author. It is part of the book's extensive apologetic apparatus, carefully crafted to establish Kircher's authority and condition the reader favorably to the work that followed. Schott paints a vivid portrait of the master at work, reflecting the scholarly persona that Kircher had forged over the preceding decades. Kircher appears as a distinctive kind of antiquary: an expert in the antiquities of Egypt and the Near East, specializing in artifacts inscribed with unusual characters and pertaining to magic or ancient mysteries, the domain that he called "recondite antiquity." (My term, "esoteric antiquarianism," is meant to evoke the Latin expressions used by Kircher, such as "antiquitas recondita" and "antiquitas arcanissima.")

---

39. Leinkauf, *Mundus Combinatus* (1993).
40. Kaspar Schott, "Benevoli Lectori," *OA* I, c4^v.

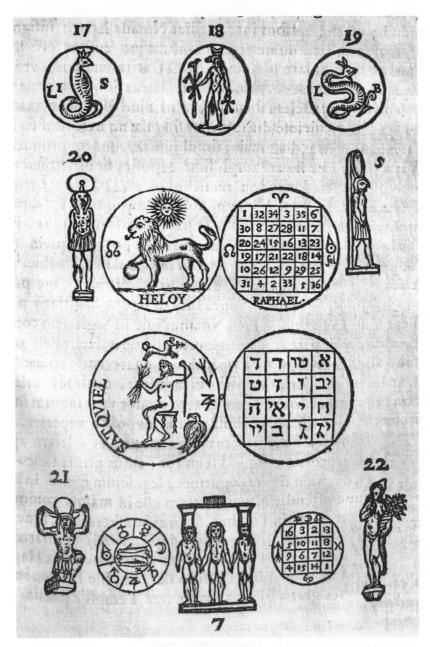

Fig. 12. "Gnostic" amulets. Kircher was renowned for his expertise in explaining ancient objects relating to magical practices. Athanasius Kircher, *Oedipus Aegyptiacus* (Rome: 1652–54), vol. 2, part 1, 465. Courtesy of Stanford University Libraries.

Kircher, writes Schott, was trained not only in Latin, Greek, and Hebrew, but also in Arabic, Chaldean, Syriac, Armenian, Coptic, and other languages, "so that for many years now he speaks not only with Greeks in Greek, with learned Hebrew rabbis in Hebrew, but with Arabs and foreigners from other provinces of Asia and Africa, of which the number here in Rome at any time is enormous, in each one's mother tongue." He goes on to describe Kircher's years of labor searching exotic literature for clues to the hieroglyphic doctrine, hunting down and transcribing texts in the libraries of Germany, France, Malta, and Italy. After studying these texts, often "half-eaten, beset by dust and cockroaches and worse, or written in nearly illegible characters," Kircher visited the piazzas, workshops, collections, and museums where Egyptian monuments were preserved. Not content with a single viewing, he returned again and again, checking to make sure that his notes and sketches "did not deviate even the least from the inscriptions." Not satisfied by Rome's bounty, he instructed correspondents to send him information about the obelisks of Constantinople and the ancient monuments in Egypt, and he appealed to scholars and aristocrats throughout Europe and the world to send him relevant material. Thereby,

> he acquired a vast quantity of hieroglyphs, idols, images, gems, amulets, periapts, talismans, stones, and similar things, which are incorporated throughout this work.... Meanwhile, he compares the noted passages of the books with other passages, hieroglyphs with hieroglyphs, entire obelisks with obelisks, the characters on statues, vases and tablets among each other and with the characters on the obelisks.[41]

These studies had made Kircher an expert in enigmatic antiquities; not only had he interpreted "the mysteries of the Egyptians," Schott specifies, but "with equal dexterity and erudition he studies and interprets the secrets of the Greeks, the amulets of the Gnostics, the arcana of the Kabbalists, the phylacteries of the Arabs, the antidotes of the Saracens, and lastly the characters, seals, delusions, superstitions, and frauds of all the deceivers." Kircher's expertise as an esoteric antiquary earned him widespread esteem, as testified by the "countless letters sent to him by princes and learned men," which Schott had seen in Kircher's archive:

> What may I say about Kircher's rare expertise in reading and interpreting inscriptions in any language and written in whatever sort of character,

---

41. Schott, "Benevoli Lectori," *OA* I, c2^v–c3^v.

however exotic, intricate, or corrupt . . . [I]t can hardly be said how many inscriptions, sacred, profane, superstitious, magical and even diabolical, carved or written on temples, palaces, cemeteries, tombs, stones, books, swords, knives, statues, coins, gems, and amulets, have been brought to him from all the parts of the world, in order to be interpreted.[42]

Although idealized, there was much truth to Schott's portrait. Many of the "countless letters" Schott mentions survive, revealing a wide circle of admirers who viewed Kircher in terms not so different from the disciple's eulogy. For example, one story Schott tells is about a nobleman who sent Kircher a wax cast of a seal with strange characters. The accompanying letter is extant in the archive of the Gregorian University and corroborates Schott's account. Its author, Philibert de la Mare, a senator from Dijon, wrote:

> In vain did I seek an interpreter among the Dutch, who for some time seem to have arrogated to themselves dictatorship of the Republic of Letters; in vain among the conceited Swedes . . . nor did I meet more success among my own people, although there are many who excel in the knowledge of foreign languages. You, then, seemed to me to be the one who could untie the knots and riddles of this Sphinx, you who have undertaken the science of recondite antiquity with hard-sought knowledge of Oriental languages.[43]

Upon receiving the object, Schott informs us, Kircher marveled that no one could decipher what was a mere trifle, as he demonstrated by promptly producing an interpretation. There are too many similar examples, Schott concludes, to recount all of them. Indeed, Kircher "himself acknowledges that he has dispatched so many interpretations to every part of the world that if they were collected together they would fill an entire volume."[44]

It is not hard to see how *Egyptian Oedipus* might be read as an apology for occult philosophy. As depicted by Schott, Kircher was preoccupied with

---

42. Schott, "Benevoli Lectori," *OA* I, c4^v–d1^r.
43. " . . . frustra tamen apud Batavos qui iampridem sibi reipublicae litterarum dictaturam arrogare videntur quaesivi interpretem, frustra apud Suecos . . . tumentes, non <felicius> denique mecum actum <est> apud nos, quamquam non pauci sint qui peregrinarum idiomatum scientia valeant: tu unus ergo mihi visus is qui Sphingis huius nodos <et> aenigmata posses expedire, quippe qui exquisita Orientalium Linguarum notitia addisti reconditoris antiquitatis scientiam." Philibert de la Mare to Kircher, Dijon, 26 January 1654, APUG 568, fol. 18^r.
44. Schott, "Benevoli Lector," *OA* I, d1^r.

esoteric lore and magical practices. He relied massively on ancient and Renaissance texts associated with occult philosophy, and the "hieroglyphic doctrine" at the heart of his investigation was simply a version of the Neoplatonists' *prisca theologia*. But the scholar described by Schott was no occult philosopher in the style of Ficino, Pico, Agrippa, or Patrizi. His principal aim was not (as Iversen would have it) "to prove the timeless universality of the [hieroglyphs'] underlying truth and at the same time to demonstrate the basic religious conformity between the Egyptian, the Greek, and his own cosmological conceptions."[45] The scholarly persona that Kircher presented to readers of *Egyptian Oedipus* was something different: an expert interpreter of the material remains of ancient civilizations, sought out by eminent collectors who required not philosophical nor theological illumination, but historical explanations of mysterious inscriptions. Kircher's virtues, as enumerated by Schott, were precisely those of the antiquary: mastery of ancient languages, skill in deciphering manuscripts in poor condition and inscriptions carved in unusual characters, the meticulous study of physical evidence, and the use of comparison to bring this diverse evidence together. Schott and de la Mare described Kircher's distinctive achievement as the combination of knowledge of "recondite antiquities" and expertise in Oriental languages, which allowed him to decipher the most difficult of ancient inscriptions. In Kircher's hands, late antique and Renaissance texts about magic and esoteric wisdom provided an explanatory framework to interpret ancient objects.

In describing the interpretation of hieroglyphic inscriptions as the end rather than the means of Kircher's project, I do not mean to suggest that he practiced pure scholarship divorced from worldly concerns, or that a work as capacious as *Egyptian Oedipus* had but a single purpose—only that the ambition of deciphering the hieroglyphs provided the driving force of Kircher's twenty-year effort. Erudite historical research had cultural prestige because of its perceived value in supporting early modern political and religious agendas. The study of Egyptian monuments, in particular, was sustained by the notion that it would confer legitimacy and authority on contemporary Catholic rulers, by establishing a connection to the Pharaonic and Roman past while simultaneously affirming the triumph of Christianity.[46] These ideological associations are the key to understanding the motivations of the popes, emperors, and other notables who supported Kircher's research. But Kircher stands out for the secondary role that religious apologetics played in

---

45. Iversen, *Myth of Egypt* (1993), 94.
46. Curran, *Egyptian Renaissance* (2007); Collins, "Obelisks as Artifacts" (2000).

his investigation of ancient religion.[47] In pursuing the hieroglyphs, I would argue, he was driven primarily by the ambition to solve a high-profile scholarly desideratum.

It is hardly possible to overestimate the role that the quest for fame played in Kircher's career, notwithstanding the Society of Jesus's insistence on modesty and self-abnegation. One of Kircher's most perceptive critics, Bishop William Warburton, hit the mark when he associated Kircher's less than rigorous handling of evidence with his excessive desire to achieve "the glory of a Discoverer."[48] As the most perplexing, and therefore challenging, ancient inscriptions known to early modern scholars, the hieroglyphs represented something like the antiquarian equivalent of the longitude problem in navigation. Or, to make a more contemporary comparison, though one with origins in Kircher's century, translating the hieroglyphs promised to confer a glory similar to that of solving Fermat's last theorem after centuries of failed attempts. By solving this great enigma, Kircher sought to confirm the reputation he had fostered as an expert in Oriental languages and a master of interpreting the most recalcitrant inscriptions, artifacts, and texts, especially those written in unusual alphabets or pertaining to magical practices.

SCHOLARS AND SYMBOLS

Recent scholarship has shown that early modern erudition's primary achievement was the development of methods for grounding the study of the past in solid evidence.[49] Philologists refined the art of textual criticism, allowing them to assess the reliability of ancient sources and debunk bogus documents, as Casaubon did with the Hermetic Corpus. And antiquaries developed parallel techniques for the study of nonliterary evidence, using ancient artifacts to fill in the gaps and correct the errors in ancient authors—for example, revising Livy's chronology of Roman consuls in the light of inscriptions discovered in the Roman Forum.[50] There are obvious ways in which Kircher may not seem to fit this mold. He based his hieroglyphic studies on ancient texts that the scholarly community had rejected as spurious, and, although he studied ancient material culture, evidently it did not lead him to sounder conclusions about Egyptian history. But these deficiencies should not blind us to how much his project shared with the

---

47. I develop this argument in later chapters, especially ch. 8.
48. Warburton, *Divine Legation of Moses* (1846), vol. 2, 196.
49. See above, introduction, "The Age of Erudition."
50. Stenhouse, *Reading Inscriptions* (2005), ch. 4.

leading erudite scholarship of his time. The basic premises and methods of Kircher's hieroglyphic studies were widely shared, especially by scholars of ancient iconography and pagan religion.

Early modern antiquaries found themselves confronted by a bewildering array of perplexing imagery. Egyptian hieroglyphs were not unique in this respect, but posed problems similar to the Greek and Roman sculptures, bas-reliefs, vases, medallions, and other ancient objects bearing images that regularly emerged from the Mediterranean soil.[51] The challenge was not only to identify what a given image depicted (for example, recognizing a god by his attributes), but also to discern what larger message a complex composition might communicate. Seeking insight into ancient iconography, antiquaries, being humanists, turned to texts for guidance. A rich body of literature that seemed to suit their evidence lay conveniently at hand. The allegorical tradition of mythography, which interpreted the stories of the ancient gods as profound poetic wisdom about nature, ethics, and theology, originated among pagan philosophers, was endorsed by some fathers of the church, and prospered in the Middle Ages.[52] The rediscovery of its most important ancient sources, especially the Neoplatonist philosophers, ensured its popularity in the Renaissance. The vast literature on emblems and other symbolic devices, which had mushroomed during the sixteenth century, offered another indispensable resource. Both traditions fueled a widespread baroque mentality that celebrated symbols as a privileged form of communication. Inevitably, antiquaries adopted symbolic interpretation as their most typical strategy for explaining complex imagery.[53] Casting a critical gaze on erudition at the turn of the eighteenth century, Joseph Addison identified the "mystical antiquary," who attributed vast wisdom to the ancients through immoderate symbolic and allegorical interpretations, as representative of antiquarianism in general.[54]

Not all symbolic interpretations were the same, however. Within the lingua franca of early modern symbolism it is useful to distinguish between two broad schools of thought. What we might call the strong theory of

---

51. Burke, "Images as Evidence" (2003); Haskell, *History and Its Images* (1993); Barkan, *Unearthing the Past* (1999).

52. Brisson, *How Philosophers Saved Myths* (2004); Seznec, *Survival of the Pagan Gods* (1995). Allegorical *interpretation* of ancient myth and literature, which presupposed that ancient authors had hidden their meaning beneath allegories, has a history distinct from that of the intentional *production* of allegorical art and literature, though of course the two traditions influenced one another.

53. Allen, *Mysteriously Meant* (1970), esp. ch. 9.

54. Addison, "Usefulness of Medals" (1721, but composed in the 1690s or 1700s), 448. See also the quotation at the top of this chapter.

symbolism asserted that the universe was filled with symbols that referred to higher things. These symbols, which could be found in nature, history, Holy Scripture, and even pagan literature, were not established by convention but possessed an ontological connection to their referents. They were one of God's chief means of communicating to men. This idea had deep roots. Originating in ancient Stoicism and Neoplatonism, it was transmitted to Christian Europe by numerous channels, most importantly through the writings attributed to Dionysius the Areopagite. To the adherent of the strong theory, the allegorical meaning of ancient myths reflected the profound wisdom of the most ancient wise men who had invented them. Homer was, truly, a theologian.[55] In a classic study, E. H. Gombrich linked the rise of emblems and allegorism in the Renaissance to the Neoplatonist theory of the mystical symbol. But Gombrich also identified a countertradition, represented by Cesare Ripa's *Iconologia* (1593), which we might call the soft theory of symbolism. Framed in terms of classical rhetoric instead of Neoplatonist metaphysics, the soft theory understood symbols in terms of Aristotle's definition of metaphor.[56] The rhetorical symbol, unlike the mystical one, was purely conventional, but it was still a powerful vehicle of communication and edification. Believers in the soft theory also engaged in allegorical interpretation, but were more likely to believe that allegorical meanings were retrospective inventions of later philosophers, and not the intention of the myths' original authors. In practice, however, it is not always easy to distinguish between symbolic interpretations rooted in the strong and soft theories.

In the sixteenth century numerous antiquaries began to apply symbolic and allegorical interpretation to the study of artifacts such as coins, statues, bas-reliefs, and funerary vases.[57] In updating Andrea Fulvio's guide to Roman antiquities, for example, Girolamo Ferrucci felt so confident in his symbolic fluency that, without recourse to a single textual authority, he explained a Mithraic relief, a classic tauroctony depicting Mithras slaying the bull, as an emblem of the virtues of farming, designed by "ancient experts in the affairs of nature."[58] At a higher lever of scholarship, the pioneering

---

55. Lamberton, *Homer the Theologian* (1986); Struck, *Birth of the Symbol* (2004). See also Ashworth, "Natural History" (1990); Copenhaver and Schmitt, *Renaissance Philosophy* (1992), 155. A powerfully evocative, but deeply flawed, account is Foucault, *The Order of Things* (1980), ch. 1; cf. Copenhaver, "Did Science Have a Renaissance?" (1992).

56. Gombrich, "Icones Symbolicae" (1985).

57. Allen, *Mysteriously Meant* (1970), ch. 9; Cunnally, *Images of the Illustrious* (1999). See also Brisson, *How Philosophers Saved Myths* (2004), 149–52.

58. Fulvio, *L'antichità di Roma* (1588), 308^v–9^v; Allen, *Mysteriously Meant* (1970), 268–70.

numismatist, Guillaume Du Choul, thought the depiction of Jupiter on an ancient coin could best be explained by a theological allegory—"superior things must be hidden from men and revealed only to the gods"—based on the "mystical and occult theology of the ancients."[59] Valeriano's *Hieroglyphica*, which made extensive use of Greek and Roman coins (and much sparser use of Egyptian inscriptions) in order to reconstruct the "transcultural, universal, and divinely inspired" language of ancient symbolism, partook of the same trend.[60] Most relevant to *Egyptian Oedipus*, in the first decades of the seventeenth century, a group of prominent antiquaries, linked to Kircher's future patrons in Rome and Aix, undertook symbolic and allegorical interpretations of ancient cult objects, both classical and Egyptian.[61]

Lorenzo Pignoria (1571–1631), a friend of Peiresc and Galileo, with close ties to the Roman scholarly circle around Francesco Barberini and Cassiano dal Pozzo, was, as we saw above, one of Kircher's important predecessors in Egyptology. Since the eighteenth century it has been common to praise Pignoria's approach to the hieroglyphs in contrast to Kircher's, above all because of his programmatic statement of method:

> I will explain the images of this table, not allegorically, but, to the best of my ability, faithfully following the ancient narratives [*ad veterum narrationum fidem*]. For I, more than anyone, hate those excessive and usually irrelevant interpretations of this kind of material, which the Platonists, departing from their master's teaching, have imposed to shore up their tottering fables. And I thought it better to confess my ignorance rather than annoy my learned reader.[62]

Pignoria made good on this promise, restricting himself to tentatively identifying most of the gods depicted on the Bembine Table, which he dated fairly accurately to the Augustan period, and he refrained from imbuing the inscriptions with profound philosophical import. Although he rejected the Neoplatonists' mythological allegories, Pignoria was still deeply interested in symbols as the key to understanding ancient paganism. He edited

---

59. Du Choul, *Discorso della religione antica* (1569), 56; Allen, *Mysteriously Meant* (1970), 257–58.

60. Curran, *Egyptian Renaissance* (2007), 234. Cf. Rolet, "Invention et exégèse symbolique à la Renaissance" (2002). On Valeriano's study of ancient coins, see Rolet, "D'étranges objets hiéroglyphiques" (2002).

61. See Miller, "Antiquary's Art" (2001); Miller, "A Philologist, a Traveller and an Antiquary" (2001); Häfner, *Götter im Exil* (2003), 84–102.

62. Pignoria, *Vetustissimae tabulae explicatio* (1605), 1^{rv}.

new editions of two of the sixteenth century's most important guides to ancient symbolism, Alciati's *Emblemata* and Cartari's *Images of the Ancient Gods*.[63] In 1628, Pignoria published a collection of letters in which he offered symbolic interpretations of images found on various ancient objects, affirming, "There is nothing so common or familiar among our ancestors, that it does not contain a mystery."[64] Trained in the Aristotelian bastion of Padua, he followed the soft theory of symbolism, and was skeptical of the Neoplatonists' claims for pagan wisdom in the deep, as opposed to the classical, past. He drew on Macrobius, the Latin Neoplatonist who was one of the most influential sources for the allegorical interpretation of pagan myth; but he seems to have treated Macrobius as a source of late antique beliefs, not more ancient ones. Pignoria's interpretation of the Bembine Table stands up better than Kircher's later effort, but even soft symbolism produced results that could seem arbitrary, and not only by modern standards. Cassiano dal Pozzo found Pignoria's symbolic interpretation of the ancient painting known as the *Nozze Aldobrandini* "so extravagant that it did not meet the case at all," and tried to hinder its republication.[65]

Some of Pignoria's Roman friends practiced symbolic interpretation more in tune with Kircher's method. In 1616 Girolamo Aleandro (1574–1629) published a treatise on an enigmatic marble bas-relief, which he interpreted using Neoplatonic allegory—especially the theory, expounded by Macrobius in the *Saturnalia*, that all the Olympian gods were manifestations of a single solar deity. Aleandro seems to have adhered to a stronger theory of symbolism than his friend Pignoria. He began his treatise by invoking the authority of Homer, the "father of all philosophers and poets," who "wrapped serious teachings in the cover of fables" as the gentle means of communicating his essentially monotheistic theology. But Aleandro hedged his bets by also citing Tertullian's opinion that such allegorical interpretations were created by later pagan philosophers embarrassed by the shameful fables of their gods.[66] Kircher borrowed Aleandro's image and interpretation in *Pamphilian Obelisk* and again in *Egyptian Oedipus* (fig. 13).[67]

No such qualms were on display in the work of Lucas Holstenius, an-

---

63. Alciati, *Emblemata* (1661); Cartari, *Imagini delli Dei* (2004; facsimile of 1647 edition).
64. Pignoria, *Symbolarum epistolicarum Liber* (1628), 13; Allen, *Mysteriously Meant* (1970), 264–65.
65. Cappelletti and Volpi, "New Documents" (1993), 278.
66. Aleandro, *Antiquae tabulae explicatio* (1616), 9, 44. See Mulsow, "Antiquarianism and Idolatry" (2005), 197–98; Häfner, *Götter im Exil* (2003), 112–13; Allen, *Mysteriously Meant* (1970), 270–72; Dempsey, "Classical Perception of Nature" (1966), 19.
67. *OP*, 236; *OA*, 206.

Fig. 13. The ancient marble bas-relief interpreted first by Girolamo Aleandro, and then by Kircher, as a representation of Apollo as solar deity, following the Neoplatonic theory of Macrobius. Athanasius Kircher, *Oedipus Aegyptiacus* (Rome: 1652–54), vol. 2, part 1, 206. Courtesy of Stanford University Libraries.

other member of the Barberini circle, whom we met earlier as Kircher's companion in Malta. Although Holstenius was primarily a classical philologist whose great passions were Neoplatonic philosophy and ancient geography, when an ancient fresco was discovered in a subterranean temple during the construction of the Barberini palace in 1629, it fell to him to produce the requisite antiquarian interpretation (fig. 14).[68] Identifying the painting as a depiction of a nymphaeum, Holstenius found its key in Porphyry's "brilliant dissertation," *The Cave of the Nymphs*, the locus classicus of Neoplatonist allegorical interpretation.[69] Following Porphyry, Holstenius found the eleven Homeric verses describing the cave where Odysseus awoke in

---

68. Holstenius, *Vetus pictura nymphaeum* (1676); Häfner, *Götter im Exil* (2003), 102–15.
69. Lamberton, *Homer the Theologian* (1986), esp. 108–33.

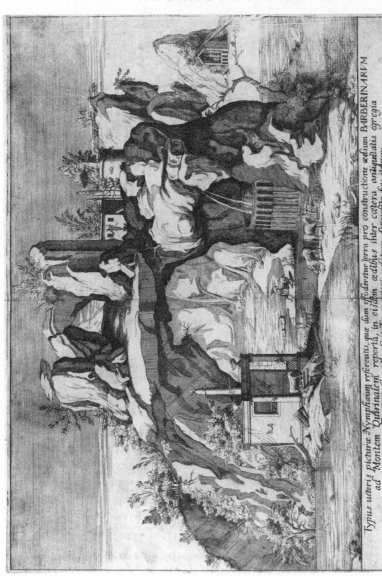

Fig. 14. The ancient fresco uncovered during construction of the Barberini Palace. Lucas Holstenius identified its subject as the Cave of the Nymphs described in Homer's *Odyssey*, and interpreted the painting as a Neoplatonic allegory of esoteric wisdom. Lucas Holstenius, *Vetus pictura nymphaeum referens commentariolo* (Rome: 1676), plate. Deutsches Archäologisches Institut, http://arachne.uni-koeln.de/item/buchseite/182731.

Ithaca as "full of secret and most arcane wisdom." Besides Porphyry, he cited Proclus's *Platonic Theology* and *Cratylus* commentary as evidence that "the secret meaning of the ancient theology was wrapped in symbols of fountains and craters." Holstenius's understanding of the fresco as a symbolic representation of the allegorical wisdom recorded in *The Cave of the Nymphs* bears comparison to Kircher's later reading of the Bembine Table as a hieroglyphic translation of the Chaldean Oracles.[70] Kircher would make use of Holstenius's collection of Platonic and Pythagorean manuscripts in completing *Egyptian Oedipus*.

The examples of Aleandro and especially Holstenius show that not only Neoplatonist symbolism and allegorism, but also recourse to the ancient theology as a historical framework were familiar elements of antiquarian practice among scholars in the Barberini circle, which Kircher would join upon his arrival in Rome in 1633. Shortly before the appearance of *Egyptian Oedipus*, Giacomo Filippo Tomasini and Fortunio Liceti, Paduan scholars in contact with Cassiano dal Pozzo in Rome, published works explaining enigmatic antiquities in a similar framework (fig. 15).[71] Even after the diffusion of Casaubon's critique, there was still much life in the idea that the most ancient pagans, including Hermes Trismegistus, Zoroaster, and Orpheus as well as Homer, had possessed profound, symbolic wisdom whether or not they had written the surviving texts attributed to them. The strong theory of symbolism and the claim of a gentile *prisca theologia* had many critics by the beginning of the seventeenth century, but also many adherents. The idea of Adamic perennial philosophy, on the other hand, still commanded a broad consensus, a reminder of the nearly universal acceptance of literalist biblical history and genealogy as the indispensable framework for understanding the origins of all civilizations.[72]

The effort to create a synthesis of "sacred" and "profane" history was the great project of pre-Enlightenment erudition, fueling research into chronology, mythography, and world religions, among other fields. By facilitating the reconciliation of gentile traditions with biblical chronology, the allegorical and symbolic interpretation of paganism sustained the literal, historical interpretation of the Old Testament. Other techniques of nonliteral interpretation served a similar function, especially Euhemerism, an ancient hermeneutic tradition that explained the myths of the pagan gods

---

70. See chapter 4, "Reading the Bembine Table."
71. Tomasini, *Manus aenae* (1649); Liceti, *De lucernis* (1652); Liceti, *Hieroglyphica* (1653).
72. Harrison, *Fall of Man* (2007); Bottin et al., *Models of the History of Philosophy* (1993); Bizzocchi, *Genalogie incredibili* (1995).

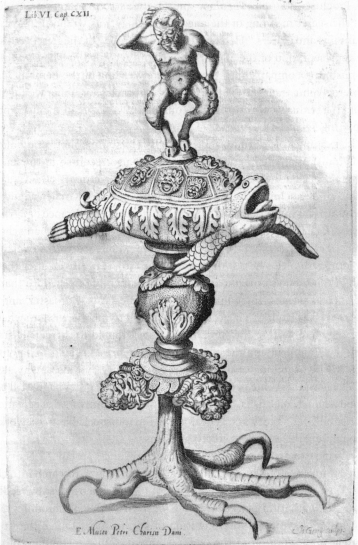

Fig. 15. An ancient bronze lamp whose image was communicated to Fortunio Liceti by Johann Rhode, a Danish professor in Padua, who also corresponded with Kircher. If the lamp had a meaning "beyond the artist's fanciful imagination," Liceti proposed that it might be "an allegory of natural things," since it could "mystically signify" the elements of air, water, and fire according to the "hieroglyphic" significance of the eagle, turtle, and satyr. "But," he concluded in good Paduan fashion, "it is uncertain that the author had this goal in mind." Fortunio Liceti, *De lucernis antiquorum reconditis* (Udine: 1652), 1169. Courtesy of Stanford University Libraries.

as imperfectly remembered stories about notable human beings. Together with etymology (in the ad hoc manner it was practiced by even the sharpest early modern philologists), Euhemerism made it possible to assimilate different pagan deities with one another and to locate their origins in biblical personalities, thereby bringing order to the chaos of polytheism's infinite pantheon. Many scholars also continued to find demonism a persuasive explanation not only for the origin of superstition, but for homologies between pagan and pious practices, which they attributed to "the Devil aping God." These methods offered alternative ways to explain any given problem, and some scholars preferred one method to another; critics of pagan wisdom, for example, tended to favor Euhemerism over allegorism. Typically, however, scholars drew on the entire arsenal, which collectively offered a powerful means not only to tame thse absurd and scandalous elements of pagan lore but, more importantly, to reconcile biblical and gentile traditions.

When not being skeptical about the hieroglyphic wisdom of ancient pagans, Pignoria defended Virgil's claim that the Trojan Antenor (a minor figure in the *Iliad*) had founded Padua, and he elaborated a theory explaining how idolatry originated with Noah's son Ham and diffused from Egypt to the entire world, including America and the Far East. Pignoria's historical narrative disagreed with Kircher's primarily in its more negative estimate of ancient paganism. More often than not, critics of the *prisca theologia* were advocates of the "plagiarism thesis." The latter differed from the former in tone more than in substance, since it similarly asserted that the best pagan thinkers depended on Moses (or earlier biblical figures), but it did so to emphasize paganism's unoriginality and tendency to corrupt the truth. By tracing pagan traditions to biblical sources, either as approximations or corruptions of revealed truth (or both simultaneously, according to the influential theory that popular polytheism coexisted with an elite esoteric monotheism), the *prisca theologia* and the plagiarism thesis offered powerful frameworks to explain ancient paganism in a way that was consistent with Christian interpretations of history.

In the seventeenth century this project produced a new body of scholarship, as the sixteenth century's manuals of pagan iconography gave way to more historical investigations.[73] The same complex of ideas found in antiquarian studies of ancient pagan imagery was at work in seventeenth-century scholarship that sought to explain the origin and progress of idolatry in the context of biblical studies. John Selden's *On the Syrian Gods*, first

---

73. Miller, "Taking Paganism Seriously" (2001); Mulsow, "Antiquarianism and Idolatry" (2005).

published in 1617, was a highly influential and innovative contribution to this genre. Inspired by Joseph Scaliger, Selden investigated the heathen gods described in the Bible in order to demonstrate how pagan sources could clarify the biblical text. He set forth a multicausal theory of the origins of idolatry, involving demons, Euhemerism, and also the idea that the ancient gods had been worshipped as "symbols" representing natural forces—a version of the theory of "physical allegory," for which he relied on the views of Porphyry as conveyed by Eusebius. On this basis, Selden claimed the existence of primitive monotheism among the pagan elite, while explaining idolatry as having originated with the vulgar masses misinterpreting the divine symbols. Selden's account of the development of pagan religion also assigned an important role to astronomy and astrology, whose invention he attributed to Adam's son Seth, although he credited Homer with the division of the zodiac into twelve signs. By endorsing the claim that Pythagoras had studied with Jews, Selden linked the Greek philosophical tradition to biblical revelation.[74]

Although Selden began with the Bible and Pignoria with mythology, they converged on the same territory: the scholarly realm of "paganology," a term coined by Jan Assmann to describe the new historical discourse about idolatry that emerged at this time. As with the allegorical interpreters of antiquities, the traditional building blocks from which Selden constructed his history of religion should not blind us to his creativity and innovation. By insisting that Holy Scripture be understood in historical context and subjecting the religion of the Bible to systematic comparison with pagan sources, Selden and other sacred philologists who followed his template—most influentially, G. J. Vossius in *On Gentile Theology* (1640)—were part of a revolution in European understanding of the Bible, with repercussions that extended far beyond their conservative, apologetic goals.[75]

By emphasizing the philosophical side of *Egyptian Oedipus* to the exclusion of the antiquarian, and assimilating Kircher's hieroglyphic studies to earlier projects of Hermetic reform, previous studies have failed to appreciate the different use that Kircher made of occult philosophy. If we place his hieroglyphic studies in a fuller context, however, by supplement-

---

74. Selden, *De dis Syris* (1629); Toomer, *John Selden* (2009), I, 211–56; Miller, "Taking Paganism Seriously" (2001), 193–200.

75. Mulsow, "John Seldens *De Diis Syris*" (2001); Miller, "Taking Paganism Seriously" (2001). See also Miller, "London Polyglot Bible" (2001); Sheehan, "Sacred and Profane" (2006); Stroumsa, *New Science* (2010); Stolzenberg, "John Spencer" (2012).

ing the Hermetic tradition of Egyptology with the archeological one, as well as with erudite studies of ancient iconography and the history of paganism—in other words, if we read his work in relation to Pignoria, Aleandro, Holstenius, Selden, and Vossius in addition to Ficino, Valeriano, Agrippa, and Patrizi—the philosophical interpretation of *Egyptian Oedipus* appears strained and unconvincing.

TAKING KIRCHER SERIOUSLY

Peter Miller has written about the complementary methodologies that Selden and Pignoria brought to the study of paganism. Selden's special weapon was his pathbreaking use of Oriental sources. But his research was almost entirely textual, in contrast to that of Pignoria, who updated Cartari by collecting material evidence, which he treated with great sensitivity.[76] Kircher was a less profound linguist than Selden and handled material evidence less subtly than did Pignoria. But his ability to combine their signature methodologies—to join knowledge of Oriental languages with expertise in "recondite antiquities"—was a rare and impressive skill set. The editor of the 1670 edition of Pignoria's interpretation of the Bembine Table found Kircher an appropriate authority to provide commentary for its appendix of images of ancient amulets (fig. 11).[77] Kircher constantly stressed that Oriental erudition had been the key to his success in deciphering the hieroglyphs and, like Selden, he offered *Egyptian Oedipus* as an example of how the study of Oriental literature could contribute to historical scholarship.

Kircher was also the inheritor of the scholarly traditions represented by Selden and Pignoria in other ways. *Egyptian Oedipus* literally fused the genres of paganology and the antiquarian atlas. Part 1, "The Temple of Isis," was a history of paganism in the spirit of Selden and Vossius. Its main theme was the Egyptian origin of superstition and idolatry and its dissemination to other peoples. After reconstructing the history of ancient Egypt and describing its gods in comparative terms, Kircher explained the reappearance of idolatry and the corruption of Adamic wisdom in Egypt after the Flood. In a long section entitled "Pantheon of the Hebrews," he investigated the idolatrous practices of various Oriental peoples, arguing that the superstitious worship of the wayward Israelites, described in the Bible, derived from an

---

76. Miller, "Taking Paganism Seriously" (2001).
77. Pignoria, *Mensa Isiaca* (1669), 86–96.

Egyptian model. Kircher then took up Selden's main theme, discussing the Syrian, Philistine, and Arabian gods. In the final section of part 1, he moved into Pignoria's territory, treating "more recently discovered Gentiles," such as the Chinese, Japanese, Tartars, Indians, and Americans. Part 3 of *Egyptian Oedipus*, "The Hieroglyphic Theater," contained images of Egyptian antiquities accompanied by Kircher's interpretations, including translations of hieroglyphic inscriptions. It belonged to a familiar genre, the antiquarian atlas, whose great exemplar was the exhaustive *Ancient Inscriptions of the Entire Roman World* (1602) compiled by Johannes Gruterus.[78] *Egyptian Oedipus* most resembled the studies of specific classes of ancient objects that became popular in the seventeenth century, such as Fortunio Liceti's comprehensive, illustrated study of ancient lamps, which overlapped with *Egyptian Oedipus* in its section on Egyptian material (fig. 15).[79]

In "Ancient History and the Antiquarian," the 1950 essay that eventually inspired the reassessment of early modern erudition, Arnaldo Momigliano referred briefly to Kircher's hieroglyphic studies, but only to dismiss their "Trinitarian nonsense." Momigliano's analysis was based on a sharp distinction between the antiquaries' "success in establishing safe rules for the use of charters, inscriptions, and coins" and their less fortunate attempts to interpret the imagery on vases, statues, bas-reliefs, and engraved gems, which "spoke a more difficult language." As Anthony Grafton comments, Momigliano "saw the central task of the intellectual historian as recovering the turning points in the development of the field" by concentrating "on those individuals and works that had made the greatest impact."[80] Consequently, Momigliano did not dwell on the antiquaries' "failure ... to produce a convincing dictionary of the figurative arts," preferring to emphasize the positive achievements of their "sober and fastidious scholarship," which "brought something of the scientific method of direct observation to historical research."[81] Symbolic interpretation, rather like etymology, was not among the methods that, in retrospect, commend the antiquaries as pioneers of modern scholarship.

But the problems that most gripped erudite scholars in the early seventeenth century (including Peiresc and his friends, the very individuals

---

78. Stenhouse, "Classical Inscriptions" (2000), 149–60.
79. Liceti, *De lucernis* (1652).
80. Grafton, "Momigliano's Method" (2007), 101. Grafton has uncovered a fascinating letter in which Frances Yates challenged Momigliano's characterization of Kircher's theory of Egyptian Trinitarianism (ibid., 115–18).
81. Momigliano, "Ancient History" (1950), 290, 300, 302–3, 309–10.

whom Momigliano described as applying Galilean empiricism to the study of the past) were, precisely, iconographic questions.[82] The "sober and fastidious" methods emphasized by Momigliano and many subsequent studies of antiquarianism were most sufficient in research on the classical (as well as the medieval) past. But they were inadequate to the task of investigating paganism's *longue durée*, which, due to its bearing on momentous debates about the meaning of Christianity and the Bible, was arguably seventeenth-century erudition's major preoccupation. In this research field, state-of-the-art historical scholarship was characterized by the coexistence of symbolism, allegorism, perennial philosophy, and biblical literalism with more modern critical and empirical methods.

In 1970 Don Cameron Allen published a history of early modern scholarship that focused precisely on the types of evidence and interpretative practices that Momigliano had downplayed. In order to make sense of ancient texts and images, the scholars in Allen's study did nothing but resort to symbolic and allegorical interpretations. (Admittedly, "the rediscovery of pagan symbolism and allegorical interpretation in the Renaissance" was the book's subject.) In this panorama Kircher appeared not as aberration but as apotheosis, receiving extensive treatment, mostly of his hieroglyphic studies.[83] While Allen's admittedly selective analysis of early modern erudition captured something important that Momigliano and others had missed, by focusing almost exclusively on those interpretative methods that were later rejected as misguided, he tended to confirm the image of antiquarianism as an eccentric byway off the highroad of intellectual history.[84] For Momigliano in 1950, taking erudition seriously meant not taking Kircher seriously. In 1970, Allen was able to discern the commonalities between Kircher and the antiquaries celebrated by Momigliano, but only by subjecting all of them to his trademark irony.

In 1994 Anthony Grafton published an influential book of essays with the avowed aim "to attack a single, general, dogma": the persistent narrative of early modern history that opposed sterile and backwards humanist

---

82. Miller, "Taking Paganism Seriously" (2001), 205; Momigliano, "Ancient History" (1950), 56–57. See also Miller, "Antiquary's Art" (2001), esp. 70–87; Dempsey, "Classical Perception of Nature" (1966), esp. 234–35.

83. Allen, *Mysteriously Meant* (1970), 119–33, 274–78. Allen's book appears to have been uninfluenced by Momigliano's famous essay.

84. To Allen's credit, this did not prevent him from recognizing, albeit in passing, the roots of the comparative study of religion and anthropology among the Renaissance antiquaries and mythographers.

scholarship to the progressive force of modern science.[85] Grafton's program—which owed something to Momigliano, who was his teacher—has been remarkably productive. In the last two decades, a growing body of research has firmly established the role of philology and antiquarian research in the emergence of modern culture.[86] In particular, Grafton and other scholars have confirmed the link, posited by Momigliano, between erudition and natural science, and pushed back the origin of many modern scholarly practices into the age of humanism.[87] Other research has explored the pervasive influence of religious apologetics on scholarship, both humanist and scientific, throughout the early modern period.[88] A new picture of seventeenth-century scholarship is taking shape in which the contradictions of Kircher's hieroglyphic studies, far from appearing as anomalies, may be seen to embody some of the essential tensions of a pivotal age.

85. Grafton, *Defenders of the Text* (1994), 1.
86. Grafton, *Joseph Scaliger* (1983–93); Herklotz, *Cassiano dal Pozzo* (1999); Pocock, *Barbarism and Religion* (1999); Stenhouse, *Reading Inscriptions* (2005); Soll, *Publishing the Prince* (2005); Edelstein, *The Enlightenment* (2010). The citations in this note and those that follow are representative but hardly exhaustive.
87. Findlen, *Possessing Nature* (1994); Bredekamp, *Lure of Antiquity* (1995); Blair, *Theater of Nature* (1997); Miller, *Peiresc's Europe* (2000); Freedberg, *Eye of the Lynx* (2002); Pomata, "Praxis Historialis" (2005); Ogilvie, *Science of Describing* (2006); Siraisi, *History, Medicine, and the Traditions* (2007).
88. Mulsow, *Moderne aus dem Untergrund* (2002); Häfner, *Götter im Exil* (2003); Sheehan, *Enlightenment Bible* (2005); Shelford, *Transforming the Republic of Letters* (2007). See also the classic study, Rossi, *Dark Abyss of Time* (1984).

CHAPTER TWO

How to Get Ahead in the Republic of Letters

Hermes was the first who erected those columns, which are called needles of the Pharaoh, and on them he carved the sciences that he discovered.
—Abenephius, *On the Mysteries of the Egyptians*[1]

COURTSHIP RITUALS AMONG THE ANTIQUARIES

Athanasius Kircher met Nicolas-Claude Fabri de Peiresc, the renowned French aristocrat and savant, at the beginning of October 1632. A professor at the Jesuit college in Avignon, Kircher found himself in Aix, Peiresc's home, while preparing a geographical survey and visiting local holy sites. In Kircher's recollection, Peiresc, being a "curious investigator of recondite things," showed great hospitality upon hearing of a young German visitor who not only was expert in Oriental languages but also knew something about Egyptian hieroglyphs. The meeting could not have been very intimate, since, in letters written just afterward, Peiresc referred to him as "Balthazar Kilner" or "Kyrner." One fact, however, left a vivid impression in Peiresc's mind, for he mentioned it in at least two letters: Kircher possessed an Arabic treatise "concerning the manner of interpreting and deciphering the hieroglyphic letters of the Egyptian obelisks," composed, as he would soon learn, by a certain "Rabbi Barachias Nephi of Babylon." Kircher promised to send him a sample of the Latin translation that he was preparing.[2]

---

1. *OP*, 45.

2. *Vita*, 42–44; Peiresc to Dupuy, Aix, 11 October 1632, in Tamizey de Larroque, ed., *Lettres* (1888–98), vol. 2, 359; Peiresc to Samuel Petit, 14 October 1632, in Peiresc, *Lettres à Saumaise* (1992), 38.

At the age of thirty, Kircher was an ambitious and energetic young scholar who had published a minor work on magnetism and won some notice among his Jesuit superiors and local German nobility. But he was still an unknown novice in the wider world of scholarship. By contrast, Peiresc, in his early fifties, was revered throughout Europe for his tireless promotion of learning (fig. 16).[3] To contemporaries as well as recent historians, he has embodied the scholar's ethos—indeed, Arnaldo Momigliano singled him out as the very archetype of the early modern antiquary.[4] From his home in Aix, Peiresc operated as an éminence grise of the Republic of Letters, using his prodigious correspondence network and collection of books and antiquities to advance scholarship in directions he deemed worthwhile. Though he published little himself, he orchestrated the work of others by means of advice, financial support, and gifts or loans of scholarly materials. Constantly on the lookout for new talent, Peiresc mobilized the scholars in his orbit according to their abilities: some were only suited to be intellectual foot soldiers, carrying out the grunt work of the Republic of Letters, while the truly gifted were tasked with the most difficult and important assignments. In the decade before his death in 1637, Peiresc was increasingly involved in promoting the study of Near Eastern languages through a circle of scholars including Jean Morin, Samuel Petit, Claude Saumaise, and several Capuchin missionaries with firsthand experience of the Orient.[5]

His scholar's antennae tingled by the description of an Arabic manuscript about Egypt and the hieroglyphs, Peiresc immediately began to cultivate a relationship with its owner. Before long, he had taken Kircher under his wing and helped launch his career. Decades later, when Kircher published his hieroglyphic studies, he professed himself Peiresc's loyal disciple.[6] Had he lived to receive the homage, Peiresc might have felt more than a little ambivalent. Before he died, he came to harbor significant reservations about Kircher's scholarship, and *Egyptian Oedipus* surely would have confirmed his fears about his protégé's lack of rigor and penchant for unsub-

---

3. On Peiresc, see especially Miller, *Peiresc's Europe* (2000) and his articles in the bibliography (many now collected in Miller, *Peiresc's Orient* (2012). See also Bresson's introduction and notes in Peiresc, *Lettres à Saumaise* (1992); Rizza, *Peiresc e l'Italia* (1965); Cahen-Salvador, *Un grand humaniste* (1951); Gravit, *Peiresc Papers* (1950). On Kircher's correspondence with Peiresc, see Fletcher, "Claude Fabri de Peiresc" (1972).

4. Momigliano, "Rise of Antiquarian Research" (1990), 54–56. For the view of Peiresc's contemporaries, see Gassendi, *Mirrour of True Nobility* (1657); and the posthumous homage: Bouchard, ed., *Monumentum Romanum* (1638).

5. Miller, "Peiresc, the Levant and the Mediterranean" (2005); Aufrère, *La momie et la tempête* (1990) Valence, ed., *Correspondance* (1892).

6. *OA* I, b3^r.

Fig. 16. The Provençal aristocrat Nicolas-Claude Fabri de Peiresc, a major figure in European antiquarian research and Oriental philology, who helped launch Kircher's career. Portrait engraved by Lucas Vorsterman, seventeenth century. Harvard Art Museums / Fogg Museum, gift of Belinda L. Randall from the collection of John Witt Randall, R4705.

stantiated speculation. Kircher was nonetheless the great antiquary's pupil, if in crucial respects a wayward one. The story of their relationship, which lies at the heart of this chapter, revolved around a series of Arabic manuscripts. It sheds light on Kircher's hieroglyphic studies by situating them in the erudite world of Oriental philology and antiquarianism that Peiresc personified.[7]

Eager to inspect Barachias Nephi's treatise personally, Peiresc invited Kircher to return to Aix at Easter. He shared his excitement with learned friends such as Samuel Petit, professor of theology, Greek, and Hebrew in Nîmes, and the natural philosopher Pierre Gassendi, whom he encouraged to be present at Kircher's visit. After Kircher sent a sample of his translation of Barachias, Peiresc told Gassendi that it made him "much more hopeful than I once was about the discovery of things that have been so unknown to Christendom for nearly two thousand years." Indeed, he was sufficiently fascinated to write to his correspondent in Tunis, the renegade Thomas d'Arcos, to inquire whether any of his Arab contacts possessed books written by "Barachias Bar Nepsi," or books that so much as mentioned him. If so, Peiresc wrote, "I would gladly pay for a copy, and would not hesitate to spend even twenty crowns, if need be."[8]

Easter came and went without a visit, but Kircher sent Peiresc a further example of his work: his so-called *Protheories* on "the explanation of the Egyptian hieroglyphs of the obelisks, extracted from an ancient Babylonian rabbi, named Rabbi Barachias."[9] After glancing at them Peiresc expressed a touch of disappointment, noting that they "were not quite what had been expected," but he remained optimistic that the Barachias manuscript would yield substantial rewards. He now expected Kircher to visit at Pentecost if not sooner. In preparation he gathered materials about Africa and Egypt that Kircher wished to examine and reiterated his invitation to Gassendi to join them.[10]

Kircher, eagerly awaited since the previous winter, returned to Aix in the third week of May 1633. In a letter composed during the visit, Peiresc

---

7. For a somewhat different interpretation of their relationship, see Miller, "Copts and Scholars" (2004).

8. Peiresc to Gassendi, Aix, 2 March 1633, in Tamizey de Larroque, ed., *Lettres* (1888–98), vol. 4, 295; Peiresc to d'Arcos, Aix, 22 March 1633, in ibid., vol. 7, 112. See also Peiresc to Petit, 18 February 1633, excerpted in Peiresc, *Lettres à Saumaise* (1992), 38.

9. A manuscript by Kircher called "Protheoriae seu Apparatus ad Hieroglyphicam explicationem" is preserved at the archive of the Gregorian University (APUG 812, fols. 1–38).

10. Peiresc to de Thou, Aix, 4 April 1633, in Tamizey de Larroque, ed., *Lettres* (1888–98), vol. 2, 488–89; Peiresc to Gassendi, Aix, 5 April 1633, ibid., vol. 4, 300–301; Peiresc to Dupuy, Aix, 16 May 1633; Peiresc to Gassendi, Aix, 5 April 1633, ibid., vol. 4, 521–22.

described how Kircher had impressed him with "very fine relations and secrets of nature," especially a marvelous weatherproof clock based on the nightshade seed's magnetic attraction to the sun. Disappointingly, Kircher postponed his demonstration of Barachias's interpretation of the hieroglyphs for a future occasion. Nonetheless, Peiresc wrote, the Jesuit was full of promise and had made an admirable impression. Kircher left Aix with an even firmer hold on the older man's imagination, as well as a chest of borrowed books to assist him in his studies.[11]

To facilitate the translation of Barachias Nephi's treatise, Peiresc resolved to bring Kircher to Aix more permanently. He lobbied to have Kircher transferred to the town's newly established Jesuit college and sought out materials that would assist his research, including the massive, just-published Arabic dictionary composed by Giggeius at the Ambrosian Library in Milan, which he ordered from Genoa. "We will neglect nothing of the midwife's art to try to make him give birth to this work," he declared.[12] But upon returning to Avignon, Kircher learned that his Jesuit superiors had other plans for him: he was being sent to Vienna, where the Holy Roman Emperor required a professor of mathematics. Peiresc persevered, dispatching letters to his powerful friends in Rome who might influence Muzio Vitelleschi, General of the Society of Jesus. He explained that a transfer to Vienna would nip in the bud Kircher's promising hieroglyphic studies, which required access to specialized resources such as could be found in Peiresc's library—in particular, certain manuscripts soon expected to arrive from Cairo.[13] The director of Aix's Jesuit college joined Peiresc's campaign, writing General Vitelleschi to praise Kircher as a great scholar whose presence justified the immediate establishment of a chair of mathematics.[14]

While waiting for a response from Rome, Peiresc urged Kircher to return to Aix with Barachias. Again he made preparations, sending the newly arrived Ambrosian Arabic dictionary to the binder and inviting learned friends to be present. But summer wore on with Kircher unable to get away from

---

11. Peiresc to Dupuy, Aix, 21 May 1633, ibid., vol. 2, 528–29. See also Peiresc to Francesco Barberini, Aix, 19 May 1633, BAV Barb. Lat. 6503, fol. 50; *Vita*, 43; Gassendi, *Mirrour of True Nobility* (1657), 85. Kircher demonstrated the clock during his subsequent visit: see Peiresc's description, *L'horologe du Pere Athanase Kircher faict avecque la graine & fleur du Solanum qu'avecque la pierre d'aymant*, 3 September 1633, BNF Dupuy 661, fol. 228^r (copy).

12. Peiresc to Dupuy, Aix, 21 May 1633, 30 May 1633, 6 June 1633; in Tamizey de Larroque, ed., *Lettres* (1888–98), vol. 2, 528–29, 533, 535.

13. Peiresc to Barberini, Aix, 12 June 1633, BAV Barb. Lat. 6503, fol. 53^r. Peiresc had already written Barberini about Kircher and Barachias on 19 May 1633; ibid., fol. 50^r.

14. Jean Louis de Revillas to Vitelleschi, no date, quoted in Romano, *La contre-réforme mathématique* (1999), 387.

Avignon, where his progress on the translation was hindered by visitors, teaching duties, preparing a new book for the press, and the intense summer heat. Frustrated by Kircher's continual delays, Peiresc lamented that the college burdened him with such mundane duties as administering student exams. As September began, Kircher resigned himself to his Austrian posting and prepared for his departure, hoping that Peiresc's orchestrations might at least allow him to stop in Rome along the way, to inspect the city's obelisks.[15] Before setting off he kept his word, stopped in Aix, and at last showed Peiresc the Barachias manuscript.

In his autobiography, completed near the end of his life, Kircher described how he translated a hieroglyphic inscription for Peiresc:

> He had a hieroglyph from an Egyptian statue fetched, and gave it to me to be interpreted, and thus, having spent a good part of the night on the interpretation, the next day I brought him the completed interpretation. When he found it to smell of the Egyptian lamp (for he was very expert in all antiquity) he spoke about my undertaking with such seriousness of words that, out of modesty, I don't consider it right to describe here.[16]

In *Pamphilian Obelisk*, published in 1650, Kircher described Peiresc's reaction to his long awaited presentation of the Barachias manuscript: "When I then showed Peiresc the codex, he was immediately filled with tremendous joy, and from that time on he left no stone unturned to promote my enterprise of restoring the hieroglyphs."[17]

In private notes recorded immediately following the September meeting, Peiresc told a different story. After making him wait nearly a year to see the manuscript, Kircher only wished Peiresc to examine a single page from the last section of the treatise, which comprised a hieroglyphic lexicon. "The bother that he made over letting me transcribe a couple of entries," Peiresc wrote, "made me suspect that he feared that I would discover that it was nothing but a kind of translation of Horapollo, which also begins with the eye on a scepter if I'm not mistaken." The visit had gotten off to a rocky start the day before, when Peiresc challenged Kircher's interpretation of a Roman obelisk in his *Protheories*. Kircher had relied on an engraving of the

---

15. Peiresc to Kircher, Aix, 3 August 1633, APUG 568, fols. 370^r–71^v; Peiresc to Petit, Aix, June 14 1633, Peiresc, *Lettres à Saumaise* (1992), 38l; Kircher to Peiresc, Avignon, 9 August 1633, BNF FF 9538, fol. 227^r; Kircher to Jean Ferrand, S. J. (in Aix), Avignon, 4 June 1633, BNF FF 9362, fol. 16^{rv} (copy).

16. *Vita*, 48.

17. *OP*, c1^r. Cf. Gassendi, *Mirrour of True Nobility* (1657), 85.

Lateran obelisk from Herwart von Hohenburg's *Hieroglyphic Thesaurus*, an atlas of hieroglyphic inscriptions published in 1610. Most of Herwart's illustrations were fairly accurate and realistic, but Kircher had chosen an exceptional image whose fanciful hieroglyphs bore no resemblance to those in the rest of the book, including another more accurate depiction of the Lateran obelisk (fig. 17). Peiresc realized at a glance that the inscriptions "were all fabricated by the imagination of the artist . . . like grotesques, which had nothing of the style of ancient Egypt." But when he pointed this out, Kircher would only acknowledge the fact with great distress, "since he had found nice, well-supported interpretations, it seemed, of all the figures there included, or most of them."[18]

Prior to the visit, Kircher had written proudly of his interpretation of this obelisk, which he planned to send to the renowned historian Michel Baudier, whom he had recently met in Avignon. Finally forced to admit his error, Kircher was, according to Peiresc, properly ashamed. The awkward mood continued when Peiresc raised the subject of the interpretation of the obelisk of Constantius, reported by the Latin author Ammianus Marcellinus, which related its inscription "to particular actions and praises of a king, much different from those first origins with which [Kircher's] Barachias amuses himself, and which he emphasized in his protheories." Peiresc had sent Kircher a copy of Ammianus in Avignon, but Kircher acted as if he had never seen it.[19] The long-awaited meeting would seem to have been a disaster, leaving Peiresc disillusioned and Kircher—if he was capable of the emotion—humiliated.

A few days after the visit, Kircher sent a farewell note from Marseilles, where he awaited the pope's galleys, thanking Peiresc for his many acts of kindness and promising to send the Barachias translation as soon as it was ready.[20] Peiresc responded by writing letters of recommendation to Cardinal Barberini, Cassiano dal Pozzo, and Pietro Della Valle, his influential friends in Rome, where Kircher planned to stop on his way to Vienna. In contrast to the tone of his private notes from the week before, Peiresc praised Kircher and called himself his "dedicated servant and admirer of his talent and worth." He did not, however, mention Kircher's hieroglyphic studies or the Barachias treatise, which had been at the focus of his earlier reports, instead referring generically to Kircher's knowledge of antiquity, the "princi-

---

18. "Note de Peiresc Après la Visite du P. Kircher," 3 September 1633, in Peiresc, *Lettres à Saumaise* (1992), 380–82.
19. Ibid.; Kircher to Peiresc, Avignon, 9 August 1633, BNF FF 9538, fol. 227^v.
20. Kircher to Peiresc, Marseilles, 6 September 1633, BNF FF 9538, fols. 228 bis^{rv}, 233.

Fig. 17. Both these images of the Lateran Obelisk appeared in Herwart von Hohenburg's 1610 atlas of Roman hieroglyphs. The first reproduces the inscriptions fairly accurately, while in the second the artist has decorated the obelisk with fanciful designs. Kircher chose the second image to try out his method of hieroglyphic interpretation. Herwart von Hohenburg, *Thesaurus Hieroglyphicorum*. Bibliothèque nationale de France.

pal languages of Christianity," and, above all, natural science.[21] In the wake of their final meeting in Aix, Peiresc's attitude cooled. Kircher's moment as his favorite new protégé was over. But Peiresc did not lose interest in Kircher, or even Barachias, for long. Although his expectations from both man and manuscript became more modest, Peiresc continued to promote Kircher's studies until his death in 1637.

BARACHIAS IN ROME

Following a calamitous sea voyage, Kircher arrived in Rome, en route from France to Vienna, in the first days of November 1633.[22] Thanks to Peiresc's letters, he quickly found himself at the center of the city's cultural life. After receiving a warm welcome from Cassiano dal Pozzo and Pietro Della Valle, he was granted an audience with the papal nephew, Cardinal Francesco Barberini, who had been informed about Kircher and the Barachias manuscript by Peiresc as early as May. Evidently, Kircher made a good impression. Barberini invited him to a second interview, during which he asked questions about his study of the hieroglyphs, Arabic literature, and the Kabbalah. The Barberini family's recent acquisition of an Egyptian obelisk, which they planned to erect in the garden of their new palace, may have contributed to the cardinal's decision to annul Kircher's assignment to Vienna so that he might remain in Rome and carry out his hieroglyphic studies. First, Kircher was to complete a Latin-Arabic edition of Barachias, along with an interpretation of the hieroglyphs on the Bembine Table as a test of Barachias's method; afterward, he was to write a book interpreting the obelisks of Rome.[23] Kircher had arrived. Not only had he been permit-

---

21. Peiresc to Cassiano dal Pozzo, Aix, 10 September 1633, in Peiresc, *Lettres à dal Pozzo* (1989), 111–12; Peiresc to Barberini, Aix, 10 September 1633, BAV Barb. Lat. 6503, fol. 60^r.

22. He arrived at Civitàvecchia, Rome's port, on 31 October: "Primo itaque octobris Genua soluimus, ac ultima eiusdem tandem serius opinione portum centum cellarum seu civitatis veteris obtinuimus integro mense in itinere commorati" (Kircher to Peiresc, Rome, 14 November 1633, BNF FF 9538, fols. 230^r–32^r). Giunia Totaro's claim that Kircher arrived some weeks earlier is based on misreadings of this letter and another by Peiresc, in which the latter described receiving letters from Rome *without* any news about Kircher. ("Il [Kircher] m'a escript de Genes du 22 septembre qu'il en partoit dans deux jours pour Rome, d'où j'ay des lettres du 22 octobre sans qu'on en mande rien." Peiresc to Gassendi, Aix, 10 November 1633, in Tamizey de Larroque, ed., *Lettres* [1888–98], vol. 4, 385.) Cf. Totaro, *L'autobiographie d'Athanasius Kircher* (2009), 63–64, 99.

23. Kircher to Peiresc, Rome, 14 November 1633, BNF FF 9538, fols. 230^r–32^r; Peiresc to Barberini, 19 May 1633; 12 June 1633, BAV Barb. Lat. 6503, fols. 50, 52–53; Kircher to Peiresc, Rome, 1 December 1633, BNF FF 9538, fol. 234^{rv}; Peiresc to Petit, 16 January 1634, quoted in Peiresc, *Lettres à Saumaise* (1992), 55; Herklotz, *Cassiano dal Pozzo* (1999), 131–32.

ted to stay in Rome, he would be conducting his chosen studies under the sponsorship of the city's most important Maecenas.

Back in the south of France, Peiresc awaited Kircher's translation of Barachias Nephi and continued to play an active long-distance role in his studies. In a letter to Claude Saumaise, one of Europe's ablest scholars of classical and Oriental literature, Peiresc described the Barachias manuscript as "very worthy of publication," and explained Kircher's reticence to show him the entire text in Aix as due to "scruples of conscience" since it likely contained "some hodge-podge of magical maxims," rather than to any effort to hide its true contents. According to Peiresc, it was not certain that Barachias was privy to Egyptian sources older than those known to Greek and Latin authors like Ammianus. But, "even should [Kircher] only give us knowledge of the succession of strange alphabets that it uses," Peiresc wrote,

> it would still be worth the attention. That's why I have all along fostered this good Father and have tried to help him with all the books at my disposal that were to his taste, in order to oblige him to keep his word to me to communicate this book and to prevent it from perishing, as is happening to great treasures in the sacking of the towns of Germany, from where the good Father carried it off.[24]

Saumaise responded that he would gladly trade half his wealth for a copy of Barachias's treatise. It was perhaps on his behalf that Peiresc asked Thomas d'Arcos in Tunis for the second time to try to locate a copy of the book, for which he was now ready to pay "its weight in gold." Peiresc also wrote the Capuchin missionary Agathe de Vendôme, asking him to seek out works similar to Barachias in the libraries of Cairo, and suggested to Cassien de Nantes that copies of Barachias and other Arabic works on the hieroglyphs might be found in the unexplored libraries of the Levant.[25]

Saumaise's enthusiasm diminished some months later, after he read a few lines of Arabic text from Barachias that Kircher had sent Peiresc, which he described as "some words taken from the Koran," with "no resemblance to the inscriptions of the ancient Egyptians." Peiresc, for all his

---

24. Peiresc to Saumaise, Aix, 14 November 1633, in Peiresc, *Lettres à Saumaise* (1992), 36–40.

25. Peiresc to Gassendi, Aix, 17 January, 1634, in Tamizey de Larroque, ed., *Lettres* (1888–98), vol. 4, 421; Peiresc to Thomas d'Arcos, Aix, 25 January 1634, in ibid., vol. 7, 117; Peiresc to Agathe de Vendôme, 22 July 1636, in Valence, ed., *Correspondance* (1892), 245; Peiresc to Cassien de Nantes, 1 November 1636, ibid., 272.

reservations, continued to promote the edition. But, increasingly skeptical of Kircher's critical powers, Peiresc encouraged him to concentrate his energies on editing and translating Oriental texts rather than writing interpretations. He worried that the hieroglyphic commentaries Kircher had undertaken for Cardinal Barberini would delay the publication of Barachias, and he repeatedly urged Kircher and his Roman patrons to reconsider their priorities.[26]

In June 1634, Kircher wrote Peiresc that he had finished his hieroglyphic commentaries—the project had grown from an interpretation of the Bembine Table to include commentaries on four Roman obelisks—which he intended to publish jointly with Barachias (fig. 18). Upon this news, Peiresc changed strategy and decided, despite his misgivings, to support the publication of the hieroglyphic commentaries as the best way to expedite the publication of Barachias. In addition to promising to send letters to Barberini, dal Pozzo, and others in support of Kircher's project, he also offered his financial assistance to pay for artists and copyists, though this would have to be kept secret from the cardinal to avoid offense.[27]

Neither the Barachias translation nor the hieroglyphic commentaries were as close to finished as Kircher had suggested. In February 1635 Kircher wrote Peiresc to express misgivings about publishing the Barachias treatise because parts of it were "superstitious" and contained descriptions of illicit magical practices. In Peiresc's mind, Kircher's new reservations confirmed his own doubts about the antiquity of the treatise's contents. "I always suspected," he wrote back, "what you never dared to confess until now, that it contained some of the author's fancies or distortions, and maybe even some rubbish and falsehood, as well as this dismal magic." Peiresc commended Kircher for his scruples but encouraged him to publish the other parts of the treatise. Just as he had an obligation to protect the public from any harmful content, Peiresc said, so he had a duty to publish everything that might be of benefit to scholars. Sounding a reproachful tone, he added that it would have proved useful now if Kircher had shown him more of the treatise back

---

26. Saumaise to Peiresc, Leiden, 10 June 1634, in Peiresc, *Lettres à Saumaise* (1992), 382; Peiresc to Petit, 16 January 1634, ibid., 55; Peiresc to Gassendi, Aix, 17 January 1634, in Tamizey de Larroque, ed., *Lettres* (1888–98), vol. 4, 421. See also Peiresc to Saumaise, Aix, 4 April 1634, in Peiresc, *Lettres à Saumaise* (1992), 55; Peiresc to dal Pozzo, Aix, 4 May 1634, 6 June 1634, 29 June 1634, in Peiresc, *Lettres à dal Pozzo* (1989), 132–34, 138–39, 140.

27. Peiresc to Kircher, Aix, 6 September 1634, APUG 568, fol. 374^{rv}. See also Peiresc to dal Pozzo, Aix, 7 September 1634; 3 November 1634; 29 December 1634, in Peiresc, *Lettres à dal Pozzo* (1989), 146–47, 157, 161; Peiresc to Saumaise, Aix, 22 September 1634, in Peiresc, *Lettres à Saumaise* (1992), 115.

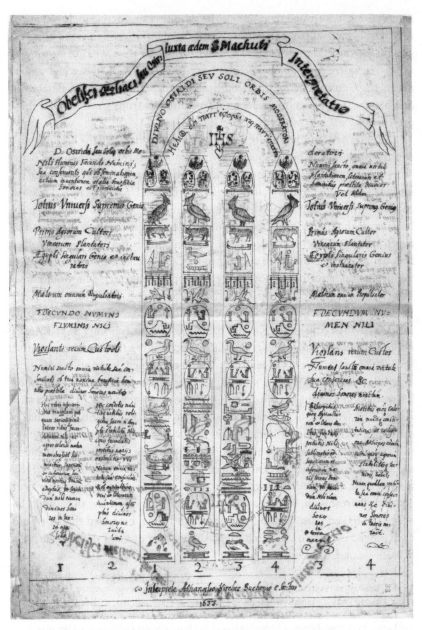

Fig. 18. An early obelisk translation by Kircher, dated 1633. It depicts the obelisk originally dedicated to Ramesses II (1279–1213 BCE), which once stood before the Church of San Macuto and now resides in the Piazza della Rotonda in front of the Pantheon. APUG 830, fol. 33. Pontificia Università Gregoriana.

in Aix, and he asked to see an unexpurgated chapter from Barachias so that he might judge for himself what was at stake in the superstitious passages.[28]

A BIBLIOGRAPHIC ENIGMA

Kircher never published his long-promised edition of Barachias. The manuscript, if it survives, cannot be located. Already the object of suspicion in the seventeenth century (see chapter 8), at the distance of almost four centuries the identity of the Arabic treatise remains obscure. For there is no known Jewish or Arabic author with a name resembling Barachias Nephi, or Abenephius, as Kircher also called him. Although Kircher scholars have paid almost no attention to the matter, the Barachias affair looms too large to be ignored in Peiresc's correspondence.[29] The most detailed treatment of the episode is found in Sydney Aufrère's study of Peiresc and the origins of Egyptology. Based on Peiresc's account of the meeting in Aix, Aufrère describes Kircher as trying to deceive Peiresc by passing off an Arabic version of the well-known *Hieroglyphica* of Horapollo as something more valuable and unique. But the accusation of fraud—although not impossible—is unsubstantiated. While Peiresc's notes and correspondence indicate profound disillusionment, they show that he found Kircher insufficiently skeptical, possessing "a great excess of good will" rather than the "bad faith" imputed by Aufrère.[30] Had Peiresc truly believed the treatise to be nothing but an Arabic Horapollo, he would not have spent the next four years actively encouraging its translation. In any case, Peiresc was mistaken in recollecting that Horapollo's hieroglyphic dictionary began, like Barachias, with the symbol of an eye on a scepter.[31]

Although Kircher allowed him to copy only a single page, Peiresc was able to skim the rest sufficiently to ascertain that the work seemed complete, except for part of the penultimate section on Egyptian history. Peiresc referred to the work as "a treatise on the history, antiquity, origins, hieroglyphic characters, religion, and obelisks of the Egyptians," and described

---

28. Kircher to Peiresc, Rome, 8 February 1635, BNF NAF 5173, fols. 25ʳ–27ᵛ; FF 9362, fols. 13ʳ–15ᵛ (copy); Peiresc to Kircher, Aix, 30 March 1635, APUG 568, fols. 364ʳ–65ᵛ.

29. See Aufrère, *La momie et la tempête* (1990), 263–87; Bresson's introduction and notes to Peiresc, *Lettres à Saumaise* (1992); Lhote's notes to Peiresc, *Lettres à dal Pozzo* (1989); and Miller, "Copts and Scholars" (2004), 103–48.

30. Peiresc to Saumaise, Aix, 14 April 1634, in Peiresc, *Lettres à Saumaise* (1992), 80; Aufrère, *La momie et la tempête* (1990), 269.

31. Peiresc may have confused Horapollo with Plutarch who discusses the hieroglyph of an eye and a scepter in *De Iside et Osiride* (1970), 133 (354 F). Cf. Macrobius, *Saturnalia* (1969), 143 (1.21.12), which gives a similar account.

the "treatise on the Egyptians" as following another text bound with it, "some discussions of morality that were not so complete." He described the book as roughly the size of an octavo, bound *à la turquesque* in blackish or reddish brown leather, and written in black ink in small handwriting on Damask paper. He judged the writing to be approximately two hundred years old.[32] Kircher claimed that he had found the treatise in the library of the Archbishop of Mainz, a plausible claim; prior to being pillaged in the Thirty Years' War, Mainz was renowned for its libraries, above all the Metropolitan Library, administered by the archbishop.[33] In a letter to Saumaise, Peiresc described the Barachias manuscript as containing "ancient obelisks, interpreted by the old rabbi according to the old traditions of the land and a countless number of figures meant for talismans, with the properties of stones, plants, and animals that are usually used with them."[34]

Peiresc's testimony leaves little doubt that Kircher possessed an old Arabic manuscript that contained images of enigmatic characters. But a more precise idea of its content is not possible from Peiresc's comments. We need not rely only on Peiresc, however. Although the manuscript that Peiresc saw is lost, Kircher cited its author frequently in his hieroglyphic studies, quoting passages in Arabic with Latin translations. By then Rabbi Barachias Nephi of Babylon had metamorphosed into "Abenephius the Arab," which may have prevented previous scholars from identifying the connection.[35] It is not evident why Kircher, who relied heavily and conspicuously on Jewish sources, would have downplayed Abenephius's Jewish identity. Barachias Nephi's epithet would refer not to the Mesopotamian city, but to Old Cairo, known from Roman times as Babylon.

As for the supposed dependence on Horapollo, eighteen quotations from Abenephius are explanations of hieroglyphic symbols, presumably taken from the section of the manuscript described as a lexicon by Peiresc.[36] Of these, five show a degree of resemblance to Horapollo, though in no case are they identical.[37] But in general, the two authors have different styles.

---

32. Peiresc, *Lettres à Saumaise* (1992), 380.
33. Peiresc to Saumaise, Aix,14 November 1633, ibid., 38-9. Serarius and Georgius, *Moguntiacarum rerum* (1722), 109-12; Jacobs and Ukert, *Beiträge zur ältern Litteratur* (1835), 5.
34. Peiresc to Saumaise, Aix,14 November 1633, in Peiresc, *Lettres à Saumaise* (1992), 38-39.
35. Already in *Prodromus Coptus* (1636), he is called "Barachias Albenephius," has lost the title of rabbi, and has gained the epithet Arab. *PC*, 254-55.
36. *PC*, 254-55; *OP*, 169, 265, 366-69, 403, 420; *OA* II.1, 110.
37. Both authors identify the ibis or stork with Hermes, the rising of the Nile with the Lion, the serpent with God, and the hawk with the soul. They both discuss the hieroglyph of a hoopoe bird with scepter, though their interpretations are entirely distinct.

Abenephius typically explains hieroglyphs as symbols of cosmological and metaphysical forces: the spirit or soul of the world; the fiery power beaming forth from the Sun; the life, motion, and fecundity of the world; the incomprehensible, inseparable, eternal nature of God; and so forth. His interpretations call to mind the Platonist interpretations of myths and rituals advanced by Plutarch, Proclus, and Macrobius (fig. 19). Horapollo, in contrast, tends to be more concrete, if not exactly down to earth. "When they wish to show a man dead from a sunstroke," reads a typical example, "they draw a blind beetle. For this dies when blinded by the sun."[38]

But hieroglyphic definitions account for a minority of the quotations that Kircher attributed to Abenephius. In other passages, Abenephius treats the hieroglyphs' relationship to other supposedly Egyptian alphabets, their mystical purpose, and the religious significance of Egyptian monuments. A number of quotations treat the history of Egypt, for the most part syncretized with a biblical narrative involving Noah's son Ham, identified with Zoroaster, who is described as the founder of magic and idolatry. For example, Ham's son Mesr or Misraim is described as the first to carve sacred mysteries on stone columns. Elsewhere, Abenephius assigns the same honor to Hermes, who in different quotations is identified with Idris, Enoch, and Osiris.[39] Many passages treat ancient Egyptian religious beliefs and rituals. In three quotations Abenephius discusses the relationship between ancient Egyptian and Jewish religion. The longest of these connects the ritual laws of Moses to the idolatry of the Egyptians and strongly resembles, as Kircher observed, a famous passage from Maimonides' *Guide for the Perplexed*.[40] Kircher also quotes Abenephius on the Egyptian division of the Zodiac, celestial influences, the twelve planetary angels that rule over different regions and elements of the world, and the manufacture of astral amulets. A few quotations treat Egyptian-Coptic vocabulary and etymologies, which had been Peiresc's great hope for the manuscript.[41]

This material resembles authentic Arabic traditions known from surviving works by other authors, such as Jalal al-Din al-Suyuti and Ibn Wahshiyya, which Kircher encountered later and cited alongside Abenephius in *Egyptian Oedipus*. Abenephius's expressions "needles of the pharaoh" to refer to obelisks, and "letters of the birds," to refer to hieroglyphs, are trans-

---

38. Horapollo, I, 41: Boas, *Horapollo* (1993), 79.
39. *OP*, 124, 353; *OA* I, 210; *OP*, 13; *OA* I, 27, 85, 177; *OP*, 163–64; *OP*, 45. See also *OP*, 167–68; *OA* I, 67.
40. *OA* I, 12, 211, 277, 335; *OA* I, 249, 259, 293.
41. *OA* II.2, 178; *OA* I, 187; *OA* I, 27, 293, II.1, 283.

Fig. 19. An Arabic interpretation of an obelisk (evidently the Villa Celimontana obelisk; see fig. 10) attributed to Barachias Abenephius, from a manuscript probably composed by Kircher in the mid-1630s. In the style of Macrobian allegory, it describes the solar deity bestowing its benefits upon the world below. In *Egyptian Oedipus* he gave the same inscription a considerably longer translation (see chapter 1). APUG 830, fol. 29. Pontificia Università Gregoriana.

lations of genuine Arabic terms. The identification of Hermes with Idris and Enoch and his role in the building of Egyptian monuments, as well as the incorporation of other biblical figures into the history of ancient Egypt, are characteristic of medieval Arabic literary traditions, as will be discussed in chapter 5. Abenephius's discussion of the construction and worship of talismans and other aspects of the Egyptians' cult calls to mind works like Ibn Wahshiyya's *Nabatean Agriculture* and Murtadi's *Wonders of Egypt*. The "strange alphabets" that Peiresc described (but which Kircher never published) are suggestive of Arabic works on the decipherment of hieroglyphs and other unusual scripts.[42]

Admittedly, reasons for suspicion remain, beginning with Kircher's odd reluctance to let Peiresc properly examine the manuscript. Several readers of *Egyptian Oedipus* have observed that the Arabic in many of the Abenephius quotations is riddled with spelling and grammatical errors, leading them to suspect that they were forged by someone lacking full command of the language.[43] Furthermore, the content of a few of the quotations is suspiciously convenient, providing precisely the evidence Kircher needed to prove a point. For example, Abenephius is the only authority to testify to Kircher's crucial claim that Hermes invented the obelisks, as opposed to the pyramids. Other dubious passages are Abenephius's Trinitarian explanation of the winged globe with a serpent and his assertion that naked human figures represent genies of the intellectual world, both key elements in Kircher's obelisk interpretations.[44] On the other hand, other quotations from Abenephius resemble Kircher's indubitably authentic sources in failing to say what would be most useful for his purpose, which one would not expect if he had invented them expressly to substantiate his arguments.

By and large, the image of Abenephius's treatise that emerges from Kircher's quotations is consistent with authentic Arabic literature on

---

42. *OP*, d2^v. See Haarmann, "Misalla" (1960–2005); Haarmann, "Medieval Muslim Perceptions" (1996), 607; Daly, *Egyptology* (2005), ch. 5.

43. See Marracci, *Prodromus* (1691), vol. 3, 170, and the comments of Mathurin Veyssiere de la Croze reported in Zorn, *Bibliotheca antiquaria* (1724–25), 134. Glasson, "Order of Jewels" (1975) relates M. Plessner's judgment that the Arabic is too grotesque to be genuine. I consulted two scholars of medieval Arabic literature. Okasha El Daly, an expert on medieval Arabic literature about ancient Egypt, believes that the errors in Kircher's Arabic quotations could be largely due to the copyist and printer. Kevin van Bladel, a student of medieval Islamic Hermetic literature, read the passages from Abenephius in *Oedipus Aegyptiacus* and reached the same conclusion as Plessner. Linguistic errors need not indicate a forgery, however. Wüstenfeld noted that an Arabic manuscript of Ibn Wasif Shah's history of Egypt contained many grammatical errors, probably indicating a Turkish editor; Wüstenfeld, "Die älteste Aegyptische Geschichte" (1861), 328.

44. *OP*, 45, 403, 414–15.

similar topics. If Kircher invented the treatise, he must have used genuine Arabic sources as his model, which would raise the question: Why attribute the material to a fictitious Arabic source rather than to real ones? Perhaps the name Barachias Nephi, or Abenephius, belonged to an obscure compiler and found its way onto a manuscript of works by another author or authors. Kircher may well have altered or fabricated some of the quotations that he attributed to Abenephius for the convenience of his argument. (This would not be out of the ordinary. Pirro Ligorio, a great scholar of Roman antiquities in the preceding century, had a nasty habit of inventing inscriptions.)[45] As chapter 5 demonstrates, he was quite capable of deliberately misrepresenting his research. But the most likely interpretation of the totality of the evidence is that he possessed a genuine manuscript, probably some kind of compendium of Arabic material related to ancient Egypt and hieroglyphic writing.[46]

DELIVERING COPTIC TO THE REPUBLIC OF LETTERS

Barachias Nephi's treatise brought Kircher to Rome; once there, he drew other Oriental manuscripts into his plans. Exploring the city's libraries, Kircher discovered valuable material on "Egyptian philosophy," including several Arabic treatises on amulets, which he described in a letter to Peiresc. Later, he acquired copies of two enigmatic Oriental inscriptions (he published both in 1636), located an Arabic manuscript on the "Saracenic Kabbalah," and projected an edition of several Arabic treatises about Egypt while also studying Jewish manuscripts in Hebrew and Aramaic.[47] Of all the Oriental texts that came into Kircher's possession, however, none was more precious than one that belonged to the Roman patrician Pietro Della Valle.

After fifteen years of traveling in North Africa and the Near East, Della Valle had returned to Rome in 1626 laden with Oriental antiquities. Of particular interest were several Coptic manuscripts he had acquired in Cairo, especially a Coptic-Arabic lexicon and grammar. Upon its discovery he wrote: "If in Rome or elsewhere, where there begins to be some knowledge of the Arabic language, one were to find someone with that ability, he could

---

45. See Stenhouse, *Reading Inscriptions* (2005). On the dialectic of scholarship and forgery more generally, see Grafton, *Forgers and Critics* (1990).

46. Aufrère, *La momie et la tempête* (1990), 276, cursorily and with no evident support identifies Barachias with the Alexandrian patriarch Eutychius. Daly, *Egyptology* (2005), 58, opines that he corresponds to Abu al-Barakat, also known as Ibn Kepir.

47. Peiresc to Kircher, Aix, 3 June 1634; 1 February 1636; 4 February 1637, APUG 568, fols. 372^{rv}, 219^r, 200^r; Kircher to Peiresc, Rome, 7 January 1637, BNF FF 9538, fol. 240^r; *PC*, 210, 250.

translate my book into Latin; and, as I will not fail to apply my industry, we could by means of the press propagate and communicate it to men of letters all over the world, [and] I would hope that by such means one could revive even this lost, dead Egyptian language of the Copts."[48]

By the beginning of the 1630s Coptic was the next frontier in Oriental studies: a lost language that promised to enrich the world of European learning as Greek, Hebrew, Syriac, and Samaritan had done before. Although some abortive efforts had been made, notably by Joseph Scaliger, the language and literature of the Egyptian Christians remained uncharted territory. "Today, we already have grammars of all the other Oriental languages," lamented Claude Saumaise in 1633, "and of this one there is nothing."[49] At the vanguard of the effort was, typically, Peiresc, whose interest had been piqued upon hearing about Della Valle's manuscripts in 1628. He set his mind to orchestrating the Coptic revival, making efforts to acquire manuscripts and casting about for a scholar worthy of so glorious but demanding an assignment. His first candidate was Jean Morin, a priest of the Congregation of the Oratory in Paris. Morin was translating the Samaritan Pentateuch for the polyglot Bible under preparation in Paris, and Peiresc had served as a go-between when Della Valle offered the use of his manuscript. In May 1631 Peiresc sent Morin three Coptic manuscripts he had recently acquired, but nothing seems to have come of Morin's Coptic studies.[50] Although some of Peiresc's Coptic texts had facing Arabic translations, real progress would require a dictionary and grammar. These only Della Valle possessed. Peiresc repeatedly wrote Della Valle asking him to send the manuscript or at least a copy, but without satisfaction.[51]

Della Valle had already chosen a local translator, the Franciscan Tomasso Obicini. Having learned Arabic and Syriac while stationed in Jerusalem, after returning to Rome, Obicini founded an Arabic school for missionaries at the convent of San Pietro in Montorio on the Janiculum.[52] Della Valle entrusted him with his manuscript, made an arrangement with the

---

48. Della Valle, *Viaggi* (1650–63), 397–98.

49. Saumaise to Peiresc, 22 January 1633, in Tamizey de Larroque, ed., *Correspondants de Peiresc, fasc. V* (1882), 26.

50. Peiresc to Dupuy, Aix, 14 July 1628, in Tamizey de Larroque, ed., *Lettres* (1888–98), vol. 1, 670; Peiresc to Holstenius, Boysgency, 24 September 1628, ibid., vol. 5, 293; Peiresc to Dupuy, Boysgency, 23 May 1631, ibid., vol. 2, 278–79. See Miller, "Peiresc and the Samaritans" (1997); Miller, "A Philologist, a Traveller and an Antiquary" (2001).

51. On Della Valle's Coptic manuscripts, see Hamilton, *Copts and the West* (2006), 201. The lexicon and grammar in question is now BAV Copt. 71.

52. Pizzorusso, "Les écoles de langue arabe" (2010), 69–72; Kleinhans, *Historia Studii Linguae Arabicae* (1930); Lantschoot, *Un Précurseur d'Athanase Kircher* (1948).

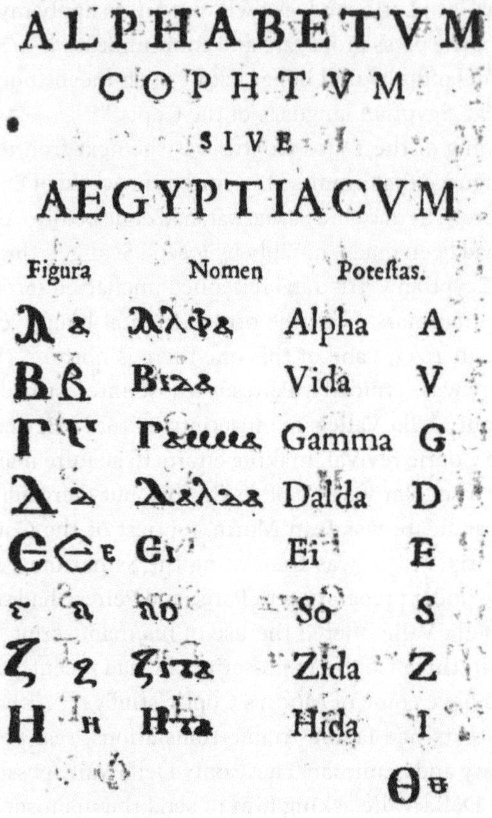

ALPHABETVM
COPHTVM
SIVE
AEGYPTIACVM

Figura	Nomen	Poteſtas.	
Ⲁ ⲁ	Ⲁⲗⲫⲁ	Alpha	A
Ⲃ ⲃ	Ⲃⲓⲇⲁ	Vida	V
Ⲅ ⲅ	Ⲅⲁⲙⲙⲁ	Gamma	G
Ⲇ ⲇ	Ⲇⲁⲗⲇⲁ	Dalda	D
Ⲉ ⲉ	Ⲉⲓ	Ei	E
ⸯ ⸯ	ⲥⲟ	So	S
Ⲍ ⲍ	Ⲍⲓⲧⲁ	Zida	Z
Ⲏ ⲏ	Ⲏⲧⲁ	Hida	I
		Ⲑ	ϴ

Fig. 20. The Coptic alphabet prepared by Tomasso Obicini and published in 1630 by the Polyglot Press of the Propaganda Fide. *Alphabetum Cophtum sive Aegyptiacum* (Rome: 1630). Bayerische Staatsbibliothek München, Signatur: L.as. 450,7.

Congregation for the Propagation of the Faith to print the completed work, and commissioned Coptic type. By 1630 the font was cut and a preliminary pamphlet was published containing the Coptic alphabet with Latin transliterations, as well as the Coptic version of Psalm 45 (fig. 20).[53] But, after this initial burst, momentum waned, as Obicini was busy with other projects.

At the beginning of 1633 Peiresc intensified his efforts, aiding two scholars who declared their desire to master Coptic: Samuel Petit, in Nîmes, and Claude Saumaise, at that moment in Leiden. Both were Huguenots widely admired for their classical learning as well as their expertise in Oriental

---

53. *Alphabetum Cophtum* ([1630]). Della Valle to Morin, Rome, 19 March 1630, *Antiquitates ecclesiae orientalis* (1682), 167.

languages. The only known Coptic manuscripts in France were the three belonging to Peiresc and a volume that François-Auguste de Thou had brought back from Egypt and left in the care of Jacques Dupuy, which contained a Coptic liturgy with facing Arabic translation. Peiresc did his best to satisfy Saumaise and Petit's competing requests for these limited resources, meanwhile seeking new manuscripts from the Levant. Then, at the end of 1632, Obicini died, and Peiresc renewed his efforts to wrest the lexicon and grammar from Della Valle, sending a letter in which he praised unnamed "worthy men" engaged in the study of Coptic on his side of the Alps. Della Valle wrote back, reporting that almost nothing of Obicini's translation had been found among his papers; since no one in Rome seemed ready to take up the task, he was, in principle, willing to send the manuscript to France. But he requested that Peiresc first send him samples of his candidates' abilities.[54]

On 16 January 1634, in Aix, Peiresc wrote to Petit requesting evidence of his expertise to forward to Della Valle.[55] He was unaware that the day before, in Rome, Della Valle had delivered his precious manuscript to Kircher, who had offered to carry out the translation, "at the request" of Peiresc and Barberini.[56] Although Peiresc later claimed credit as the promoter of Kircher's Coptic studies, he most definitely had not intended this turn of events. In Peiresc's plans, Kircher was to translate Arabic texts relevant to Egypt and the hieroglyphs, while Petit and Saumaise carried out the more urgent and important investigation of Coptic. Peiresc learned how his design had come undone in a letter from Jean-Jacques Bouchard, who visited Della Valle in February to plead on Peiresc's behalf for the transmission of the lexicon.[57]

Did Peiresc feel betrayed? His own protégé, who was only in Rome and in contact with Della Valle through Peiresc's good offices, had undermined his patiently nurtured plans, and at precisely the moment when the manuscript he had been "courting for ten years" seemed at last within his grasp. Or perhaps Peiresc appreciated the situation's irony. The previous September he had again written Della Valle, pressing him to send the manuscript. As before, he artfully avoided identifying the Protestant colleagues to whom he

---

54. Della Valle to Peiresc, 12 July 1633, BNF FF 9542, fols. 144ʳ-45ʳ.

55. Peiresc to Petit, Aix, 16 January 1634 and Aix, 10 February 1634, in Peiresc, *Lettres à Saumaise* (1992), p. 55, n. 11; p. 51 n. 4.

56. "Diario di Pietro della Valle di alcune cose memorabile," ASV Fondo Della Valle–Del Bufalo 186: "15 Gennaio. Io diedi al Pre Athanasio Kircher Giesuita il mio libro originale manoscritto antico delle grammatiche e vocabulario egitto in < ... > Arabo < ... > versione Latina per < ... > come si era offerto di fare di istanza del Sig. Peiresk, et anche del Sig. Card. Barberino..."

57. Jean-Jacques Bouchard to Peiresc, Rome, 11 February 1634, in Tamizey de Larroque, ed., *Correspondants de Peiresc, fasc. III* (1881), 26.

intended to entrust the translation, but he referred by name to Athanasius Kircher, S. J., who might pass through Rome on his way to Vienna, and whose edition of Barachias might benefit from consulting the dictionary. It would hardly be surprising if Della Valle associated Kircher with Peiresc's "worthy men."[58] Given Della Valle's publishing arrangement with the Propaganda Fide, it is hard to imagine that either of Peiresc's Calvinist candidates would have made an acceptable translator. If Della Valle had sent the manuscript to France for translation, it would have to have been published there as well.[59] In the transalpine race to deliver Coptic to the Republic of Letters, Rome stayed pure with the Jesuit Kircher replacing the Franciscan Obicini, while Catholic France, as in the Thirty Years' War then raging, threw its lot in with the Protestants, as Peiresc supplied his manuscripts to Saumaise.

Kircher was now in possession of the only known Coptic dictionary and grammar in Europe, coveted by the leading Oriental philologists. This placed him at great advantage to go to press with the first Latin version of the text, satisfying a major scholarly desideratum. But advantage was not yet victory. Peiresc continued to campaign behind the scenes in favor of Petit and Saumaise, expressing doubt that Kircher would be able to carry out the translation, and renewing his requests to Della Valle for the manuscript. Petit nonetheless withdrew from contention, graciously proclaiming his wish that Kircher should promptly publish the Coptic dictionary, as well as Barachias. Saumaise, who foresaw the glory to be won by delivering Coptic to the Republic of Letters, was less easily dissuaded. Peiresc encouraged him with the hope that Della Valle would take the manuscript back from Kircher, whom he described as irresolute, easily distracted, and bad at setting priorities.[60] At the same time he wrote Della Valle, suggesting that all in all, it would make more sense to send the manuscript to Peiresc's men in France to make the Latin translation, after which it could be returned to Rome for printing. (By now Peiresc's representations to Della Valle bordered on blatant deception.)[61] Meanwhile, he had three agents in the Levant seek-

---

58. Peiresc to Della Valle, 7 September 1633, ASV Fondo Della Valle-Del Bufalo, 52; Della Valle to Peiresc, Rome, 12 July 1633, BNF FF 9542, fol. 144ʳᵛ.

59. See Peiresc to Saumaise, Aix, 11 September 1635, in Peiresc, *Lettres à Saumaise* (1992), 168–70.

60. Peiresc to Petit, Aix, 21 March 1634, ibid., p. 51, n. 4; Petit to Peiresc, April 1634 (no day), in Tamizey de Larroque, ed., *Corréspondants de Peiresc*, fasc. XIV (1887), 57; Peiresc to Saumaise, Aix, 4 April 1634 and 10 April 1634, in Peiresc, *Lettres à Saumaise* (1992), 50–56, 80–82.

61. Peiresc to Della Valle, 6 April 1634, ASV Fondo Della Valle–Del Bufalo 52. See also Peiresc to Della Valle, 9 February 1634, in the same folder, where Peiresc's circumlocutions about the identity of his collaborators reach an apex.

ing out Coptic manuscripts on his behalf. The latter effort paid off: in October 1634 he reported to Saumaise the discovery of another Coptic-Arabic lexicon and grammar, which was being transcribed in the Levant and would soon be sent to France.[62]

Peiresc was playing all the angles. Once it became clear the Kircher was resolved to translate Della Valle's manuscript, Peiresc encouraged him and wrote letters to his patrons urging them to bring the work to completion as soon as possible. At the same time he supported the competing effort by Saumaise, who was his favorite. All Peiresc's efforts to acquire new Coptic texts were carried out for Saumaise's benefit—he even requested relevant material from Kircher and Barberini in Rome, without explaining that they might be used to facilitate a rival project. After the new exemplar from the Levant reached him in May 1635, his letters to Rome began to include vague warnings about other men working on a similar publication.[63] He advised Barberini that if the Roman edition did not appear soon, it might be preempted—without revealing that he himself was orchestrating the competition. In September, after learning that Kircher was only prepared to publish a preliminary study of Coptic rather than an edition of the grammar and lexicon, Peiresc turned blunt, informing Barberini that he possessed his own exemplar of the Coptic-Arabic text: until now he had dawdled for the sake of Barberini's honor, but the Romans had to publish the lexicon soon, for he could not go on making excuses to worthy scholars eager to work on its translation. A few days later, he sent Saumaise the lexicon and grammar (he had been waiting over the summer for a copy to be made) along with other newly acquired Coptic manuscripts.[64]

---

62. Agathange de Vendôme to Peiresc, Cairo, 18 March 1634, in Valence, ed., *Correspondance* (1892), 24–26; Peiresc to Gilles de Losches, Aix, 22 December 1633, 20 May 1634, ibid., 14–15, 52; Peiresc to Cassien de Nantes, Aix, 29 September 1635, ibid., 193; Peiresc to Saumaise, 2 October 1634, in Peiresc, *Lettres à Saumaise* (1992), 134.

63. Peiresc to Petit, Aix, 7 April 1634, in Peiresc, *Lettres à Saumaise* (1992), p. 51, n. 4. Peiresc to Kircher, Aix, 3 June 1634; 30 March 1635; 10 June 1635, APUG 658, fols. 372^r–73^v; 354^r–55^v; 194^{rv}; Peiresc to dal Pozzo, Aix, 29 June 1634, 29 December 1634, and 3 May 1635, in Peiresc, *Lettres à dal Pozzo* (1989), 140, 161–65, 184; Peiresc to Kircher, Aix, 14 August 1635, 8 October 1635, and 1 February 1636, APUG 568, fols. 362^{rv}, 369^r, 219^r; Peiresc to Saumaise, Aix, 4 April 1634, in Peiresc, *Lettres à Saumaise* (1992), 55–56; Saumaise to Dupuy, Leiden, 5 June 1634, in Saumaise, *Epistolarum liber* (1656), 96; Saumaise to Peiresc, Leiden, 1 June 1635, in Tamizey de Larroque, ed., *Correspondants de Peiresc, fasc. V* (1882), 34. On the arrival of the dictionary, see Peiresc to J.-A. de Thou, 14 May 1635, Valence, ed., *Correspondance* (1892), 131.

64. Peiresc to Barberini, Aix, 17 June 1635 and 7 September 1635, BAV Barb. Lat. 6503, fols. 135^r, 148^r; Peiresc to Saumaise, Aix, 11 September 1635 and 25 September 1635, in Peiresc, *Lettres à Saumaise* (1992), 160–63, 197–200.

For all his wile, one cannot impugn Peiresc's motives. He simply wanted what was best for the Republic of Letters, that an edition of the Coptic lexicon appear as soon as possible. Saumaise may have been his favorite, but Peiresc never tried to impede Kircher's progress. On the contrary, he did what he could to see that Kircher completed his edition expeditiously.

At the beginning of September 1636 Kircher's *Coptic Forerunner* issued from the Propaganda's press. As foreseen, it was not a Latin edition of the lexicon and grammar, but a preliminary study in anticipation of that work. After Peiresc received a copy, he encouraged Saumaise to continue his translation. Bitter at having lost the chance to publish the first study of Coptic, Saumaise undertook a critique of *Coptic Forerunner* (Peiresc dissuaded him from carrying it through) and continued his plan to publish his own edition of the dictionary and grammar.[65] But the easily distracted Jesuit turned out to have greater staying power. After Peiresc's death in 1637, Saumaise abandoned his Coptic studies while Kircher went on to publish *Egyptian Language Restored* (Rome, 1643), containing a Latin translation of the Coptic-Arabic lexicon and grammar.[66] Together with *Coptic Forerunner*, it earned Kircher an enduring reputation as the founder of European Coptic studies.[67]

KIRCHER'S COPTIC

For Kircher, Coptic was above all a tool for understanding ancient Egyptian history and the hieroglyphs. *Coptic Forerunner's* central argument was that Coptic was closely related to the language spoken by the ancient Egyptians who produced the hieroglyphs. Disputing earlier claims that Coptic and Egyptian were related to languages such as Chaldean, Syriac, Ethiopian, Arabic, and Armenian, Kircher claimed they were instead most closely related to Greek. With words he would soon regret, he suggested that Greek derived from Egyptian and that in ancient times the two languages had been mutually intelligible, being as similar as Italian and Spanish or French and Latin.[68] His account was not entirely so naive as such statements suggest: he specified that Coptic resembled Greek only with respect to vocabulary, not grammar or syntax. In that case, however, one wonders how they could have been as close as Spanish and Italian.

---

65. Peiresc to Saumaise, Aix, 22 September 1636, 29 November 1636, and 15 December 1636, ibid., 319, 329–31, 352.
66. Kircher, *Lingua Aegyptiaca restituta* (1643).
67. See chapter. 8, "Oedipal Conflicts among the Copticists."
68. *PC*, 171–72.

The similarities and filiations among different nations, their languages, and their alphabets was the dominant theme of *Coptic Forerunner*. Kircher argued that in early Christian times the Coptic-Ethiopian Church founded colonies throughout the Near East and as far away as India and China. As evidence he presented the Sino-Syriac monument, a bilingual stele discovered in 1625 by Jesuits in China, which testified to the early presence there of Christianity.[69] Kircher went on to describe the routes supposedly used by the apostles to reach distant regions of Asia. By showing the extent of intercourse between the Mediterranean and the Near and Far East, he claimed that "Egypt was the nursery not only of the progress of the Christian religion in distant lands, but also of all the superstitions propagated to the world at every time," carried by Egyptian priests along the same routes later used by the apostles.[70]

Kircher here laid the groundwork for *Egyptian Oedipus*, contending that the Indians and other Asians received their wisdom from the Egyptians, as demonstrated by the common worship of heavenly bodies and animals in ancient Egypt, India, China, Japan, Tartary, Cathay, and even America.[71] The book was thus a forerunner not only to Kircher's forthcoming *Egyptian Language Restored*, containing the grammar and lexicon, but also to *Egyptian Oedipus*. This was explicit in its subtitle, which promised to reveal "not only the origin, age, vicissitudes, and changes of the Coptic or Egyptian language ... but also the restoration of hieroglyphic literature." The final chapter of *Coptic Forerunner* provided a preview of the method by which Kircher promised to translate the hieroglyphs. After first discussing inscriptions that seemed to mix "Egypto-Coptic" writing with hieroglyphic symbols, Kircher presented the long-promised exhibition of his technique— a "specimen of hieroglyphic interpretation"—in which he translated a fragment from the Bembine Table (fig. 21).[72] Kircher never lost sight of the fact that Coptic was a means to a greater end: the investigation of the hieroglyphic doctrine that would take full form in his magnum opus. The final appendix to *Coptic Forerunner* contained a detailed preliminary outline of *Egyptian Oedipus* and an open appeal to the Republic of Letters to send him relevant material.

Kircher's belief in the affinity between Coptic and Greek demonstrates

---

69. Billings, "Jesuit Fish in Chinese Nets" (2004); Mungello, *Curious Land* (1985), 164–72.
70. *PC*, 86, 119–20.
71. *PC*, 121.
72. *PC*, 238–77, discussed in Godwin, *Athanasius Kircher's Theatre of the World* (2009), 61–63.

Fig. 21. A "hieroglyph" from the Bembine Table. Kircher's interpretation in *Coptic Forerunner* served as a preview of the method he would use in *Egyptian Oedipus*. *Prodromus Coptus* (Rome: 1636), 239. Courtesy of Stanford University Libraries.

his limited comprehension of the language whose modern study he is famous for founding. Two "Coptic" poems that he composed in 1637 or 1638 for inclusion in a posthumous tribute to Peiresc, though written with Coptic characters, more closely resembled classical Greek.[73] Kircher had rushed to press in 1636, and *Egyptian Language Restored*, published in 1643, demonstrated deeper knowledge, including a retraction of the claim that Greek derived from Coptic. Nonetheless, even the more mature work contained serious linguistic errors, not just in the Coptic text but in Arabic as well.[74]

The section of *Coptic Forerunner* most indicative of Kircher's scholarly limitations had nothing to do with Coptic. Kircher devoted half of the eighth chapter—ostensibly concerned with the "usefulness of Coptic"—to an inscription, carved in a mysterious alphabet that had been discovered by Tomasso Obicini at the foot of Mount Horeb in the Sinai (fig. 22). The inscription had eluded all prior efforts at decipherment and Kircher declared

---

73. The poems appeared in Bouchard, ed., *Monumentum Romanum* (1638), 96. See Chaine, "Une composition oubliée du Père Kircher" (1933–34), 207–8.

74. *LAR*, 507. See also *OP*, 142–45. For a critical assessment of Kircher's Coptic studies, see Hamilton, *Copts and the West* (2006), 195–228.

Fig. 22. The Mount Horeb inscription, discovered in the Sinai by Tomasso Obicini. Against Peiresc's protestations, Kircher published his interpretation, which claimed that ancient Jewish sages foretold the virgin birth. *Prodromus Coptus, sive Aegyptiacus* (Rome: 1636), 204, 207. Courtesy of Stanford University Libraries.

that although it did not pertain to Coptic or anything Egyptian, he would be remiss not to discuss it. He described how he used his "usual combinatory method," trying to match the inscription's characters with three Oriental languages, Hebrew, Samaritan, and Syriac, until he achieved success. He concluded that the inscription was written in a form of ancient Chaldean, which some call Assyrian and "Targumic," others Aramaic, Lebanese, or Babylonian, and which was used in the time of the kings of Israel and the Babylonian captivity. It had been difficult to identify, he asserted, because it had been carved in the style of a monogram in order to conceal mysteries from the masses. With a bold leap of epigraphic imagination, Kircher transformed the characters into their supposed Hebrew analogues, revealing the inscription to read ייי עלמתא יבטן, or, in Kircher's translation, *Deus Virginem concipere faciet* (God will make a virgin conceive). In a final burst of creativity, he argued that the final word also encoded the phrase ותבטן בן, *Et illa pariet filium* (And she shall bring forth a son). According to Kircher, the form of the inscription demonstrated that it had been composed several centuries before Christ. Thus it was evidence that ancient Hebrew wise men had predicted the virgin birth, a claim he supported by appealing to kabbalistic literature.[75]

KIRCHER AGONISTES

A defensive undercurrent runs through all of Kircher's works. In his autobiography he wrote of a backlash that supposedly greeted his precocious success: "During this time I came to be persecuted in no small way, when learned men, considering the enviable novelty of my difficult argument, along with my young age . . . not only questioned my trustworthiness, but also made false accusations, calling me Thraso and impostor."[76]

In truth, there was more than jealousy behind the doubts that met Kircher's work. From the start of his career he was the object of both accolades and questions about his critical powers, sometimes from the same individuals. After receiving criticism from Peiresc and a warning about the possibility of a rival edition of the Coptic lexicon appearing in France, Kircher responded in the following terms:

---

75. *PC*, 200–218. An earlier, shorter manuscript of the text is extant: "Athansii Kircherii e soc: Jesu / Scripturae mirabilis et toto oriente celebratissimae in monte Sinai rupi cuidam incisae / Interpretatio Nova et Antehac a nemine adhuc enodata. / Ad Eminentiss: et Reverendiss: S. R. E. Cardinalem Franciscum Barbarinum Maecaenatem munificentissimum," BNF Dupuy 448, fols. 161ʳ–63ʳ.

76. *Vita*, 54.

I am hurt that you have so small an opinion of me that you judge that I presumptuously wish to undertake a matter of which I am ignorant; indeed you waver about my good faith. Surely pretense, falsehood, and whatever is contrary to true and genuine sincerity are so foreign to me that I would prefer that all my scholarly labors perish, rather than commit such a crime in the Republic of Letters. I do what I can; accepting or rejecting what I have composed is for my superiors; I am ready for and indifferent to either. You may think that my works proceed from some sort of compulsive desire to write, or from vainglory and appetite for praise, which I despise as diametrically opposed to piety; but I only undertook these matters lest I seemed to fail in my duty with respect to the talents granted to me by God in his infinite goodness.[77]

Peiresc's letter does not survive, but elsewhere he always delivered criticism gently. Kircher's defensive response surely expressed awareness of criticism from other quarters. In a subsequent letter Peiresc assured Kircher that he had never doubted his good faith and tried, delicately, to restate his concerns. He explained that it was not Kircher's trustworthiness but that of his sources that worried him. He brought up the example of Herwart von Hohenburg, who had allowed inaccurate information to discredit his hieroglyphic studies. "It is better to proceed more slowly and do a better job," he advised, "especially given the importance of the things you treat."[78] Peiresc trusted Kircher's good intentions, but not his critical acumen, and admonished him to establish his claims by citing sources and evaluating them carefully. Wary of Kircher's penchant for speculation, he repeatedly advised him to avoid writing long commentaries on his Oriental texts and to con-

---

77. "Ad Copticum dictionarium quod attinet, et de cuius fideli versione tàm sollicita est, ut etiam me usque ad alterius, nescio cuius in galliam recèns allati proditionem expectare velit. Doleo ego me D. V. tam exiguam opinionem habere, ut eam rem quam ignorem, praesumptiosiùs me aggredi velle arbitretur; imò de fide meâ vacillet; certè fucus, fallacia et quidquid verae et germanae sinceritati contrarium est, à me procul absunt; ut meas omnes lucubrationes perire malim, quàm huiusmodi scelus in Rempublicam literariam comittere. Facio quod possum, superiorum meorum arbitrio quae composui vel admittendo vel reprobando, ad utrumque paratus sum et indifferens; nec enim opera mea ex nescio quo scripturientis animi pruritu, aut ex vanâ gloriâ et aestimationis appetitu quam veluti religioni è diametro contrariam detestor, procedere putet; sed solùm me ea aggredi, nè talentis à Deo opt. max. prò suâ infinitâ bonitate mihi concessis defuisse videar." Kircher to Peiresc, Rome, 8 February 1635, BNF FF 9362, fols. 13^v–14^r; NAF 5173, fol. 25^v (copy).

78. Peiresc to Kircher, Aix, 30 March 1635, APUG 568, fol. 365^v. See also Peiresc to dal Pozzo, Aix, 4 May 1634, in Peiresc, *Lettres à dal Pozzo* (1989), 132, in which he expresses his worry that Kircher's attempt to explain the hieroglyphs will lead to an embarrassment like the "benedetta calamità" of Herwart.

centrate instead on making accurate translations. Above all, Peiresc was skeptical about Kircher's interpretations of hieroglyphs, doubting the very possibility of ever arriving at a convincing solution. He cautioned Kircher numerous times not to rush to print, lest he publish something that he would later regret. He likewise tried to dissuade Kircher from publishing his interpretation of the Mount Horeb inscription, which he feared would damage Kircher's reputation.[79]

Kircher, heedless, forged ahead, publishing his interpretation of the inscription and, eventually, his expositions of the hieroglyphs. But despite his characteristic displays of self-assurance, he knew that his conclusions were vulnerable and his authority contested. In an apologetic pattern repeated throughout his publications, Kircher appealed to external authorities in order to establish the reliability of his claims. To appease doubts about his Christological decipherment of the Mount Horeb inscription, for instance, he adduced the names of eminent linguists who supposedly accepted his interpretation.[80] The plaudits by experts in different languages at the beginning of *Coptic Forerunner* were similarly tendered as testimony to the credibility of his use of sources in those languages. This strategy reached its apogee at the beginning of *Pamphilian Obelisk*, where Kircher, anticipating his critics, printed attestations not only from linguistic experts testifying to the reliability of his quotations from Oriental sources, but also from the curators of the museums and cabinets of antiquities whose objects Kircher depicted and used as evidence.[81]

A CAREER IN ORIENTAL STUDIES

Despite its shortcomings, *Coptic Forerunner* won praise and marked a milestone in Kircher's career, bringing him to the attention of scholars throughout Europe as an investigator of Oriental languages and antiquities. Eminent students of Hebrew and Arabic established correspondence with him after reading *Coptic Forerunner*, including the great Jesuit chronolo-

---

79. Peiresc to Kircher, Aix, 30 October 1636; 4 February 1637, APUG 568, fols. 217^r–18^v; 200^r–201^v; Peiresc to dal Pozzo, Aix, 29 April 1636, in Peiresc, *Lettres à dal Pozzo* (1989), 237; Peiresc to [Jacques] de Saint-Sauveur Dupuy, Aix, 22 April 1636; 13 May 1636, in Tamizey de Larroque, ed., *Lettres* (1888–98), vol. 3, 474; 484. See also Peiresc to Gabriel Naudé, 5 June 1636 and Peiresc to Saumaise, Aix, 2 June 1636, in Peiresc, *Lettres à Saumaise* (1992), 284.

80. *PC*, 160. In the 1670s, Antonio Baldigiani, a younger professor at the Collegio Romano, claimed that Kircher had the habit of fabricating such testimonies without authorization; Findlen, "Living in the Shadow of Galileo" (2009), 236.

81. *PC*, ††1^v; *OP*, f4^r–g3^r.

ger Denis Petau in Paris, as well as Protestants like Johannes Buxtorf the younger in Basel, and Johannes Elichmann in Leiden. Peiresc, for all his reservations, did not hesitate to identify himself with Kircher's publication, proudly claiming to have been its instigator.[82]

After the publication of *Coptic Forerunner* in fall 1636, Kircher's correspondence with Peiresc returned to the Barachias translation and his hieroglyphic studies. Peiresc restated his long-standing opinion that Kircher should first publish translations of Barachias, the Coptic lexicon, and other Oriental texts, and that only after, when the time would be ripe—in Peiresc's private thoughts this likely meant never—should he complete his work on the hieroglyphs. At first, to Peiresc's consternation, Kircher reiterated his opinion that the Barachias treatise could not be published on account of its superstitious magical content. He declared his intention instead to incorporate passages from Barachias and other Arab authors in *Egyptian Oedipus*, arguing that framing them in a larger interpretative work would render them acceptable to the ecclesiastical censors. But shortly thereafter he changed plans, writing that he would collaborate with the Maronite scholar Abraham Ecchellensis on a translation of selections from treatises by Barachias and two other Arabic authors.[83] This encouraging news was the last Peiresc heard on the subject before he died on 24 June 1637. By then, Kircher was on Malta.

Kircher eventually became famous as a "polymath," celebrated for the encyclopedic breadth of his studies, not least his investigations of natural philosophy. This image has dominated modern studies, which subsume Kircher's investigation of the hieroglyphs within his larger quest for "universal knowledge."[84] This approach offers a valuable perspective from which to discern the common ground beneath his various studies of nature, philosophy, and history. A recently discovered manuscript, probably composed shortly after his arrival in Rome, indicates that Kircher once conceived of his hieroglyphic studies as part of a massive "Universal History of the Char-

---

82. Petau to Kircher, Paris, 15 February 1643; 3 August 1645, APUG, 567 fols. 210^r-11^v, 69^{rv}; Buxtorf to Kircher, Basel, 1 August 1637; 1 May 1639, APUG 557, fols. 364^r-65^r, 322^v; Elichmann to Kircher, Leiden, 29 July 1638, APUG 557, fols. 366^r-67^v; Peiresc to de Losches, 23 December 1636, in Valence, ed., *Correspondance* (1892), 268–69; Peiresc to Cesaire de Roscoff, 27 December 1636, ibid., 303.

83. Kircher to Peiresc, Rome, 3 December 1636, BNF FF 9538, fols. 236^r-37^r; Kircher to Peiresc, Rome, 7 January 1637, BNF FF 9538, fol. 240^{rv}. See also Peiresc to Kircher, Aix, 4 February 1637, APUG 568, fols. 200^r-201^v.

84. Most importantly, Leinkauf, *Mundus Combinatus* (1993). See also Schmidt-Biggemann, *Topica Universalis* (1983), 176–86; Daxemüller, "Welt als Einheit" (2002); Englmann, *Sphärenharmonie und Mikrokosmos* (2006); Vercellone and Bertinetto, eds., *Athanasius Kircher* (2007).

acters of Letters and Languages," which would also have included treatises on the origin and relationship of all known languages, cryptography and steganography, the Kabbalah, magic and divination, and the combinatory art of Ramon Llull.[85] But the interpretation of Kircher as a "universal scholar" obscures other aspects of his career. Kircher quickly abandoned his proposal for the universal history (although he eventually returned to many of its components in later studies). Instead, as this chapter has shown, during his first years in Rome he narrowed his vision. It proved a successful strategy. Capitalizing on his possession of two rare Arabic manuscripts, Kircher first won fame as an Oriental philologist.

It is not an exaggeration to say that Barachias Nephi, though possibly imaginary, made Kircher's career. Arriving in Provence a young, unknown scholar from Germany, Kircher parlayed his possession of a mysterious Arabic manuscript into the patronage of Peiresc and Barberini, two of Europe's most influential promoters of learning, catapulting himself onto the center stage of the Republic of Letters. Peiresc's correspondence leaves no doubt that he took an interest in Kircher because of the Barachias manuscript. Prior to Kircher's undertaking the translation of the Coptic-Arabic dictionary, Peiresc rarely if ever mentioned him without referring to the Arabic author. Even afterwards, the treatise remained the touchstone of their relationship and the subject of their last surviving letters. Because he feared that Kircher would be unable to complete his edition of Barachias in Vienna, Peiresc appealed to Barberini, who brought Kircher to Rome for that express purpose. Without Barachias, Kircher would have gone to Vienna to teach mathematics and might have spent the rest of his life in Austria.

From the point of view of Peiresc scholarship, the Barachias episode, like the story of the Coptic lexicon and grammar, demonstrates the gulf between Peiresc's rigor and judiciousness and Kircher's indiscriminate, perhaps unscrupulous, methods. There is no denying the differences between their scholarly styles. Peiresc was undoubtedly right to intuit that the Arabic treatise did not contain older, more authentic traditions about Egypt and the hieroglyphs than those already known from Greek and Latin literature, an insight that would have precluded Kircher's interpretation of the hieroglyphs had he not so adamantly refused to accept it. But from the point of view of Kircher scholarship, which has paid little attention to Peiresc's influence or to Kircher's study of Near Eastern sources, these stories hold a different lesson. They show the extent to which Kircher's studies were

---

85. "Characterum literarum linguarumque totius universi historia universalis," BAV, Barb. Lat. 2617, fols. 33–34. See Stolzenberg, "Universal History" (2011/2012).

shaped by Peiresc and his world. In the first half of the seventeenth century, the critical philology and antiquarian scholarship associated especially with French and Dutch savants was not so clearly differentiated from the encyclopedism that flourished most luxuriantly among central European polymaths.[86] As subsequent chapters demonstrate, baroque Rome provided a fertile environment in which these strains of erudition could mingle.[87]

86. Häfner, *Götter im Exil* (2003); Mulsow, *Moderne aus dem Untergrund* (2002); Rossi, *Logic and the Art of Memory* (2000); Schmidt-Biggemann, *Topica Universalis* (1983); Grafton, "World of the Polyhistors" (1985).

87. Arguably, the balance of scholarly styles shifted over time, and in the last decades of his life Kircher's work had an increasingly central European character; see Findlen, "Living in the Shadow of Galileo" (2009), 237.

CHAPTER THREE

Oedipus in Rome

Nearly twenty years have passed, kind Reader, since, during the pontificate of Urban VIII, I was called to Rome from France in order that I might demonstrate a certain model of hieroglyphic interpretation on the Roman obelisks; for rumor reported me to be toiling at the restoration of this hitherto unknown literature from an ancient Arabic manuscript.
—Athanasius Kircher[1]

CATHOLIC COSMOPOLIS

Upon returning to Rome from Malta in spring 1638, Kircher was eager to return to the unfinished edition of the Coptic-Arabic lexicon and his ever-expanding hieroglyphic studies. As he brought both projects to completion in the following years, he encountered numerous obstacles, which he overcame only with the assistance of new and powerful patrons and by harnessing the exceptional resources that made Rome, even in the reactionary climate of the second third of the seventeenth century, a vital center of scholarship, especially in the fields of antiquarian and Oriental studies. Kircher's investigations of Egypt and the hieroglyphs, culminating in the monumental *Egyptian Oedipus*, were products of a network of institutions, collections, and individuals centered in Rome and extending throughout Europe and the world.[2]

---

1. *OP*, a4^r.
2. On Roman science and scholarship, see Donato and Kraye, eds., *Conflicting Duties* (2009); Romano, "Rome, un chantier" (2008); Romano, ed., *Rome et la science moderne* (2008); Romano, "Sciences, activités scientifiques, et acteurs" (2002); Romano, "Il mondo della scienza" (2002); Rietbergen, *Power and Religion* (2006); Burke, "Rome as Center of Information" (2002); Grafton, ed., *Rome Reborn* (1993); Baldini, "Christoph Clavius" (1983).

When Kircher had learned in 1635 that a rival edition of the Coptic lexicon was underway in France, he wrote Peiresc to defend his enterprise. Stung by his mentor's lack of faith, he argued that, being in Rome, he was uniquely qualified to carry out the translation. "Know that I am not relying only on my translation," he wrote,

> for [the work] has been reviewed by men with the greatest knowledge of Arabic. Thus I would judge that no one could carry it out more correctly elsewhere. Whoever there is in France who may wish to attempt the same thing, surely let him know that I consider that if he shall not have the support of Coptic and Arabic books, which I have in abundance, if he has not been instructed with a decent knowledge of Coptic, if he shall not have experts in every kind of language to whom he can turn regarding the doubts that constantly occur in this sort of work, he will achieve nothing, and shall not arrive at the true and genuine translation.[3]

Kircher may have underestimated the talent and ingenuity of his French rival, Claude Saumaise, but he made an irrefutable point about the advantages that his location afforded him.

Rome had been at the forefront of Oriental studies since the Renaissance. The city's preeminence in the realm of textual materials reached back to the founding of the Vatican Library in the fifteenth century, when Oriental manuscripts, including gifts from the Coptic Patriarch, were among its original holdings. The library's collection of such materials gradually expanded during the sixteenth century and received a major boost in 1622, when Catholic forces took Heidelberg and General Tilly awarded the famous Palatine Library as booty to Pope Gregory XV. Transported to Rome under the direction of Leone Allacci and later catalogued by Lucas Holstenius, the *Fondo Palatino* dramatically augmented the Vatican's assets, above all in the realm of Greek and Latin manuscripts, but also contributed valuable Oriental works, including manuscripts that had once belonged to

---

3. "Sciat igitur D. V. me non tantùm versioni meae, utpote à versatissimis viris ac arabicae linguae peritissimis iam revisae, non confidere, sed ità confidere, ut neminem alibi veriùs eam expedire existimem; ponamus enìm in Galliâ esse, qui simile quid tentare velit, certò tamen sciat, eum si non habuerit subsidia librorum et Coptorum et Arabicorum quibus ego abundo, si non mediocro linguae Coptae cognitione fuerit imbutus, si non habuerit viros omnigenis linguis peritos, ad quos in dubiis quae in huiusmodi opere infinita occurrunt, recurrat, nihil efficet, nec ad veram et genuinam interpretationem perveniet." Kircher to Peiresc, Rome, 8 February 1635, BNF FF 9362, fol. 14^r; BNF FF 5173, fols. 25^v–26^r (copy).

Guillaume Postel.[4] By Kircher's time the library possessed an extraordinary collection of Oriental texts, probably the best in Europe, reflected in the abundance of references in his publications to Arabic, Hebrew, Chaldean, Syriac, Samaritan, Armenian, and Coptic manuscripts consulted at the Vatican.

Rome was also Europe's leader in Oriental typography. The city's supremacy dated to the *Typographia Medicea*, founded in 1584, the first publishing enterprise devoted to Arabic texts.[5] In the seventeenth century the tradition was carried on by the Congregation for the Propagation of the Faith, established in 1622 to oversee Catholic missionary programs. The Propaganda promoted linguistic studies in order to train missionaries, and its printing office, the famous Polyglot Press, published grammars, dictionaries, and devotional materials in the languages of potential converts. Drawing on an incomparable, continually expanding collection of typefaces, the Propaganda's signature publications were a series of pamphlets displaying exotic alphabets, printed both as didactic aids and as showpieces of the Church's resources and power to transmit its message universally. Within a decade of its founding, the press issued works using Arabic, Syriac, Chaldean, Estrangelo (an archaic Syriac alphabet), Georgian, Coptic, Ethiopian, Persian, and Samaritan characters.[6] Not the least of Kircher's advantages in his bid to go to press before Saumaise was the Coptic font, cut in 1630 to print Obicini's alphabet, which waited at his ready (see fig. 20). *Coptic Forerunner* was published by the Propaganda Fide, and with its Coptic, Syriac, Arabic, Estrangelo, Samaritan, Armenian, Hebrew, and Ethiopian text, it amounted to a catalog of the Polyglot Press's typefaces.

In addition to texts and typography, Rome offered Kircher access to Egyptian and Oriental antiquities on a scale that no European city could rival. Most spectacular was the open-air museum of obelisks scattered about the city, which provided the foundation of Kircher's hieroglyphic project. In addition, Rome was home to numerous museums and cabinets of antiquities which included Oriental artifacts. Kircher himself amassed a significant collection of such material in his museum at the Collegio Romano. (Its Egyptian holdings eventually formed the nucleus of Turin's Egyp-

---

4. Levi della Vida, *Ricerche* (1939), 290–337; Hamilton, "Eastern Churches" (1993); Bignami Odier, *Bibliothèque Vaticane* (1973), 107–8, 114, 123–24; Bepler, "Vicissitudo Temporum" (2001); Proverbio, "Alle origini delle collezioni" (2010).

5. Jones, "Medici Oriental Press" (1994); Vervliet, "Robert Granjon" (1967).

6. Pizzorusso, "L'indagine geo-etnografica" (2005); Pizzorusso, "I satelliti di Propaganda Fide" (2004); Pizzorusso, "Antipodi di Babele" (2000); Henkel, "Polyglot" (1971).

tological Museum.) His location at the hub of the Society of Jesus's global information network facilitated the steady flow of information and artifacts from missionaries passing through the college on their way to and from distant lands, while Jesuits at European courts played a similar role, conveying Kircher representations of Oriental antiquities from local collections.[7]

Above all, Rome possessed unique human resources. Manuscripts and typefaces were of little value without scholars capable of reading and writing Oriental languages. A city uncommonly open to foreigners, at the center of the expanding and increasingly interconnected Catholic world, Rome was home to a diverse community of linguistic talent. This was due in particular to Gregory XIII (1572–85), whose policy of outreach to Eastern churches created an enduring legacy of Orientalist expertise, including Europe's first regular chair in Arabic at the University of Rome.[8] Most important were two colleges that Gregory founded for Oriental students. The College of Neophytes, established in 1577, offered instruction to new Christians, primarily converts from Judaism, but also those from Islam and occasionally from dissident Eastern churches.[9] The College of Maronites, founded in 1584, served the dual purposes of educating Arab Christians from the Lebanon (the Maronites were formally in communion with the Roman Catholic Church) who would propagate Tridentine Catholicism in their homeland, and of bringing linguistic experts to Rome, who would educate missionaries. After 1627, the Collegio Urbano, founded by Urban VIII to train missionaries under the auspices of the Congregation for the Propagation of the Faith, also brought students from the Near East.[10] The colleges served as anchors of permanent, if small, communities of Jewish and Moslem converts as well as Oriental Christians, who taught languages, worked in the Vatican Library, and enabled relatively easy commerce between Rome and the Levant. They also housed small but significant libraries. Kircher made use of Hebrew manuscripts from the College of Neophytes in his study of

---

7. On Roman collections, see Stenhouse, "Visitors, Display, and Reception" (2005); Herklotz, *Cassiano dal Pozzo* (1999); Findlen, *Possessing Nature* (1994); Whitehouse, "Towards a kind of Egyptology" (1992), 74–79; Roullet, *Egyptian* (1972). On Kircher's museum, see Mastroianni, "Kircher e l'Oriente" (2001); Leospo, "Collezione egizia" (2001), 125–30; and references above, Introduction, n. 27.

8. Russell, ed., *"Arabick" Interest* (1994), 4. Although Guillaume Postel taught Arabic at the Collège de France from 1538 to 1542, instruction did not continue; see Toomer, *Eastern Wisedome* (1996), 26.

9. Levi della Vida, *Ricerche* (1939), 405–11; "Neofito" (1840–61).

10. Grégoire, "Costituzioni" (1977), 176–80; Heyberger, *Chrétiens du Proche-Orient* (1994), 408–9; Pizzorusso, "Les écoles de langue arabe" (2010), 248–56.

the Kabbalah.[11] At the College of Maronites he found his favorite Syriac source, the *Philosophia* of Mor Isaac, as well as Arabic texts that aided his study of Coptic.[12]

Kircher developed important working relationships with two associates of the Neophyte and Maronite colleges. Giovanni Battista Iona, a converted Jew from Safed and erstwhile jeweler at the Polish court, who worked as a *scriptor* at the Vatican Library and taught Oriental languages at La Sapienza and the Neophyte College, assisted Kircher with Hebrew and biblical Aramaic.[13] Even though Hebrew had been institutionalized in much of Christian higher education by the seventeenth century, scholars wishing to delve deeply into Jewish literature still depended on the guidance of "native informants." Sometimes practicing Jews supplied this service; Kircher acknowledged two Roman rabbis who served him as research assistants.[14] But in the age of the ghetto, forced preaching, and the yellow star, the presence of educated converts in Rome made Jewish learning considerably more accessible.

Kircher's most significant relationship was with the remarkable Maronite scholar Ibrahim al-Haqilani, known in Europe as Abraham Ecchellensis. Born in the Lebanon in 1605, Ecchellensis first came to Rome in 1620 to study at the College of Maronites, afterwards taking a degree in philosophy at the Collegio Romano. In 1631 he left for the Levant and began a diplomatic career in the service of the Druze Emir Fakhr al-Din. His assignments included business in Tuscany with the emir's Medici allies, and, following Fakhr al-Din's execution by the Ottomans in 1635, he taught briefly at the University of Pisa. In 1636 Ecchellensis was called by the pope to teach Arabic and Syriac at the Sapienza University and participate in the Arabic translation of the Bible that had been underway since the

---

11. Besides Moses Cordovero's *Pardes Rimmonim*, discussed in ch. 5, Kircher explicitly cites at least two other manuscripts from the Neofiti library: a "Commentarius vetus Hebraicus M.S." (*OP*, h3^r) and "Bet Melchisedech" (*OP*, 23; see also *OP*, 194, *OA* II.2, 145).

12. *OP*, c1^r. Kircher never realized his intended edition of Mor Isaac, though a short Latin excerpt survives as APUG 561, fol. 90^r. He quoted the manuscript abundantly as evidence of "Syrian" beliefs: e.g., *PC*, 272; *OP*, 25, 158, 320; *OA* I, 171, 257, 320; *OA* II.1, 7, 407, 417–18, 424; *OA* II.2, 148. Assemani, *Bibliotheca Orientalis* (1719), 461–62, refuted Kircher's attribution of this manuscript to the Nicene father, Isaac Syrus or Ninivita. For manuscripts consulted in aid of his Coptic studies, see *PC*, 186, 195.

13. Kircher's relationship to Iona went back at least to 1636, when Iona contributed a "rabbinic rhyme" to *Prodromus Coptus* (*PC*, †††2^v). For Iona's linguistic assistance, see *OP*, g2^v; *OA* I, b2^r. See also *OA* II.2, 238, for Iona communicating a Hebrew book with information on Egyptian astrology. A collection of notices on Iona's life and works can be found in the introduction of Carmignac, ed., *Evangiles* (1982).

14. *OA* I, b2^{rv}.

1620s.[15] Like Kircher with Barachias beneath his arm three years earlier, Ecchellensis came to Rome bearing Oriental manuscripts. Even before his arrival, rumors of his chest of Arabic and Syriac books circulated among Roman scholars. Writing to Peiresc, Holstenius singled out a particularly noteworthy item in Ecchellensis's cargo: an "ancient history of Egypt," which he and Kircher "passionately awaited."[16]

Kircher wasted no time to examine the intriguing manuscript, a treatise by an author called Gelaldinus (Jalal al-Din al-Suyuti), which Ecchellensis had acquired in North Africa while on a mission to ransom slaves. (Kircher promptly sent a brief outline to Peiresc, who forwarded it to the Dutch Arabist Jacob Golius and wrote to the Levant in search of another copy.)[17] Kircher and Ecchellensis quickly formed a collegial relationship, as testified by the Syriac and Arabic poems that Ecchellensis contributed to *Coptic Forerunner* in time for its publication in September 1636 (fig. 23). By the beginning of 1637 they had announced a plan to collaborate on several translation projects, including the Coptic-Arabic lexicon and grammar as well as several Arabic treatises on Egypt, including Gelaldinus and Barachias. Kircher described their working arrangement in a letter to Peiresc, recounting how every day Ecchellensis came to his study in the Collegio Romano, where the two sat side by side, making suggestions and correcting one another as they translated.[18]

Ecchellensis and Iona were the most prominent of a group of Oriental informants in Rome that included Maronites, Copts, Ethiopians, Armenians, and converted and unconverted Jews who assisted Kircher in his studies. When he needed guidance with Ethiopian texts, for example, he queried the

---

15. Heyberger, ed., *Orientalisme, Science et Controverse* (2010), especially Heyberger, "Abraham Ecchellensis" (2010). See also Rietbergen, "Maronite Mediator" (1989); Levi Della Vida, "Abramo Ecchellense" (1960).

16. Holstenius to Peiresc, 2 May 1636, in Holstenius, *Epistolae* (1817), 495. See also the letters of 4 June and 6 September, ibid., 504, 271.

17. Thomas d'Arcos to Peiresc, 30 June 1633, in Tamizey de Larroque, ed., *Thomas d'Arcos* (1889), 27; Peiresc to Golius, Aix, 29 November 1636, in Peiresc, *Lettres à Saumaise* (1992), 340–43; Peiresc to Cassien de Nantes, 1 November 1636, in Valence, ed., *Correspondance* (1892), 272.

18. "Et quamvis ego in hoc negotio primus detulerim D. Abrahamo, noluit tamen ipse id onus in se suscipere solus. Sed voluit ut id faret in mea praesentia, me coadiuvante, coniudicante, et coexplicante, sic enim sperabat futurum, ut unus alterius defectum supplendo, mutua ingeniorum incitatione opus illustrius fieret, et solidus: Opus igitur aggredimur bonis avibus, conveniente dicto D. Abramo quotidie meum musaeum, ut ibi ea qua decet diligentia consumetur." Kircher to Peiresc, Rome, 7 January 1637, BNF FF 9538, fol. 240^r. See also Peiresc to Kircher, Aix, 4 February 1637, APUG 568, fol. 200^r.

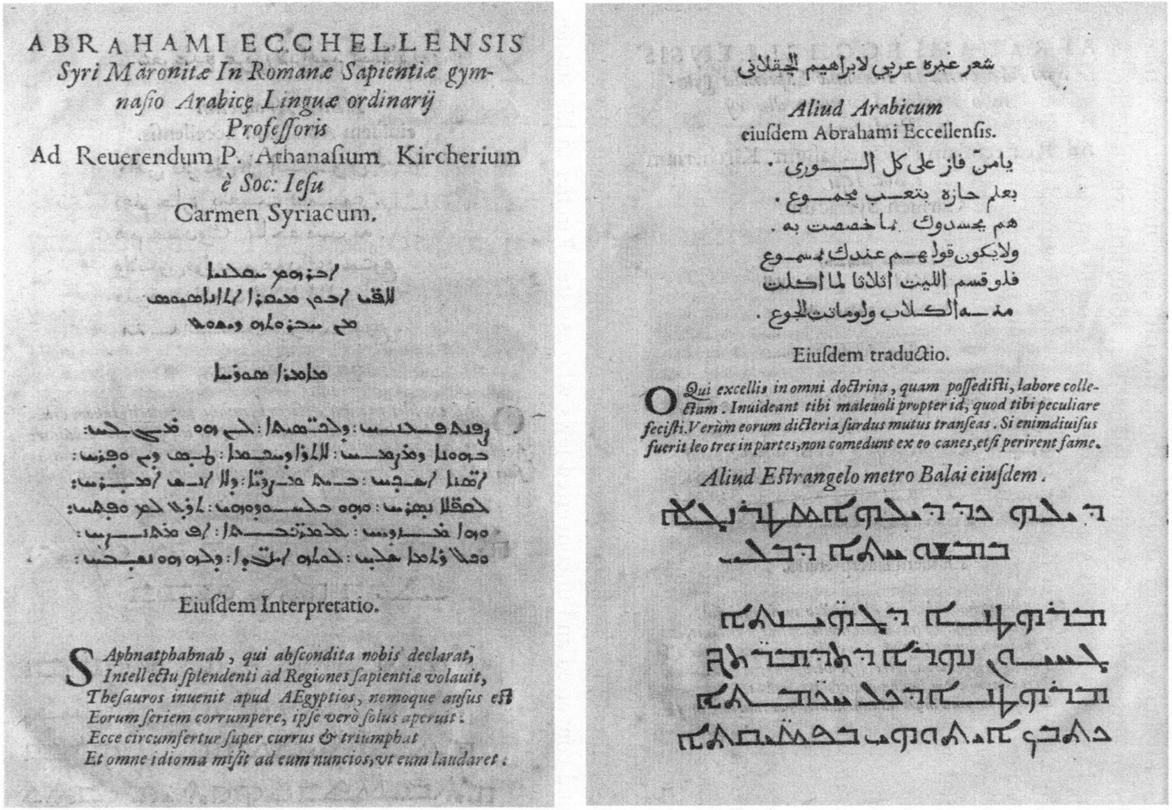

ABRAHAMI ECCHELLENSIS
Syri Maronitæ In Romanæ Sapientiæ gymnasio Arabicæ Linguæ ordinarij Professoris
Ad Reuerendum P. Athanasium Kircherium è Soc: Iesu
Carmen Syriacum.

[Syriac text]

Eiusdem Interpretatio.

S *Apbniatphabnab, qui abscondita nobis declarat,*
Intellectu ostendenti ad Regiones sapientiæ volauit,
Thesauros inuenit apud Ægyptios, nemoque anßus est
Eorum seriem corrumpere, ipse verò solus operuit:
Ecce circunfertur super currus & triumphus,
Et omne idioma misit ad eum nuncios suos vt eum laudaret.

[Arabic text]

Aliud Arabicum
eiusdem Abrahami Eccelensis.

[Arabic text]

Eiusdem traductio.

O *Qui excellis in omni doctrina, quam possedisti, labore collegisti. Inuideant tibi maleuoli propter id, quod tibi peculiare fecisti. Verùm eorum dicteria surdus mutus transfeas. Si inimici tui fuerit leo tres in partes tum comedans ex eo canes, etsi perirent fame.*

Aliud Estrangelo metro Balæi eiusdem.

[Syriac text]

Fig. 23. Poems in Syriac (using two different alphabets) and Arabic, composed by the Maronite scholar Abraham Ecchellensis to celebrate Kircher's Coptic studies. Athanasius Kircher, *Prodromus Coptus, sive Aegyptiacus* (Rome: 1636), ††4r/v. Courtesy of Stanford University Libraries.

Abyssinian priests with whom he frequently conversed. (Sometimes even native speakers came up short: about a difficult passage in a Vatican Armenian manuscript, Kircher reported that not even his Armenian associates in Rome could explain it.)[19] When Kircher set about completing his edition of the Coptic-Arabic lexicon in the 1640s, Ecchellensis had gone to Paris. But he readily found new linguistic assistants: two Copts residing in Rome, Iuhana Kozi and Michael Schatta, the latter of whom he employed as an amanuensis. After returning to Egypt, Schatta sent Kircher materials to use in *Egyptian Oedipus*. "It can hardly be believed," Kircher wrote about preparing the lexicon, "how much work I expended translating the names of things, and truly I would have given up if not for the assistance of two Copts and of others, especially Abraham Ecchellensis, a man celebrated for knowledge of many things and Oriental languages, who supported me in many ways and compared everything to the autograph."[20]

KIRCHER, COURTIER

It was not enough, however, just to be in Rome. Kircher's ability to exploit the city's resources depended on his relationships to powerful protectors. Early modern antiquarian research was closely tied to the world of the wealthy patricians who collected the materials upon which scholars relied. Rome was an important center, not only because of its incomparable archeological density but also because of its aristocratic culture, distinguished by the multiplicity of courts belonging both to cardinals and the indigenous aristocracy. In general, patrician collectors tended to be more interested in the aesthetic dimension of antiquities, especially esteeming ancient sculpture as opposed to the often less sensuous fare that scholars valued as evidence. In the wake of the Counter-Reformation, however, many Catholic collectors felt a need to justify their passion for pagan antiquities, which encouraged them to support work that would demonstrate their collections' utility. Furthermore, while recent scholarship has overturned the image of Rome after 1633 as a scientific wasteland, there can be no doubt that the constraints imposed on natural philosophical pursuits in the wake of the Galileo affair left a vacuum that erudite research expanded to fill.[21]

---

19. *PC*, †††3ʳ, 47, 130.
20. *LAR*, **2ʳ, **4ᵛ, 500, 527; *OA* II.2, 204, 414, *OA* III, 544.
21. Herklotz, "Excavations, Collectors and Scholars" (2004); Stenhouse, "Classical Inscriptions" (2000); Claridge, "Archaeologies" (2004); Romano, "Rome, un chantier" (2008), 115–19.

During the pontificate of Urban VIII, Rome, more than ever, became a mecca for antiquarianism.[22] In particular, the desire to affirm the Church's power to master the pagan past led the Barberini circle to foster investigations of ancient religion, especially allegorical interpretations of paganism as a monotheistic solar cult.[23] After 1623, when his uncle Maffeo ascended the papal throne and made him cardinal, Francesco Barberini became one of Rome's wealthiest, most powerful men and its preeminent collector of antiquities and patron of culture. An avid bibliophile, he amassed an extraordinary collection of books and manuscripts, housed in the library of the Barberini palace, while also directing the Vatican Library.[24] The cardinal's *primo maestro di camera*, Cassiano dal Pozzo, was an influential aristocratic patron in his own right; his famous "Paper Museum," an archive of natural history and antiquarian illustrations, provided scholars a unique and invaluable research tool.[25] During his first Roman years, Kircher was blessed by the favor of the cardinal nephew, with dal Pozzo often serving as an intermediary. (Dal Pozzo played a similar role between Kircher and the general of the Society of Jesus.)

After accepting Kircher as his protégé in November 1633, Cardinal Barberini instructed his librarian, Joseph Maria Suarès, to give the new arrival a tour of the Vatican and Barberini libraries and all the important antiquities—obelisks, pyramids, ruins, and statues—that might aid him in his studies. Kircher formed friendly relationships with members of the cardinal's learned entourage, including the French-born Suarès, fellow German émigré Lucas Holstenius, and the Greek Leone Allacci, all three influential scholars who worked not only for Barberini's household but at the Vatican Library as well. Thanks to the cardinal, Kircher found the doors open to the city's private and ecclesiastical collections and was relieved of academic duties so that he could devote himself entirely to scholarly pursuits.[26] Barberini's largesse was sustained by regular epistolary prodding from his old

---

22. Onori, Schütze, and Solinas, eds., *I Barberini* (2007); Rietbergen, *Power and Religion* (2006); Herklotz, *Cassiano dal Pozzo* (1999); Fosi, *All'ombra dei Barberini* (1997); Hammond, *Barberini Patronage* (1994); Völkel, *Römische Kardinalshaushalte* (1993).

23. See above, ch. 1, "Scholars and Symbols"; Häfner, *Götter im Exil* (2003), 81–116; Mulsow, "Antiquarianism and Idolatry" (2005).

24. The literary scene around the Barberini pontificate was documented in media res by Allacci, *Apes Urbanae* (1633).

25. Herklotz, *Cassiano dal Pozzo* (1999); Solinas, ed., *Cassiano dal Pozzo* (1989).

26. Kircher to Peiresc, Rome, 1 December 1633, BNF FF 9538, fol. 234^{rv}; Peiresc to Kircher, Aix, 3 June 1634, APUG 568, fol. 372^r; Peiresc to dal Pozzo, Aix, 6 June 1634, in Peiresc, *Lettres à dal Pozzo* (1989), 138–39. Kircher is not listed as teaching at the Collegio Romano until the Triennial Catalog of 1639: ARSI Rom. 57, fol. 153^v.

friend, the tireless Peiresc. For example, at a time when access to the Vatican Library was restricted and books were chained to their reading tables, Peiresc wrote the cardinal on behalf of Kircher, who had complained about the library's limited hours and the inconvenient walk (less than two miles) from the Collegio Romano.[27] Kircher's important work with Vatican manuscripts would be greatly facilitated, Peiresc explained, if only he were permitted to take books home, "to work with them there continuously and without the interruptions that occur when he is required to go to the Vatican, from so far away and only on certain days and during certain hours, not being able to have access at his pleasure . . . especially when the weather is bad, either because of the heat or the rain or anything else."[28] While the outcome of this extraordinary request is unknown, the previous chapter demonstrated how much Kircher's completion of the reputation-making *Coptic Forerunner* was due to Barberini patronage, strategically directed from afar by Peiresc. This good fortune was not lost on Kircher, who, looking back in 1636, wrote: "For three years now in Rome I spend my time occupied with thoroughly exotic and foreign languages under the auspices of his Eminence Cardinal Barberini."[29]

When Kircher returned from Malta, he found himself in a changed environment. Under financial strain, Cardinal Barberini had informed him earlier that publication of the Coptic lexicon and grammar would need to be funded with proceeds from the sale of *Coptic Forerunner*, an unlikely revenue stream.[30] Without Peiresc acting as his guardian angel, Kircher never regained the cardinal's favor. He had feared such a turn of events from the moment news of Peiresc's demise reached him on Malta. As he put it

---

27. Kircher's request (in a lost letter of April 29 1635) is mentioned in Peiresc to Kircher, Aix, 10 June 1635, APUG 568, 194^v; see also Peiresc to Gassendi, Aix, 3 June 1635, in Tamizey de Larroque, ed., *Lettres* (1888–98), vol. 4, 511.

28. "Mà credo che sarebbe necessario in questo caso che gli si potesse confidare qualche volume da transferire nel suo Musaeo, per attenderci di continuo et senza l'interruptioni che vi occorrono quando conviene andare alla Vaticana da tanto lontano et solamente a certi giorni, et à certè hore, non potendo egli comè credo <di> disponere à suo bene placito, ne d'un compagno ne forzi della sua persona propria, ogni volta che gli sarebbe necessario, per questo <estento> massimo quando s'incontrano cattivi tempo così di caldo, come di pioggia o altri." Peiresc to Barberini, Aix, 17 June 1635, BAV Barb. Lat. 6503, fol. 135^v.

29. Athanasius Kircher [to Philippe Bebius], [Rome], 14 October 1636 (draft), APUG 561, fol. 83^r.

30. Kircher to Peiresc, Rome, 7 January 1637, BNF FF 9538, fol. 240^{rv}. On sales of *Prodromus Coptus* and Barberini's financial difficulties, see Holstenius to Peiresc, Rome, 4 December 1636, in Holstenius, *Epistolae* (1817), 275–76; Peiresc to Holstenius, Aix, 5 November 1636, in Tamizey de Larroque, ed., *Lettres* (1888–98), vol. 5, 464; Peiresc to Kircher, 4 February 1637, APUG 568, fol. 200^v; dal Pozzo to Peiresc, 7 January 1637, BNF FF 9539, fols. 83^r–84^r.

in a letter, "hateful death has taken from me the man whom I had served as the unique patron of my studies." But rather than succumbing to grief, Kircher recounted, he promptly scanned the horizon for a suitable replacement, setting his sights on Cassiano dal Pozzo.[31] While dal Pozzo would continue, as before, to offer occasional assistance, he declined to become Kircher's patron. Pietro Della Valle may have been another prospect, since he too possessed the social status to serve as an intermediary to the cardinal and was keen to see his Coptic manuscript finally in print. Unfortunately for Kircher, however, Della Valle had fallen out with the Barberini after killing one of the pope's lackeys in a scuffle, and had gone into exile.[32]

Kircher felt the loss of Barberini's support profoundly. "For seven years," he confided to a correspondent in 1640,

> I remained in Rome, called by a certain eminent prince in order to prepare the publication of an unprecedented work. Under his patronage I served for some time not unhappily. But by some fate it happened that the esteemed prince not only dropped me from his memory, but also abandoned responsibility for all the studies and books that he had promised, notwithstanding that he was urged on by letters from learned men from every part of the world in favor of the works' preparation and publication.

Deprived of support for his Orientalist research, Kircher resumed his teaching duties and turned his attention to mathematics and natural philosophy.

---

31. "Cum praeterlapsis aliquot ab hinc septimanis mors praestantissimi, & nulla commendatione sat celebrandi viri D. Nicolas Fabricii de Peresc unici bonarum omnium artium fatoris et Patris percrebuisset hic Melitae, dici vix potest, quam incredibilis dolore animique maestitia percussus sim, cum in hoc uno studia mea veluti in solidissima quaedam columna fundata videbantur; huius enim consilio auxilioque Aegyptiaca opera Romam vocatus agressus sum, huius continuâ sollicitatione magnam quoque eorum partem conseci; huius denique industria & incomparabili zelo quae feliciter inceperam, feliciter eram finiturus. Sed heu inanes spes hominum et propositarum rerum inconstantem exitum; rapuit mihi invidiosa mors eum, quem unicum literarum mearum patronum adseriveram, et incorruptem Iudicem. . . . Cum itaque, uti dixi, de morte tanti veri certior factus, varius omnino essem animo. continuo circumspexi quis in defuncti locum sua centuratus in operibus incaeptis persequendis patrocinator esse posset et promotor, et ecce inter quamplurimos alios variae conditionis Amicos, Ill.ma D.tio V.ra sola occurrit, cuius authoritate quae apud plurisque Romanae Urbis proceres pollet maxima, et per sua singularis eruditione at zelo erga Remp: Literariam inconcusso, interruptum studiorum meorum filum continuare posset. Pro ea itaque qua ill.mam D.m Vm semper fui confidentia modo ad eam confugio, orans et obstans ut studiorum meorum patrocinium apud Em.m Cardinalem suscipere non dedignetur." Kircher to Cassiano dal Pozzo, Malta, 15 August 1637, BISM ms. H 268, fol. 9^{rv} (facsimile consulted at the Accademia Nazionale dei Lincei).

32. Ciampi, *Della vita* (1880), 118–20; Della Valle to Kircher, Gaeta, 26 June 1638, APUG fol. 291^{rv}.

According to Kircher, some people erroneously attributed his failure to complete his Coptic and hieroglyphic studies to a lack of ability, rather than to its true cause.[33]

Kircher's patronage crisis, which produced an uncharacteristic five-year gap between books, resolved itself spectacularly in 1640. Far from Rome, two friends, Johannes Marcus von Marci and the Grand Burgrave Bernard Martinic, had been advancing Kircher's case in the court of the Holy Roman Emperor. Primed by Martinic, Ferdinand III granted Marci an audience and responded favorably to the proposal of becoming the Jesuit scholar's long-distance protector. The emperor offered not only to finance Kircher's publications but also to provide him with an annual salary and funds for an assistant. The first product of what would be a decades-long relationship between Kircher and the Habsburgs, *Lodestone; or, On the Magnetic Art*, was printed in Rome in 1641 with a dedication to the emperor.[34] But Ferdinand made it known that his paramount desire was for Kircher to complete his unfinished studies of Coptic and the hieroglyphs.[35]

RESTORING THE EGYPTIAN LANGUAGE

Kircher returned first to the Coptic-Arabic lexicon and grammar. Although he and Ecchellensis had completed much of the Latin translation in 1637, the manuscript languished for several years in the office of the Propaganda Fide.[36] Thanks to the emperor's support, Kircher, assisted by Schatta and Kozi, put the finishing touches on the work, sending it at last to the press in 1643.[37] *Egyptian Language Restored* began with the Latin translation of a Coptic-Arabic grammar, which ran to fewer than forty pages, followed

---

33. "Noverit ni fallor R.V me ad septennium ferè Romae morari ab Em.mo Principe quodam ad operis hucusque intentati editionem expediendam vocatum; sub cuius quidem patrocinio per aliquod tempus non infeliciter merui; at nescio quo fato contigerit, ut notus Princeps mei non tantum memoriter deposuerit; sed et omnem quoque studiorum librorumq[ue], promissorum curam abiecerit, <sed> non obstante quod omni mundi parte literis doctissimorum virorum ad operum promotionem evulgationemq[ue] instanter sollicitaretur." Kircher [to Johann Gans, Rome, c. 1640], APUG 561, fols. 88^r–89^r (draft). Cf. Johann Marcus von Marci to Kircher, Regensburg, 3 August 1640 and Prague, 17 September 1640, APUG 557, fol. 124^r; 127^r.

34. Kircher, *Magnes* (1641).

35. Marci to Kircher, Regensburg, 3 August 1640, APUG 557, fol. 124^r. See also Martinic to Kircher, Prague, 1 September 1640 and 17 September 1640, APUG 556, fol. 285^r, 127^r; Fletcher, "Johann Marcus von Marci" (1972).

36. Kircher to the Congregation for the Propagation of the Faith, [Rome, no date], APUG 561, fol. 62^{rv}. See also *LAR*, **1^v–**2^r.

37. Kircher, *Lingua Aegyptiaca restituta* (1643).

by a Coptic-Arabic-Latin lexicon of almost five hundred pages.[38] Kircher concluded the work with ten appendices dealing with Egyptian and Coptic matters.

In the first appendix, Kircher retracted his claim about Greek's dependence on Coptic, concluding, "It is settled, therefore, that except for the alphabet, Egyptian is as different from Greek as possibly can be." Other chapters dealt with the institutes of the Coptic Church; Coptic names of God; the Coptic calendar; Coptic names of stars, plants, and animals; Egyptian cities; and Egyptian philosophical terminology. Kircher printed a catalog of "the books of the Egyptian Arabs" in the library of the Duke of Tuscany, which Giovanni Battista Raimondi had brought back from Egypt in the previous century. He also provided a catalog of Coptic books that were conserved in the libraries of Cairo according to a list Peiresc had sent him. Throughout the work Kircher emphasized the relevance of Coptic literature for understanding the hieroglyphs, and whetted his readers' appetites for the forthcoming *Egyptian Oedipus*. As Ecchellensis put it in an attestation printed at the front of the book, Kircher's work on Coptic was "like a key by which that unknown and thoroughly decayed literature of hieroglyphic wisdom [could be] unlocked."[39]

With *Egyptian Language Restored*, Kircher cemented his reputation as an Oriental philologist. In 1644 the College of Cardinals appointed him to a committee of leading Orientalists charged with reviewing an Arabic translation of the Bible.[40] He later served on another prestigious committee that prepared a Latin translation of the *plomos* (lead tablets) of Granada, controversial religious texts written in archaic Arabic characters and suspected to be heretical forgeries.[41] A token of Kircher's reputation in the wider Republic of Letters can be found in his contribution to Jean Plantavit de la Pause's Hebrew-Aramaic thesaurus, published in Lodève, France in 1644. In the style of the day, the work was prefaced by tributes to the author from an international roster of scholars, in which Hebrew and Syriac poems by Kircher appeared alongside testimonies by the convert scholar Philippe d'Aquin, professor of sacred languages at Paris, the Jesuit theologian Denis

---

38. To be precise, it contains editions and translations of two distinct dictionaries and two grammars from BAV Vat. Copt. 71. See Hamilton, *Copts and the West* (2006), 210.

39. *LAR*, 509, 512, **4ᵛ.

40. See , "Congregationes Cardinalices" (1886): 887–89; Pizzorusso, "Filippo Guadagnoli" (2010), 257.

41. "Interpretatio Laminarum Granatensium Romana, Ordine Pontificio Fideliter Facta," 15 July 1669, BAV Vat. Lat. 7509, fols. 1–184. On the *plomos*, see Harris, "Forging History" (1999): 945–66; García-Arenal and Rodríguez Mediano, *Un Oriente español* (2010).

Petau, the Maronites Gabriel Sionita and Vittorio Scialac, and the Rabbi Salomon Azubi.[42]

THE PROS AND CONS OF ROMAN ERUDITION

In the next century, as knowledge of Coptic progressed, *Egyptian Language Restored* would be subjected to considerable criticism.[43] But for its time the work was a considerable achievement, which Kircher, despite mediocre Arabic and less Coptic, accomplished by mobilizing Rome's resources. At the beginning of the project, when Kircher was locked in a race with Saumaise to publish the first study of Coptic, Peiresc privately made disparaging comments about Roman scholarship.[44] Above all he criticized the "jealousy" with which the Romans guarded their materials, and the sluggish pace at which they executed their plans. As the contest wore on, he also criticized the quality of Roman erudition. In Peiresc's opinion the most capable scholars were on his side of the Alps, and Della Valle's obstinacy was preventing the translation from being done in the best possible manner.[45]

There is no doubt that Saumaise was a more able linguist than Kircher. And there were real differences in the style of scholarship practiced in different parts of Europe and among members of different churches. Nonetheless, Peiresc's criticisms of Rome should be taken partly with a grain of salt. Many of his negative comments were elicited by his frustration in failing to obtain Della Valle's manuscript and his impatience to see the work published as soon as possible. Peiresc was fully aware of Rome's strengths. He knew that the Vatican's Coptic holdings were unrivaled and he eagerly sought information about their content, just as he impatiently awaited for Barberini to send him the latest exotic alphabets from the Propaganda's press. After initially warning the Capuchin missionary Gilles de Losches about the dangers of Roman inertia, Peiresc arranged for him to go to Rome to oversee the printing of Ethiopian books. At one point he even suggested to Saumaise that Kircher's encounter with a Maronite who was to help him

---

42. Plantavit de la Pause, *Planta Vitis* (1644).
43. See chapter 8, "Oedipal Conflicts among the Copticists."
44. By this time Peiresc had a long and frustrating history with Roman erudition. On the mutual suspicions between French and Roman scholars, see Miller, "A Philologist, a Traveller and an Antiquary" (2001).
45. Peiresc to Saumaise, Aix, 14 November 1633, in Peiresc, *Lettres à Saumaise* (1992), 31–32; Peiresc to Petit, Aix, 16 Jan 1634, ibid., 50; Peiresc to de Losches, Aix, 10 July 1634, in Valence, ed., *Correspondance* (1892), 65; Peiresc to Bouchard, Aix, 15 December 1633, in Tamizey de Larroque, ed., *Lettres* (1888–98), vol. 4, 94–95.

with the "African and Nubian" vocabulary in the Coptic lexicon indicated that Rome was a superior place to carry out that enterprise.⁴⁶

The Congregation for the Propagation of the Faith sponsored Kircher's studies because it believed that knowledge of Coptic would advance its mission to promote Catholicism among infidels and heretics. Such religious motivations were characteristic of Oriental studies in general, among both Catholics and Protestants. The Calvinist Saumaise, for example, became interested in Coptic because he hoped it would prove useful in a theological polemic against Catholics. Having heard that some Capuchin missionaries claimed that the Copts agreed with Rome about Christ's real presence in the sacrament, Saumaise borrowed Dupuy's bilingual manuscript of the Coptic liturgy. He translated some lines of the Arabic version, which he thought proved that the Coptic teaching—and by implication the teaching of the primitive Church—was in fact similar to the Calvinist position.⁴⁷

In this respect Coptic followed the pattern of the study of other Near Eastern languages. In the wake of the Reformation, controversies over scriptural interpretation and the doctrines of the early Church led scholars to explore linguistic traditions that preserved versions of biblical texts and early Christian literature, such as Syriac, Samaritan, Arabic, Armenian, and Ethiopian. Protestants were also curious about the national churches of eastern Christendom because of their tradition of independence from the papacy. The Catholic Church, by contrast, increasingly resigned to the loss of much of Europe to Protestantism, hoped to recoup some of its loss through reconciliation with "dissident" Oriental churches.⁴⁸ With stakes like these, Oriental philology was always on the verge of turning into a sectarian battleground, as happened in Kircher's day in the controversies that erupted around the translation of the Samaritan Pentateuch.⁴⁹

Religion was the primary motor of Oriental studies, but it was accompanied by an independent current of humanist and historical interest. In his surviving letters to the Catholic Peiresc, Saumaise never mentioned the debate over the Eucharist. Instead, their discussions revolved around mat-

---

46. Peiresc to Barberini, Boysgency, 14 March 1631 and Aix, 3 May 1635, BAV Barb. Lat. 6503, fols. 30ʳ, 131ᵛ–32ʳ; Peiresc to Saumaise, 22 September 1634, in Peiresc, *Lettres à Saumaise* (1992), 114.

47. Saumaise to M. Daille, Grigny, 22 October 1631, in Saumaise, *Epistolarum liber* (1656), 69–72.

48. Russell, ed., *"Arabick" Interest* (1994); Toomer, *Eastern Wisedome* (1996); Hamilton, Van Den Boogert, and Westerweel, eds., *Republic of Letters and the Levant* (2005); Pizzorusso, "Filippo Guadagnoli" (2010).

49. See *Antiquitates ecclesiae orientalis* (1682).

ters such as Coptic's value for understanding the origin of languages and for the study of antiquity more generally. Not that they ignored religious matters—to do so would have been impossible. All available Coptic literature was ecclesiastical, and Peiresc was deeply interested in Coptic versions of the New Testament. Indeed, his initial interest in Coptic may have been to include such texts in the Paris polyglot Bible. But unlike the liturgy's description of the Eucharist, the Coptic gospels did not obviously threaten to raise contentious theological questions. Saumaise's wish to use Coptic to defend Calvinist doctrine is emblematic of the central role of sectarian polemics in early modern philology. At the same time, his fruitful interaction with Peiresc reveals a simultaneous desire for scholarly intercourse across confessional lines, equally characteristic of the time. In order to collaborate rather than debate, Catholic and Protestant scholars developed a lingua franca that bracketed controversial theological issues.

Kircher stands out for the relatively small role that sectarian polemics and theology played in his study of Coptic. To the extent that he wrote about Coptic's value for supporting Catholic doctrine, he did so at the prodding of Della Valle, who asked him to revise an early draft of *Coptic Forerunner* that did not sufficiently emphasize this matter.[50] In a chapter entitled "The Usefulness of the Coptic Language," Kircher dutifully explained how knowledge of the Coptic liturgy would help refute heretics. But he wrote passionately about the priceless knowledge of pre-Christian antiquity that awaited discovery in Coptic literature, hitherto obscured by linguistic barriers and hidden away in Egyptian libraries.[51]

During the seventeenth century, Rome lost its primacy in Oriental studies to Paris, Leiden, and eventually Oxford and Cambridge. But in the larger panorama of European scholarship, the Catholic capital remained a vital site, offering scholars of Near Eastern languages unique resources through its libraries, museums, printing facilities, educational and ecclesiastical institutions, as well as the communities of scholars and collectors associated with them. Even the dominant influence of the Church—lest one forget, the Coptic saga unfolded in the aftermath of Galileo's 1633 condemnation—brought advantages as well as disadvantages. On the negative side, there were significant constraints on intellectual freedom, and relevant scholarship by Protestants could be difficult to access. But Counter-Reformation

---

50. See Della Valle to Kircher, 6 October 1634, ASV Fondo Della Valle–Del Bufalo 53 (copy of undated letter, APUG 557, fols. 440^r–41^v). Compare *PC*, 185 to Della Valle, *Viaggi* (1650–63), 399.

51. See especially, *LAR*, 511–13.

policies materially sustained Orientalist research and other kinds of erudition, which were understood as essential tools for the Church's mission, while Rome's concentration of manuscripts, antiquities, and linguistic expertise kept the city at the center of international scholarly networks.[52] Della Valle explained the trade-off succinctly in a letter to Jean Morin regarding some Persian histories that he planned to publish. Publishing such works in Rome was difficult, he wrote, on account of "the excessively scrupulous censorship" of the authorities; but, on the other hand, he judged that in no other city could Persian characters be printed so easily.[53]

OEDIPUS TRIUMPHANT

In 1643, Kircher once again floated the idea of a trip to Egypt, this time seeking imperial sponsorship, but nothing came of the idea.[54] Instead, following the publication of *Egyptian Language Restored*, he returned again to mathematical sciences, composing his first in-folio publications, both dedicated to Habsburg princes: *The Great Art of Light and Shadow* (1646), on optics and sundials, and *Universal Musurgia* (1650), on the science of sound and music.[55] Meanwhile, Kircher's hieroglyphic studies had grown so massive that printing them would require an extraordinary financial outlay, above all for illustrations and Oriental typefaces. These technical matters were important enough to be discussed in the initial negotiations with the emperor undertaken by Marci on Kircher's behalf. Immediately after receiving Ferdinand's pledge of support in 1641, Kircher and his agents at the imperial court contacted the Grand Duke of Tuscany to negotiate the use of the elegant Arabic font, cut by Robert Granjon, that had been sent to Florence after the demise of the Medicean Press. Kircher also wrote to Munich, requesting to borrow the original copper plates from Herwart's *Hieroglyphic Thesaurus*, but unfortunately the author's heirs had recycled them to make domestic utensils.[56] Even with imperial patronage, however, unreliable funding

---

52. See Miller, "Making the Paris Polyglot" (2001); Miller, "A Philologist, a Traveller and an Antiquary" (2001).
53. Della Valle to Morin, undated (circa August/September 1630), in *Antiquitates ecclesiae orientalis* (1682), 182–83.
54. Marci to Kircher, 19 September 1643, Prague, APUG 557, fol. 90; Fletcher, "Johann Marcus von Marci" (1972), 106.
55. Kircher, *Ars magna lucis* (1646); Kircher, *Musurgia universalis* (1650).
56. Marci to Kircher, Regensburg, 3 August 1640, APUG 557, fol. 124^r; Johann Gans to Kircher, Regensburg, 10 February 1641, 30 April 1641, 25 June 1641, APUG 561, fols. 132^{rv}, 134^{rv}, 118^r; Ferdinando II de Medici, Grand Duke of Tuscany to Kircher, Livorno, 26 March 1641, APUG 556 fol. 67^r; Georg Spaiser to Kircher, Munich, 13 May 1644, APUG 567, fol. 228^r.

continued to delay the completion of *Egyptian Oedipus*, perhaps due to the strains of war on the emperor's purse.[57]

Relief came from an unexpected quarter. Giambattista Pamphili's election as Innocent X in 1644 did not seem especially auspicious for Kircher, as the new pope was not noted for the devotion to scholarship that had characterized his predecessor. After donning the tiara, however, Innocent undertook the renovation of the Piazza Navona, site of the Pamphili palace. The centerpiece of this urban renewal project was to be a great ornamental fountain, fed by the pure waters of the Acqua Vergine and crowned by an ancient Egyptian obelisk. The obelisk, which lay in five pieces at the site of the ancient circus of Maxentius, along the Via Appia outside the city walls, required extensive restoration. As a renowned expert on Egyptian antiquities, Kircher was placed in charge of excavating its fragments. In collaboration with Gian Lorenzo Bernini, who won the competition to design the fountain and oversaw the heavy work of transportation and assembly, Kircher designed reconstructions of inscriptions that were missing from the damaged stones.[58] He may also have influenced the design of Bernini's masterpiece, the Fountain of the Four Rivers, whose statues suggest Kircherian motifs, such as the geohydraulic theories later published in *Underground World* (fig. 24).[59]

This was civic architecture on a grand scale. The obelisk's transportation alone was estimated to cost twelve thousand scudi, and the entire restoration of Piazza Navona eighty thousand. In the midst of a food shortage, Innocent X raised funds by levying a new tax, arousing popular outcry expressed in pasquinades. A satirical verse pasted to the recumbent obelisk declared:

Not obelisks and fountains, we want something else instead:
Bread, we want bread, bread, bread!

Another lampoon repurposed the words of the Vulgate Bible: "Command that these stones be made into bread." As the obelisk went up, Kircher's Jesuit colleagues distributed food to Rome's poor, but the hungry crowds lost control, crushing several people to death. When Bernini's spectacular fountain was finally unveiled on 12 June 1651 (the obelisk itself had gone up in

---

57. See *OP*, a4^v.
58. Cancellieri, *Il mercato* (1811), 35–45; Iversen, *Obelisks in Exile* (1968), 83–89.
59. See Rowland, "Th' United Sense of the Universe" (2001); Rivosecchi, *Esotismo in Roma barocca* (1982), 117–38; Marrone, *I geroglifici* (2002), 85–100; Fehrenbach, *Compendia Mundi* (2008).

Fig. 24. Roman society admiring the Pamphilian obelisk atop Bernini's masterpiece, the Fountain of the Four Rivers, in Piazza Navona. Pope Innocent X funded the immensely expensive project by a special tax during a time of food shortages, provoking public outcry and vandalism. Giovanni Iacomo Rossi, "Obelisco Panfilio, eretto dalla santita di N S Innocentio X," 1651. Typ. 625.55.390, Houghton Library, Harvard University.

August 1649), guards had to be posted to arrest protesters who threw objects in the fountain and committed other acts of vandalism.[60]

Kircher, happily, occupied a higher station on baroque Rome's social pyramid. The pope's unpopular tax not only gave him the opportunity to collaborate with Rome's greatest artist but, more importantly, it financed the publication of the first volume of his hieroglyphic studies. According to Kircher, Innocent X summoned him and spoke these words:

> Father, I have resolved to erect an obelisk, a massive structure of stone. It shall be your massive job to give it life with an interpretation. Thus I wish for you to dedicate yourself earnestly to the work that I have put in your charge on account of the gift granted you by God, so that all who gaze on the massive structure in wonderment, due to its strange characters, by means of your interpretation may grasp their secret inner mysteries and meaning.

Kircher enthusiastically set about his assignment, the pope having instructed the Jesuit General Vincenzo Caraffa to provide him with all necessary resources, including an assistant.[61] The result, *Pamphilian Obelisk*, appeared during the jubilee year. Ostensibly, the book, which ran to more than five hundred pages in-folio, was a study of the obelisk erected by Innocent X in Piazza Navona. But the interpretation of the obelisk occupied only the last of five parts, while the rest consisted of material of a more general nature. As such, *Pamphilian Obelisk* offered a synopsis of Kircher's views on the "hieroglyphic doctrine," whose full expression he would give in the two-thousand-page *Egyptian Oedipus*. But it may better be thought of as the first volume of the larger work. Most of the content of *Pamphilian Obelisk* was originally intended to form part of *Egyptian Oedipus*, and the book contained essential parts of Kircher's argument that were not treated in the later work.[62]

After the publication of *Pamphilian Obelisk*, Kircher's project stalled again. Then, in December 1650, he received word from the imperial court:

---

60. Gigli, *Diario* (1958), 323, 326, 334–35, 341, 356–57, 385–86, 409. The original pasquinades read: "Noi volemo altro che Guglie, et Fontane, / Pane volemo, pane, pane, pane"; and "Dic ut lapides isti Panes fiant" (Matthew 4:3). See also Cancellieri, *Il mercato* (1811), 42–60, which collects various rare and unpublished materials.

61. *Vita*, 57–58.

62. The unity of the two works is indicated by the fact that the judgments and testimonials at the beginning of *Obeliscus Pamphilius* were meant to cover *Oedipus Aegyptiacus* as well; see *OA* I, d1ᵛ. The main hieroglyphic dictionary that Kircher uses to justify his translations in *OA* III appears in *OP*, book 4.

Ferdinand III would provide him all necessary means to complete *Egyptian Oedipus*. The emperor insisted that Kircher cease other activities, and asked how much money would be needed to put the work in print. In the end, Ferdinand pledged the enormous sum of two thousand scudi, sent in three payments, allowing Kircher to spend the next four years completing his magnum opus. With funds available again for an amanuensis (all the more necessary since Kircher now suffered from a tremor in his writing hand), Kaspar Schott, a former pupil and old friend from Würzburg, joined him in Rome from 1652 until 1655.[63]

Kircher had attained the pinnacle of worldly success as a scholar. Ferdinand and Innocent were the most illustrious of his princely admirers, but they were joined by many lesser luminaries from the firmament of European court society, as testified by the profusion of princes, dukes, and cardinals with whom Kircher corresponded and to whom he dedicated sections of *Egyptian Oedipus*. By now he was also well known to European scholars, many of whom had long anticipated the appearance of *Egyptian Oedipus* (fig. 25). When his hieroglyphic studies finally hit the bookshops, they met with immediate commercial success. Before *Egyptian Oedipus* was even ready, Kircher worried that the print run of one thousand copies for *Pamphilian Obelisk* would be insufficient to meet demand, as a single Dutch bookseller snatched up all five hundred remaining copies.[64]

KIRCHER'S WEB

At the beginning of *Pamphilian Obelisk*, discussing the sources and methods of that book as well as the forthcoming *Egyptian Oedipus*, Kircher offered the equivalent of the acknowledgment section of a modern academic monograph, adducing "the libraries and collections of antiquities, and the support of learned men, by whose advice and aid the author was helped." It reads like a map of the resources and networks whose essential role in Kircher's studies has been demonstrated throughout this chapter. "With many years of labor," he relates,

> here in Rome especially, I procured all the authors who mention anything of Egyptian matters, whether in manuscript or printed, in Latin,

---

63. See Johann Maximilian Lamberg to Kircher, Vienna, 3 December 1650, APUG 557, fol. 301^r; Johann Gans to Kircher, Vienna, 29 January 1651, 18 March 1651, APUG 561, fols. 124^r, 116^r. *Vita*, 61, gives the figure as three thousand scudi.

64. Kircher to an unidentified bookseller [Rome, circa 1652–55], APUG 561, fol. 79^{rv} (draft).

P. ATHANASIVS KIRCHERVS FVLDENSIS
è Societ: Iesu Anno ætatis LIII.

Honoris et observantiæ ergò sculpsit et D.D.C.Bloemaert Romæ 2 Maij A.1655.

Fig. 25. The official portrait: Kircher in 1655, the year he published *Egyptian Oedipus*, engraved by Cornelis Bloemaert (perhaps based on the painting in fig. 7). This was the first printed portrait of Kircher, an indication of his fame. Modified slightly to represent aging, it would be reprinted many times, disseminating Kircher's likeness throughout the Republic of Letters. Courtesy of the Smithsonian Institution Libraries, Washington.

Greek, Hebrew, Arabic, Chaldean, Egyptian, Syriac, Armenian, and Ethiopian, as well as those published in the languages used in Europe, hidden away in the libraries of the Vatican, the Angelica, the Barberini, the Sforziana, the colleges of the Society of Jesus and other private museums, and I read them carefully and with discernment . . .[65]

He went on to enumerate, at great length, the textual authorities on which his work relied, singling out Oriental works he had consulted in Roman libraries, such as the Vatican's Ethiopian and Coptic manuscripts, the *Pardes Rimmonim* from the College of Neophytes, al-Suyuti's history, which Ecchellensis had brought to Rome from Egypt, and Mor Isaac's *Syriac Philosophy* from the College of Maronites.[66] Having based his investigation on the premise that the hieroglyphic doctrine must be patched together from countless fragments dispersed in different languages, Kircher argued that no one before him could explain the Egyptian inscriptions because those who had tried had not devoted themselves to the Oriental literature in which the secret hid. In the passage above, Kircher made explicit the role played by Rome's great wealth of libraries and museums in allowing him to carry out this heroic task of historical reconstruction.

Next, Kircher turned to the Egyptian objects and other antiquities that facilitated his study, reeling off a list of Roman museums that included the collections of Cardinal Barberini, Francesco Angeloni, Giovanni Pietro Bellori, Francesco Gualdo, Giovanni Battista Casali, Marzio Milesi, Giovanni Battista Romano, Pietro Stefanoni, Hippolito Vitelleschi, Pietro Della Valle, and Cassiano dal Pozzo.[67] Kircher's debts extended beyond Rome, however, and he also acknowledged materials received from Johann Rhode in Padua, the Duke of Tuscany, the Holy Roman Emperor's collections in Naples and Vienna, Carlo di Ventimiglia in Palermo, D. Abela in Malta, the Duke of Bavaria, and, above all, Peiresc in Aix. Kircher also paid tribute to fellow scholars who had provided help, including Leone Allacci and Lucas Holstenius (citing in particular the latter's collection of manuscripts pertaining to the Platonists and Pythagoreans) and Gabriel Naudé in Paris. To situate his hieroglyphic studies within the field of contemporary scholarship, he invoked leading lights of European erudition (mostly northern European and

---

65. *OP*, b2^r, b4^r.
66. *OP*, b4^r–c1^r.
67. Discussions of most of these collections may be found in Herklotz, *Cassiano dal Pozzo* (1999). See also Findlen, *Possessing Nature* (1994); Stenhouse, "Visitors, Display, and Reception" (2005).

Protestant) like Selden and Greaves in England; Scaliger, Saumaise, Heinsius, Vossius, and Boxhorn in the Netherlands; Bangius, and Bartholin in Denmark; and Petau in France.[68] Finally, Kircher acknowledged the assistance of two men closer to home: Jacques Viva, his amanuensis, and Leone Santi, the prefect of studies at the Collegio Romano. "With these resources, provided to me from nearly all of Europe," Kircher concluded, "by combining each and all appropriately among themselves, I have derived wonderful light for untangling the most secret sphinxes."[69]

Large-scale antiquarian projects required substantial collaboration, which Rome facilitated with its cosmopolitan intellectual community and central position within global information channels.[70] *Egyptian Oedipus* and the studies leading up to it could not have been less the work of a lone scholar. Kircher relied on an extensive network of informants, assistants, and other supporters. His published acknowledgments, though generous, only hinted at the extent of his debts. *Egyptian Language Restored*, for example, named Athanasius Kircher as its sole author, but the evidence of unpublished correspondence has revealed that most of the translation was carried out in partnership with Abraham Ecchellensis. Translations of Arabic texts that Kircher published in *Egyptian Oedipus* were likely fruits of the same collaboration, just as his study of Hebrew and Aramaic relied on Jewish assistants.

Too often, modern treatments have extracted Kircher's studies from the world he described in his acknowledgments, which this chapter has sought to recover. While they inevitably invoke the great names of Renaissance occult philosophy and hieroglyphic symbolism (Ficino, Patrizi, Horapollo, Valeriano, and so forth), one searches them in vain for the names of Kircher's Oriental collaborators like Abraham Ecchellensis and Giovanni Iona; or of the Arabic sources, such as al-Suyuti and Abenephius, that he claimed were the key to his breakthrough. Even the better-known contemporaries from whom Kircher took inspiration, like Scaliger, Selden, and Vossius, have rarely been considered relevant to understanding his work. Such omissions amount to a failure to understand Kircher in relation to the scholarship of his own time and place rather than of previous centuries. Attention paid to the making of Kircher's hieroglyphic studies leads to a new image of his

---

68. See *OP*, c1^v, where these authors are among those singled out by Kircher for the aid that their works provided him.

69. *OP*, b4^v–c2^r.

70. Cf. Stenhouse, *Reading Inscriptions* (2005), 149–60, on Gruterus's compilation of Roman inscriptions.

most famous work. *Egyptian Oedipus* was a product of seventeenth-century Rome's distinctive scholarly ecosystem with its human and institutional resources and connections to Europe and the Near East. Its author was deeply involved in contemporary trends in Oriental philology and antiquarianism. This image, which has emerged in the first part of this book by reconstructing the social world in which Kircher's studies unfolded, finds confirmation in the following chapters, which examine the content of *Egyptian Oedipus* and *Pamphilian Obelisk*.

CHAPTER FOUR

Ancient Theology and the Antiquarian

Rome shews in whole and parcels all the rubble
Of wasted Aegypt giving pleasant trouble,
And most sweet rack to witts, to know, and see
The mangled parent of Antiquitie. . . .
No longer shall it be so. For their *Sphinx*
We have found an OEDIPUS, doth solve the links
Of chayn'd mysterious emblemes, holy rites,
Close riddles, obscure symbols; Aegypt's nightes;
Scarce having other darkenesse. KIRCHER'S he,
That whylome gave a proofe of masterie
O're such concealed wisdome . . .
—James Alban Gibbes, M.D.[1]

KIRCHER'S OEDIPUS COMPLEX

For Kircher, the enigmatic Egyptian inscriptions came first. His determination to explain them created exigencies. Starting from the standard theory that hieroglyphs were a symbolic language encoding the secrets of ancient Egyptian priests, he turned to occult philosophy because it seemed to provide an appropriate framework in which he could explain hieroglyphic monuments and other antiquities pertaining to ancient religion, philosophy, and magic. This framework had three distinct components: doctrinal, historical, and semantic. The first component gave Kircher access to the content of "the hieroglyphic doctrine," the teachings recorded on the obelisks and other monuments. It was to be found through the study of texts

[1]. OA I, a3^r.

in known languages prior to interpreting the hieroglyphic inscriptions. By his own description, Kircher could translate the hieroglyphs because he already knew what they said. The historical component provided a genealogy of knowledge that allowed Kircher to argue that the doctrinal component in fact corresponded to the teaching of ancient Egyptian priests, had survived the vagaries of time, and could be recovered. It was based on the concept of the *prisca theologia*, which posited a tradition of primordial wisdom among "ancient theologians." Finally, the semantic component provided a theory and practice of symbolic communication, which enabled Kircher to convert hieroglyphic inscriptions into Latin descriptions of teachings that resembled the doctrinal component. This Neoplatonic semantics, introduced in chapter 1, will be further examined in chapter 7. This chapter concerns the historical and doctrinal components.

Kircher insisted emphatically on the truth of the historical component of his investigation—his narrative about the transmission of occult philosophy by ancient theologians, like Hermes Trismegistus—which was crucial for his claim to have accurately translated the hieroglyphs. By contrast, his treatment of the doctrinal component was primarily descriptive. To explain the inscriptions on Rome's obelisks required knowledge of the culture that produced them. Believing that knowledge to be found in the testaments of occult philosophy, Kircher needed familiarity with the Hermetic Corpus, the Chaldean Oracles, and the Kabbalah, just as he needed expertise in paleography and Oriental languages. To use the terminology of nineteenth-century scholarship, one might say that Kircher treated occult philosophy as a *Hilfswissenschaft*—an auxiliary science, like numismatics or archeology, that the scholar of antiquity must master in order to make correct sense of his sources and bring the past to light. Alternatively, one can think of Kircher as a scholar whose subject matter was occult philosophy, but whose treatment of ancient documents of esoteric wisdom and magic, in contrast to that of his fifteenth- and sixteenth-century predecessors, was more historical (concerned with describing the past) than philosophical (concerned with defending true propositions about the nature of things). The success of *Egyptian Oedipus* did not depend on persuading the reader to accept the truth of the "hieroglyphic doctrine." It depended on persuading the reader that ancient Egyptian priests had believed in its truth.

Against this claim, one might argue that Kircher adopted the descriptive mode of historical discourse pragmatically, in order to gain the freedom to discuss controversial doctrines.[2] There is no question that Kircher held

---

2. For studies along such lines, see references above, ch. 1, "The Hermetic Philosopher."

some views that went beyond the bounds of legitimate opinion as defined by the Society of Jesus, and that he tested the limits of what a Jesuit scholar could discuss in print.[3] Perhaps, as Ingrid Rowland ingeniously argues, he intended his story about Hermes Trismegistus as an allegory of his own life and work, alerting discerning readers that, like Hermes, he had concealed his true meaning beneath symbols and allegories. (Rowland has in mind Kircher's unorthodox natural philosophical opinions, especially his anti-Aristotelian astronomy, rather than his treatment of occult philosophy and the hieroglyphs.)[4] In other books, notably his studies of magnetism, Kircher espoused a cosmology and metaphysics with Neoplatonic elements, including the idea that the universe was structured in hierarchical chains of influence, which made magic possible, if dangerous.[5] In *Egyptian Oedipus* he enthusiastically endorsed aspects of the hieroglyphic doctrine, not least the claim that ancient pagan wise men possessed knowledge of the Trinity.

But there is no reason to interpret Kircher's avowed antiquarian agenda as a ruse to propagate heterodox ideas. The underlying structural logic of his hieroglyphic studies confirms their primarily historical, rather than philosophical, ambition. After laying the historical and theoretical groundwork, *Egyptian Oedipus* and *Pamphilian Obelisk* culminate with images of hieroglyphic inscriptions and other Egyptian antiquities, accompanied by Kircher's interpretations. As Kircher explained, the first of *Egyptian Oedipus*'s three parts demonstrated the links between the religion of ancient Egypt and other pagan nations, which made it possible to recover the hieroglyphic doctrine by "collating" Egyptian material with traditions from other cultures.[6] The second part assembled those traditions and explained the principles of symbolic communication, while the final part used symbolic hermeneutics to produce translations of Egyptian inscriptions. Similarly, *Pamphilian Obelisk* began with the history of the transmission of the hieroglyphic doctrine (books 1 and 2), followed by an exposition of its content based on Platonic interpretations of the Egyptian gods, a primer on the allegorical and Euhemeristic interpretation of myth (book 3), a dictionary of hieroglyphic symbols (book 4), and finally a translation of the Pamphilian obelisk (book 5).

3. See below, ch. 6.
4. Although she describes Kircher as a clandestine disciple of Giordano Bruno, Rowland's Bruno was no Hermetic magus, à la Yates, but rather a prophet of modern science. Rowland, "Kircher Trismegisto" (2001); Rowland, *Giordano Bruno* (2008). See also Rowland, "Athanasius Kircher, Giordano Bruno" (2004); Rowland, "Th' United Sense of the Universe" (2001).
5. Leinkauf, *Mundus Combinatus* (1993), esp. 110–29; Stolzenberg, "Connoisseur of Magic" (2001).
6. *OA* I, 421.

Because Kircher drew the essential building blocks of his hieroglyphic studies from texts and ideas associated with Renaissance Neoplatonism, *Egyptian Oedipus* may seem like an encyclopedia of occult philosophy, comparable to the sixteenth-century projects of Cornelius Agrippa or Francesco Patrizi. But Kircher used those building blocks to construct something different. By examining how his narrative about the propagation of ancient philosophy, magic, and religion functioned within the overall structure of *Egyptian Oedipus* and *Pamphilian Obelisk*, this chapter shows how Kircher repurposed occult philosophy into a historical and theoretical framework for explaining antiquities.

THE HIEROGLYPHIC DOCTRINE

Having equated "the Encyclopedia of the Egyptians" with "the secret wisdom of the ancient Hebrews, Chaldeans, Egyptians, Greeks, and other Oriental peoples,"[7] Kircher argued that the content of the hieroglyphic doctrine could be reconstructed from the skillful reading of the documents of Egypt's various cultural offspring. This was the burden of the massive second part of *Egyptian Oedipus*, entitled "Egyptian Gymnasium." More than any other part of Kircher's hieroglyphic studies, it gives the impression of a compendium of Renaissance occult philosophy—a "phenomenology of the occult" or "*Summa Magiae*," as R. J. W. Evans puts it.[8] Spread over two volumes and more than one thousand pages, it contains twelve sections (Kircher called them "*classes*"), each a substantial, stand-alone treatise. Since all the traditions from which he would reconstruct the hieroglyphic doctrine had to do with symbolic wisdom, Kircher devoted the first section to a general discussion of symbolism, encompassing the arts of emblems and *imprese*, military insignia, enigmas, riddles, parables, fables, and adages. There followed treatises on the origin, development, and mystical powers of languages and alphabets; the esoteric sayings of ancient pagan sages; the Hebrew Kabbalah; Arabic magic; Egyptian cosmology; hieroglyphic mathematics (including numerology, cosmic harmony, and astrology); hieroglyphic mechanics; hieroglyphic medicine; hieroglyphic alchemy; hieroglyphic magic; and hieroglyphic theology.

In the third section, entitled "Mystagogic Sphinx," Kircher examined Greek texts associated with the *prisca theologia*: the Chaldean Oracles, which, following Patrizi and Gemisthus Pletho, he attributed to Zoro-

---

7. *OA* II, title page.
8. Evans, *Making of the Habsburg Monarchy* (1979), 440–41.

aster (although elsewhere he clarified that they were actually composed by Hermes Trismegistus); the Orphic hymns; the Pythagorean verses; gnomic utterances, largely extracted from Horapollo, that he attributed to the Egyptians themselves; and a series of statements, derived from Proclus, Pico, and Ficino, pertaining to the allegorical interpretation of classical mythology. According to Kircher, the ancient sages used "enigmatic sayings" (*effata enigmatica*) to protect their profound teachings about God, angels, the world, and the soul from the masses. In reprinting these texts side by side, along with glosses, Kircher followed the example of Patrizi, who, as part of his program to replace Aristotelianism with Neoplatonic occult philosophy, had assembled the Chaldean Oracles, the Orphic hymns, and the Hermetic Corpus in an appendix to his *New Philosophy of Everything* (1591).[9]

Although Kircher did not reprint the *Corpus Hermeticum* or *Asclepius* in their entirety, he cited them throughout his work and devoted the final section of "Egyptian Gymnasium" to an exposition of their teachings, described as "Hieroglyphic theology." Following discussion of Hermetic teachings about subjects such as the Trinity, the world spirit, demons, the origin of souls, and the influence of divine ideas on lower things, Kircher confronted Egyptian theology's darker side: theurgical practices involving sacrifices, rituals, and incantations. He acknowledged that, as Hermes Trismegistus had been a pagan guided only by natural light, his theology contained errors, such as the doctrine of the transmigration of souls.[10] But the dominant tenor of Kircher's discussion was positive, attributing to Trismegistus prophesies of the decline of ancient religion, the rise of a new faith, the advent of Christ, the future judgment, the glory of the blessed, and the punishment of sinners. Kircher—or was it his sources?—wavered between assigning Hermes Trismegistus some kind of revelatory knowledge and stating that he was dependent only on natural light. Although he acknowledged that Saint Augustine was unsure whether Trismegistus knew these things by knowledge of the stars or by revelation, he noted that Lactantius did not hesitate to number him among the sibyls and prophets. In the final pages of *Egyptian Oedipus*, Kircher asserted:

> Hermes Trismegistus, the Egyptian, as the first founder of the hieroglyphs, was rightly the founder and parent of all Egyptian theology and philosophy and the first and most ancient of all those among the

---

9. Patrizi, *Nova de universis philosophia* (1591); Leijenhorst, "Francesco Patrizi's Hermetic Philosophy" (1998).
10. *OA* II.2, 497–546.

Egyptians who thought correctly about divine things; and he carved his opinion about these matters on the stones and giant masses of rock that would endure forever, as has been thoroughly proven previously. Not only the pagan commentators of the ancient philosophy, but also the orthodox memorials of the Holy Fathers demonstrate that everything that Orpheus, Musaeus, Linus, Pythagoras, Plato, Eudoxus, Parmenides, Melissus, Homer, Euripides and other praiseworthy men proffered concerning God and divine things followed after him. And Trismegistus in the *Pimander* and *Asclepius* asserted first that God is one and good in essence; afterward the rest of the philosophers followed him.[11]

The "Egyptian" cosmology, which Kircher identified as the basis of many beliefs and practices associated with the hieroglyphic doctrine, resembled the classic Neoplatonic universe, consisting of intellectual, celestial, and terrestrial worlds emanating from a transcendent divinity.[12] To explain Egyptian magic and its offshoots, which made use of sympathies and antipathies to influence good and bad genies or intelligences, he relied on the theories of Proclus, in *On Sacrifice and Magic*, and Iamblichus, in *On the Mysteries of the Egyptians*, texts made known to early modern Europe through Marsilio Ficino's translations.[13] Kircher also made extensive use of Ficino's original work. His account of "hieroglyphic medicine," based on the stars' influence on terrestrial bodies, was largely derived from *On Living One's Life According to the Heavens*, Ficino's famous primer on astral-medical magic.[14] Kircher insisted that such material was meant only to shed light on antiquity. It was, as the Loompanics catalog might have put it, "for informational purposes only."[15] But his detailed descriptions of magical and divinatory practices made *Egyptian Oedipus* serviceable as an encyclopedia of magic, not unlike Cornelius Agrippa's notorious *On Occult Philosophy*, which treated a similar range of material within a similar Neoplatonic framework, and from which Kircher borrowed a good deal.[16]

---

11. *OA* III, 568.
12. *OA* II.1, 401–40.
13. *OA* II.2, 435–96.
14. *OA* II.2, 345–86; Cf. Ficino, *Three Books on Life* (1989).
15. In the 1980s and 1990s the catalog of the libertarian-anarchist book distributor Loompanics Unlimited, which offered instructional books on subjects such as tax evasion and bomb making, carried the disclaimer: "Certain of the books and papers in this catalog deal with activities and devices which would be in violation of various Federal, State and local laws if actually carried out or constructed. Loompanics Unlimited does not advocate the breaking of any law. . . . All titles are sold for informational purposes only."
16. Agrippa von Nettesheim, *De occulta philosophia* (1992).

Since the pioneering studies of Giovanni Pico, Johannes Reuchlin, and Agrippa in the late fifteenth and early sixteenth centuries, many European scholars viewed the Jewish Kabbalah as an important font of occult philosophy, compatible with the teachings of gentile ancient theologians. Kircher affirmed an especially close relationship between Egyptian and Hebrew wisdom: "The Hebrews have such an affinity for the rites, sacrifices, ceremonies, and sacred disciplines of the Egyptians," he wrote, "that I am fully persuaded that either the Egyptians were Hebraicizing or the Hebrews were Egypticizing."[17] According to Kircher, the true Kabbalah preserved the same Adamic wisdom that Hermes Trismegistus had encoded in the hieroglyphs, while the "rabbinic superstitions" found in many kabbalistic treatises were closely related to Egyptian idolatry.[18] The fourth section of "Hieroglyphic Gymnasium" was a substantial treatise on the "Kabbalah of the Hebrews," which examined hermeneutic methods based on the manipulation of the Hebrew alphabet, the names of God and their use in mystical prayer, the doctrine of the ten *sefirot* or divine numerations, and the so-called "natural Kabbalah"—which, as with the other divisions, contained both a true doctrine and a false one, the latter including what Kircher called kabbalistic magic and astrology.[19] Kircher made frequent use of kabbalistic sources in other parts of his work as well, notably in the section nominally devoted to "Egyptian philology," in which he described the mystical powers of the Hebrew alphabet and argued that a form of Hebrew had been the original, universal language spoken before Babel.[20]

KIRCHER'S HERMETIC HISTORY

Kircher justified using sources like the Chaldean Oracles and the Kabbalah to explain Egyptian hieroglyphs by constructing a narrative about the transmission of ancient culture. According to Kircher, two traditions lay at the root of human civilization: first Adam founded the arts and sciences necessary for the preservation of mankind, and then his son Cain invented black magic and idolatry. These two traditions existed side by side during the first generations of mankind, with Adam transmitting pristine wisdom to his pious son Seth, from whom it passed to Enoch and Noah, while Cain

17. *OA* I, b^{1v}. Kircher's phrasing echoes a famous remark by Numenius, in his commentary on the *Timaeus*, about how "Plato Pythagorizes" and "Pythagoras Platonizes." I owe this observation to Anthony Grafton.
18. *OA* II.1, 359.
19. *OA* II.1, 209–360.
20. *OA* II.1, 42–122.

instructed his offspring in superstition. Eventually, Noah's cursed son Ham, known to some descendants as Zoroaster, combined the Adamic wisdom he had learned from Enoch's son Methuselah with the superstitious magic of the Cainites. After the Flood he passed this tainted legacy on to his children, the founders of different pagan civilizations.[21]

Since its purpose was to support his interpretation of the hieroglyphs, Kircher's history focused on Egypt. After the universal deluge, he argued, Egypt, which had known kingdoms in antediluvian times, was repopulated by the descendants of Ham's son Misraim, who spread their progenitor's corrupt teachings.[22] After several generations, Hermes Trismegistus appeared on the historical stage to revive the pure wisdom of the antediluvian patriarchs by untangling it from the superstitions introduced by Ham. According to Kircher, he was the second of two historical figures correctly to be identified as Hermes, although others mistakenly were also called by that name. (In sorting out the mass of conflicting testimonies about such ancient figures, Kircher had recourse to two complementary strategies. On the one hand he made claims of identity, arguing that the same individual had different names among different peoples, as in the case of Ham-Zoroaster. He also claimed that certain names came to be applied incorrectly: for example, Hermes Trismegistus was mistakenly called Zoroaster by people for whom the latter name was synonymous with great wisdom.)[23] The first Hermes had lived before the Flood and was none other than the biblical Enoch, known to the Arabs as Idris; according to some reports, he built pyramids in Egypt that were destroyed by the Flood. The second Hermes Trismegistus was not a native Egyptian, but rather descended from Canaan, the fourth son of Ham, and was born in Asia. He flourished approximately three hundred years after the Flood, was a contemporary of Abraham, and probably lived for one hundred and ninety years.[24]

The second Hermes cultivated the wisdom of the antediluvian patri-

---

21. The elements of this narrative are scattered and repeated in numerous places in *Obeliscus Pamphilius* and *Oedipus Aegyptiacus*. In particular, see *OP*, 2–32, 79–88, and *OA* II.1, 42–80, II.2, 142–50.

22. *OA* I, 65–68. On the significance of these antediluvian dynasties, see Grafton, "Kircher's Chronology" (2004).

23. On this basis, Kircher made Trismegistus the author of the "Oracles of Zoroaster" (i.e., the Chaldean Oracles); see *OP*, 107. In his discussion of the Oracles in *OA* II.1, 129–30, however, he declined to state that Hermes Trismegistus was their author, asserting only that the Hermetica and the Zoroastrian Oracles were so close that they must come from the same source, and that that the Oracles were nothing but hieroglyphs taken from the columns of Hermes. See also *OA* I, 94.

24. *OP*, 20–31, 45–48.

archs, testing his knowledge through experience, and traveled widely. When the Orient grew crowded as the postdiluvian world repopulated, he emigrated to Italy where he soon achieved godlike fame. After internecine wars broke out among the Italians, however, he moved to Egypt, then ruled by Misraim, a lover of wisdom, who made Trismegistus his supreme scribe and counselor. After Misraim's death the Egyptian throne passed to Mesramuthisis, during whose reign Hermes ruled as regent for thirty-nine years, spreading the hieroglyphic doctrine through all Egypt. To preserve the purified teachings that he had recovered for posterity, Trismegistus invented a new medium on which to record them: the obelisk. (Previously, the impious doctrines of Ham had been written on pyramids, but, according to Kircher, their excessive obliquity rendered them impractical). To protect the sacred teachings from the profane masses, he invented a new form of writing intelligible only to the priestly elite: the hieroglyphs.[25]

Alas, related Kircher, Hermes Trismegistus's reforms did not take root and, "with the passage of time, as is accustomed to happen, [the hieroglyphic doctrine] was corrupted by the priests and turned into the nursery of every superstition." Thus, the hieroglyphic legacy Kircher attempted to unravel was a mixed one. Pure Adamic wisdom was preserved on the oldest obelisks, erected by Trismegistus and his immediate successors. But the majority of surviving obelisks, constructed by later Egyptian priests, recorded a mixture of the original Hermetic teachings with superstitious magic and idolatry. From Egypt, Hermes' teachings were passed to other cultures, both by foreigners who visited Egypt, such as the founders of Greek philosophy and religion, and by Egyptian priests who traveled abroad, especially to the Orient. Through similar channels, the nations of the world received the impious rituals of Egyptian magic and idolatry, which Kircher described as the template of all pagan religion.[26]

For Kircher, Egypt was the birthplace neither of true religion nor of idolatry and superstition; both traditions could be traced to the first generations of man and the figures of Adam and Cain. But it was the postdiluvian crucible in which the antediluvian traditions were revived and intermixed, and out of which they passed to the rest of humanity. Egypt was the fountainhead of the *prisca theologia*, and Hermes Trismegistus, whose reputation as a pious Egyptian sage had been propagated by Renaissance Neoplatonists since Ficino, was the father of gentile philosophy and "theosophy."[27]

25. *OP*, 45–8, 92–98.
26. *OP*, 46, 104–13; *OA* I, 168 and passim.
27. *OA* II.2, 497.

THE TRANSMISSION OF CULTURE AND THE HISTORY OF LEARNING

Despite the central role he assigned to the *prisca theologia* and Hermes Trismegistus, Kircher's purpose was different from that of the Renaissance occult philosophers on whom he so heavily relied. For Ficino and his Christian Neoplatonist successors, the point of the *prisca theologia* was to legitimize Platonism and related philosophical and magical traditions. By establishing the existence of a pious tradition of gentile theology that paralleled the Mosaic tradition, they would present Christianity as the fulfillment of both. The wisdom of ancient gentile theologians like Hermes Trismegistus might have an independent origin, either in rational speculation or divine inspiration, or it might derive from Moses or earlier Hebrew patriarchs. The difference between these alternatives was theologically significant, but it was not of utmost importance to occult philosophers like Ficino, who were primarily concerned to validate the Platonic tradition by appealing to church fathers who had acknowledged the existence of the *prisci theologi*. Consequently, the invocation of the ancient theology typically had been more a phenomenological than a historical assertion. The resemblance of different systems of thought was the main point, not the precise nature of their transmission over time. For Ficino, the *prisca theologia* was simply a roster of a half-dozen ancient pagan thinkers that served as a symbol of the existence of a tradition of pious pagan wisdom.[28] By contrast, Kircher used the genealogy of the *prisca theologia* to establish the antiquity of the hieroglyphic doctrine and the continuity of its transmission, regardless of its truth. The esoteric antiquary was an expert in occult philosophy, but his aim was different. If the hieroglyphs encoded the teachings of Hermes Trismegistus and later Egyptian priests, Kircher needed to demonstrate that those teachings, though fragmented and dispersed, had survived to the present day, thus making it possible to reconstruct them. In producing a transmission history that would perform this function, Kircher built on the skeleton of the *prisca theologia* by drawing on other models.

The concept of the ancient theology asserted the existence of a tradition of pious pagan wisdom; but Kircher's investigation concerned falsehood—"the errors of the ancients"—as much as truth. His history of the hieroglyphic doctrine thus wove together the genealogy of the ancient theology and a parallel history of superstitious magic and idolatry. Both strands had

---

28. Allen, *Synoptic Art* (1998), 24–26. See above, introduction, "Occult Philosophy," for further references.

polemical origins. While the doctrine of the *prisca theologia* was meant to demonstrate the piety of select pagan philosophers, the legend of Ham was created to discredit magic by identifying Noah's cursed son as its founder.[29] These traditions originally spoke not to historical concerns, but rather to philosophers and theologians preoccupied with the probity of knowledge. The concept of the *prisca theologia* was inherently a tool of syncretism, and thus a means to establish connections between different traditions and cultures. But Kircher did something more, using the story of Hermes Trismegistus and the spread of Egyptian knowledge as the basis of a universal history that situated all known civilizations in a single, coherent, chronological narrative.

As Kircher well knew, some skeptics asserted that the meaning of the hieroglyphs had been irrevocably lost and that his project was doomed from the outset. The investigations of ancient religions, arts, and sciences, largely of non-Egyptian provenance, which formed the bulk of *Egyptian Oedipus* were meant to refute this pessimistic claim by demonstrating the dissemination of ancient Egyptian traditions to other nations who had preserved them piecemeal in their scattered memorials. In the second book of *Pamphilian Obelisk*, "Apparatus for Interpreting the Hieroglyphs," after explaining how Hermes Trismegistus had founded the hieroglyphic doctrine and later Egyptian priests had corrupted it, Kircher chronicled the vagaries of its subsequent transmission. He began by demonstrating, with citations to Maimonides and Philo of Alexandria, that the Egyptian sciences had first flourished long before Moses. After the exodus from Egypt, they were fostered by a succession of pharaohs, so that when the first Greek philosophers, Orpheus, Linus, and Amphion, voyaged to Egypt they were able to study the ancient wisdom recorded in the hieroglyphic inscriptions. The allegories of the Orphic hymns described the hieroglyphic mysteries so precisely, Kircher wrote, that they seemed to have been written by an Egyptian priest rather than a Greek poet. When Cambyses conquered Egypt, the invaders had not, as some would claim, destroyed all traces of the hieroglyphic doctrine. While Egypt was under Persian and then Greek rule, Pythagoras, Plato, Democritus, Thales, Eudoxus, and even Aristotle had visited and imbibed its ancient wisdom. So Kircher pronounced, citing numerous classical authorities.[30]

Kircher then turned to the dissemination of the hieroglyphic doctrine in the Near East as testified by the circulation of the Chaldean Oracles, which,

29. Flint, *Rise of Magic* (1991), 333–38.
30. *OP*, 104–6.

although attributed to Zoroaster, were really composed by Hermes Trismegistus.[31] Hermetic wisdom flourished among the Hebrews, Chaldeans, Arabs, and other nations, although, as in Egypt, it was soon corrupted by star worshipers. Kircher discussed various ancient Oriental authorities on Egyptian literature, whose opinions were preserved in the works of later authors such as Eusebius, Clement of Alexandria, and Origen, devoting special attention to the ancient Phoenician writer Sanchuniathon. Continuing into the Christian era, Kircher presented a long list of witnesses to the survival of knowledge of ancient Egyptian literature, including Maimonides, Salamas, al-Suyuti, Abenephius, and Ibn Wahshiyya among Arabic writers, and many Western authors such as Plotinus, Porphyry, Iamblichus, Dionysius the Areopagite, Justin Martyr, Origen, Tertullian, Clement, Eusebius, and Irenaeus, "who weighed the writings of the ancients accurately and presented the Egyptian doctrine, although fragmentarily, nevertheless with clarity." Then, after a long interval, Kircher concluded, Pierio Valeriano and Nicolas Caussin had taken up the mantle, followed by Lorenzo Pignoria and Herwart von Hohenburg.[32]

Kircher's final examples seem especially weak evidence for the preservation of ancient Egyptian literature. By his own description, the *Hieroglyphicae* of Valeriano and Caussin did not really treat Egyptian matters, Pignoria's interpretation of the Bembine Table lacked profundity, and Herwart's was wide of the mark. But his motive for bringing the lineage up to his own era is clear. "From all these [testimonies] that have been brought forth, perhaps more profusely than was necessary," Kircher concluded, "it is perfectly clear that not only were the hieroglyphic disciplines not destroyed, but they were propagated even until today through the memorials of the foresaid authors, although they neglected the marrow of the symbols and passed on to posterity only the shell, without the inner meaning."[33] At the same time that he claimed to have recovered knowledge lost for millennia, Kircher needed to establish that the ancient Egyptian teachings had been preserved—although broken up into fragments and hidden beneath symbols—continuously from antiquity to the present. Throughout these chap-

---

31. On the attribution of the Chaldean Oracles to Zoroaster, see Dannenfeldt, "Pseudo-Zoroastrian Oracles" (1957); Stausberg, *Faszination Zarathushtra* (1998). On Kircher's attribution to Trismegistus, see above n. 23.

32. *OP*, 106–13, quotation at 113. Kircher also enumerated many "historians and philologists" whose memorials supplied scattered evidence of Egyptian wisdom, such as Athanasius, Epiphanius, Lactantius, Augustine, and Jerome, among the holy fathers; and Ammianus Marcellinus, Martianus Capella, Pliny, Aelian, Solinus, and Macrobius, among the pagans.

33. *OP*, 114.

ters he intoned this refrain: the hieroglyphic doctrine and Egyptian literature had not perished, but survived scattered in the memorials of different nations. By demonstrating the preservation of Egyptian wisdom through the millennia, Kircher sought to establish the link—absolutely essential to his antiquarian endeavor—between the literary records at his disposal and the inscriptions on the ancient Egyptian obelisks.

The final product was a history that resembled less the schematic genealogies of Renaissance occult philosophers than the encyclopedic histories of learning known as *historia litteraria* and *historia philosophica*. The former was an erudite genre that emerged in the sixteenth century, which aspired to narrate the history of all knowledge, from the beginning of the world to the present, based on the premise of the identity of writing and culture.[34] Kircher's debt to the tradition of *historia litteraria* is evident from the central historical role he assigned to writing and literary technologies such as inscriptions, pens, ink, papyrus, parchment, and codices, including in the chapter in *Pamphilian Obelisk* devoted to the vexed question of whether writing and books had existed before the Flood. (Kircher answered affirmatively.)[35] In the seventeenth century *historia litteraria* gave rise to *historia philosophica*, the first histories of philosophy. The key works in this new genre, by Georg Horn, Thomas Stanley, and G. J. Vossius, appeared in the same decade as *Egyptian Oedipus*. An important function of these histories of philosophy was to synthesize Judeo-Christian and pagan traditions within the framework of sacred history. In this respect they resembled the erudite histories of idolatry that proliferated in the same period, and it is no coincidence that Vossius made important contributions to both genres. It was standard practice in such works to begin with Adam, regularly styled as the first philosopher, and to describe the transmission of true wisdom as well as its evil twin, superstition, via Adam's progeny and Noah and his sons. It was also standard practice for early modern histories of philosophy to include the wisdom of the Egyptians, Chaldeans, and other "barbarians," represented by figures like Hermes Trismegistus and Zoroaster, who were described as the teachers of the first Greek philosophers. The genealogy of occult philosophy thus became a normal feature of seventeenth-century histories of philosophy and learning, even if their authors were not usually

---

34. Kelley, "Writing Cultural History" (1999). See also: Bottin et al., *Models of the History of Philosophy* (1993).
35. *OP*, "Liber I. De Litterarum, & Obeliscorum Origine," 1–91. *OA* II.1, "Classis II. Grammatica. seu De Philologia Aegyptiorum, & de primaeva artium ac scientiarum institutione," 42–122.

proponents of Neoplatonic doctrines.³⁶ Despite the doubts that had been cast on the authenticity of specific texts, the historicity of the *prisca theologia* remained a scholarly commonplace in the 1650s.

These early modern histories of learning were motivated by the encyclopedic ideal of universal knowledge and, in most cases, by belief in Adamic perennial philosophy. They aimed to reform knowledge by sifting truth from falsehood through an investigation of the history of doctrines; in effect, applying the method of ecclesiastical history to philosophy. Ecclesiastical history, which reemerged in the early modern era as an important tool in theological debates among Protestants and Catholics, sought to separate orthodox doctrine from heresy through the study of the history of the Church.³⁷ Kircher's hieroglyphic studies were based on the similar premise of an original, perfect teaching that had been obscured over time and could be recovered through the study of primary sources. "Egyptian literature never perished," he affirmed, "but . . . always lay hidden, scattered here and there in the memorials of ancient authors."³⁸

With its relentless leitmotif of the corruption of once-pure doctrines by sinister forces, Kircher's history of pagan wisdom (like many *historiae philosophicae*) was reminiscent of Protestant ecclesiastical histories: the *prisca theologia*, like the Protestants' true Church, had no institutional protector during its long dark ages. But Kircher was no Protestant. If the corruption of doctrine over time was preeminently a Protestant historical trope, Kircher's understanding of the process of corruption was distinctly Catholic and voiced in the language of the Counter-Reformation. Sublime truths were defiled and distorted, according to Kircher, whenever "someone, by his own judgment and for the pleasure of his reeling mind, has his hands full explaining, correcting, and applying the divine and secret hidden meanings of sacred scriptures and twisting them to illegitimate uses."³⁹ At the heart of his argument was the characteristically Catholic notion that the reliability of his interpretation of the hieroglyphs was vouchsafed by

---

36. Bottin et al., eds., *Models of the History of Philosophy* (1993). On Stanley, see Haugen, *Richard Bentley* (2011), 18–29. Disciplinary histories of fields such as mathematics, astrology, and alchemy also frequently invoked the *prisca theologia*, but typically in a perfunctory manner, without argument or sustained development. See Popper, "Abraham, Planter of Mathematics" (2006).

37. See Momigliano, *Classical Foundations* (1990), 131–52; Ditchfield, *Liturgy, Sanctity and History* (1995), esp. part 3.

38. *OP*, b4ʳ.

39. *OA* II.1, 210. Kircher referred here to the corruption of the Hebrew Kabbalah, but the point would have applied generally.

a continuous transmission of doctrine. The Catholic Church asserted its legitimacy against Protestant critics by claiming an unbroken chain of doctrine and authority going back to the apostle Peter, who had been deputized by Christ himself. (Protestant and Catholic ecclesiastical histories were respectively attempts to refute or defend this continuity.) Although Kircher's hieroglyphic doctrine was not preserved en masse by a single institution like the Catholic Church, his account of its history appealed to a notion of doctrinal continuity similar to the one that vouchsafed Catholic claims to religious authority.[40]

For Kircher the omnipresent, entropic force of corruption and the dichotomy between esoteric truth and vulgar superstition were historical constants, allowing him to make sense not only of Egyptian teachings, but of the religious and philosophical traditions of the entire world. By setting non-Christian traditions within a comprehensible pattern, the historical framework of *Egyptian Oedipus* resembled a *historia litteraria* or *philosophica* in which the history of paganism took center stage and the *prisca theologia* was given an outsize role.

READING THE BEMBINE TABLE

Kircher's antiquarian mobilization of occult philosophy, based on the historical framework described above, can be seen in action in his interpretation of the Bembine Table. A bronze tablet decorated with silver, copper, and niello inlay, depicting Egyptian cult scenes centered on the figure of Isis, the Bembine Table (also known as the *Mensa Isiaca*) was the most famous Egyptian artifact in Renaissance Europe.[41] Having first come to notice in the 1520s, when it was acquired by Cardinal Pietro Bembo, its fame spread following Enea Vico's printing of an engraving in 1559. Kircher knew the table only from reproductions, such as those printed by Lorenzo Pignoria and Herwart von Hohenburg, the latter of which served as the model for the plate printed in *Egyptian Oedipus* (fig. 26). First commissioned by

---

40. On stylistic similarities between Kircher's *China illustrata* and traditional ecclesiastical histories, see Grafton, *Footnote* (1999), 154. For further discussion of Kircher's assertions about the continuity of tradition, see below, ch. 7, "The Divorce of Mercury and Theology."

41. Modern Egyptologists consider the table to be an Egypticizing product of Imperial Rome rather than a genuine Egyptian artifact, though it may well have been used in an Italian Isiac temple. Its images are not, in fact, hieroglyphs. It is now housed in the Egyptological Museum of Turin. Leospo, *Mensa Isiaca* (1978). See also Whitehouse, "Towards a kind of Egyptology" (1992), 69–70; Iversen, *Myth of Egypt* (1993), 55–56; Curran, *Egyptian Renaissance* (2007), 231–33.

Fig. 26. The Bembine Table, or *Mensa Isiaca*, was the most famous Egyptian antiquity in early modern Europe and the object of the most significant early efforts to interpret hieroglyphs. The table is not in fact Egyptian, and its images, which depict cult scenes, are not hieroglyphic. It was produced in Imperial Rome, probably for use in the Temple of Isis. Athanasius Kircher, *Oedipus Aegyptiacus* (Rome: 1652–54), vol. 3, fp. 78–79. From a reprint in Manly P. Hall, *The Secret Teachings of All Ages* (1928).

Cardinal Barberini in 1633 as a test of Barachias Nephi's method, Kircher's eighty-page interpretation of the Bembine Table initiated the series of hieroglyphic translations in the final volume of *Egyptian Oedipus*.[42]

Kircher framed his interpretation in distinctly antiquarian terms: he would explain one of the most celebrated ancient inscriptions, which previous interpreters, like Pignoria and Herwart, had been unable to account for in a satisfactory manner.[43] He began by considering the table's provenance, determining that it was of Egyptian rather than Roman construction, since in the latter case, he supposed, it would contain some admixture of Latin. He then addressed its function, arguing that it had been a sacred altar in the temple of Isis. To support this thesis Kircher cited numerous ancient sources—including Arnobius, Clement of Alexandria, Justin Martyr, and Lucian, as well as the Hebrew Old Testament and its Aramaic paraphrase—on the importance of Egyptian temple sanctuaries as centers of learning and on the use of images inscribed on their walls, tables, and altars. This led him to an excursus on ancient sacrificial altars in general, drawing on classical sources such as Festus, Arnobius, Diogenes Laertius, and Cicero. An Arabic quotation from Ibn Wahshiyya succinctly established that ancient Egyptian altars were inscribed with images and letters.[44]

Having established to his satisfaction that the Bembine Table had been a sacred altar in an Isiac temple, Kircher proceeded to the meaning of its inscriptions. The Egyptians, he explained, citing Iamblichus, used symbols to represent the cosmos and the "workshop of the gods," as well as to indicate sacred mysteries. According to Kircher, the table adumbrated the entire "Egyptian theosophy," which he understood in terms of a Neoplatonic metaphysics of archetypal, celestial, and elementary worlds, animated by the radiating power of the transcendent divinity.[45] In order to explain this system and show how the symbols on the *Mensa Isiaca* represented it, he drew on virtually all the traditions that defined his elaboration of the hieroglyphic doctrine in previous volumes of *Egyptian Oedipus* and in *Pamphilian Obelisk*.

The heart of Kircher's interpretation was a freely rendered paraphrase of the Chaldean Oracles.[46] To demonstrate that the table's figures represented such teachings, he deployed Neoplatonic hermeneutic methods (discussed

---

42. *OA* III, 79–160.
43. On the interpretations of Pignoria and Herwart, see above, ch. 1, "Renaissance Egyptology."
44. *OA* III, 81–84.
45. *OA* III, 85–87.
46. On the Chaldean Oracles, see Majercik, *Chaldean Oracles* (1989); Klutstein, *Marsilio Ficino et la theologie ancienne* (1987); Stausberg, *Faszination Zarathushtra* (1998).

in detail in chapter 7), identifying individual signs by reference to sundry authorities on ancient symbolism, and assigning them collective meaning based on their location in the composition.[47] Using the terminology of the Chaldean Oracles, as mediated through the commentaries of Psellus, Kircher explained that the center of the table represented the triad of the *fundus paternus* or supreme divinity, surrounded by other triads belonging to the archetypal world, while the smaller figures along the perimeter represented the ideal forms in the mind of God.

To explain the table's symbolism, Kircher also drew on many other primary sources of occult philosophy, including the *Corpus Hermeticum*, the Orphic hymns, and pseudo-Aristotle's *Egyptian Theology*. He made extensive use of the *Zohar*, dilating on the parallels between kabbalistic teachings concerning the "three mothers" and the Chaldean Oracles' enumerations of divine and archetypal triads. Kircher stated that the words of the *Zohar*'s reputed author, Simeon ben Yochai, were so close to the meaning of the Bembine Table that they almost seemed to explain them, and he later suggested that ben Yochai might have copied from the Chaldean Oracles.[48] In addition, he cited many secondary authors associated with the preservation of the ancient theology, especially Iamblichus, Proclus, and Porphyry (the ancient Neoplatonists most influenced by the Chaldean Oracles) as well as Plotinus, Clement of Alexandria, Eusebius, Horapollo, and Abenephius. Drawing on Iamblichus, he described how ancient Egyptian priests applied the metaphysical system described on the Bembine Table to sacrificial, theurgical rituals.[49]

Kircher did not explicitly discuss the transmission of the hieroglyphic doctrine in his chapter on the Bembine Table; but the historical account set forth in earlier sections of *Egyptian Oedipus* and in *Pamphilian Obelisk* established the justification for bringing together the Chaldean Oracles, *Corpus Hermeticum*, Orphic hymns, Greek Platonic philosophers, the Kabbalah, Arabic writers like Ibn Wahshiyya, and so forth in order to unravel the meaning of an ancient Egyptian inscription.

DIVULGING THE SECRETS OF ANTIQUITY

Kaspar Schott left a vivid description of Kircher at work deciphering the *Mensa Isiaca*:

---

47. See Griggs, "Antiquaries and the Nile Mosaic" (2000) for an analysis of Kircher using similar methods to interpret a Hellenistic mosaic.
48. *OA* III, 109–10.
49. *OA* III, 155–58.

I recall, when once he gazed attentively at the Bembine Table, which he held spread out before the table at which he sat absorbed in studying and writing, so clear and uncommon a light appeared to him (as he reported to me) that by a single precise intuition he thereupon clearly knew the whole mystery of the table and said that by no account did he doubt but that the meaning which he had proposed in his exposition was the very meaning of the Egyptian priests.[50]

Schott's depiction of Kircher's sudden illumination has the air of a mystical revelation. Such an association of erudition and mysticism was not in itself unusual. The Florentine scholar Giovanni Pico della Mirandola, for example, famously pursued tireless philological research in the service of self-transcendence and unmediated, Seraphic knowledge of God. But in Schott's vignette, the relationship between scholarship and gnosis was reversed: divinely granted intuition led Kircher to historical knowledge. His eureka moment did not bring profound insight into God or the cosmos, but confirmation of the historical accuracy of his interpretation of an ancient artifact.

The frontispieces of *Pamphilian Obelisk* and *Egyptian Oedipus*, drawn by the distinguished painter Giovanni Angelo Canini and engraved by the master printmaker Cornelis Bloemaert, with imagery surely devised by Kircher, present a similar picture of his enterprise.[51] In the background of the first, the Pamphilian obelisk lies broken and decayed beneath the scythe of Saturn, god of time (fig. 27). Fame stands chained in manacles, her eyes downcast and her trumpet lowered to indicate the oblivion into which the hieroglyphic doctrine has fallen. A winged Kircher sits in the foreground with pen in hand, recording his interpretation of the obelisk, while Hermes, messenger of the gods and personification of hermeneutics, reveals its meaning by unfurling a scroll with hieroglyphic inscriptions. The closed end of the scroll lies in the hands of Harpocrates, the Egyptian god, who holds a finger to his lips and rests a foot upon a crocodile; both symbols, following Plutarch, of sacred silence.[52] But even as Kircher gazes at the god above, his writing arm is supported by the classical and Oriental sources on which he based his work. For all its allegorical flourish, the scene is not far from Schott's description of Kircher studying the Bembine Table.

The frontispiece of *Egyptian Oedipus* depicts Kircher as Oedipus before

---

50. Schott, "Benevoli Lectori," *OA* I, c4^v.
51. On Kircher's frontispieces in general, see Mayer-Deutsch, "Iconographia Kircheriana" (2001); Godwin, *Athanasius Kircher's Theatre of the World* (2009), 23–46.
52. *De Iside*, 68.378c, 75.381b; Plutarch, *De Iside et Osiride* (1970), 225, 237.

Fig. 27. Kircher, winged, translates the Pamphilian obelisk with help from Hermes and in disregard of the infant god, Harpocrates, who calls for silence. Athanasius Kircher, *Obeliscus Pamphilius* (Rome: 1650), frontispiece. Courtesy of Stanford University Libraries.

the Sphinx, declaiming his answer to her ancient riddle (see fig. 1). It, too, evokes the idea of sacred revelations, inspired by Plutarch's account of the Egyptians placing sphinxes before their temples in order to intimate "that their teaching about the gods holds a mysterious wisdom."[53] Kircher's discovery of those mysteries, however, is depicted as the result of "Sense and Experience" and "Reason," the foundations of human, as opposed to revealed, knowledge according to the Thomist formula. Their angelic personifications hover over Kircher-Oedipus, holding an open book that lists the many languages of Kircher's sources. Ten seals represent Egyptian wisdom, Phoenician theology, Chaldean astrology, Hebrew Kabbalah, Persian magic, Pythagorean mathematics, Greek theosophy, mythology, Arabic alchemy, and Latin philology. As Kircher explained elsewhere, the Oedipus who would unravel the meaning of the Egyptian hieroglyphs first must understand the secret philosophy and theology of the ancients through the study of all these traditions: "You see, therefore, how much preparation was necessary, how many foreign memorials had to be penetrated."[54]

Kircher made the most of the aura of mystery and profundity that surrounded his subject matter. In fashioning himself as an expert on "recondite antiquity," he effectively fused the persona of the "professor of secrets," who divulged choice arcana based on knowledge of natural magic, with that of the antiquarian-philologist able to decipher the meaning of objects from the past.[55] In his hieroglyphic studies, as in other investigations both literary and scientific, Kircher envisioned scholarship as the revelation of secrets, employing the "dialectic of concealment and disclosure," which Kocku von Stuckrad describes as inherent to the concept of the *prisca theologia* and discourses of esoteric knowledge.[56] The antiquary's quest to reassemble the fragmented remains of the past shared a deep structural and emotive affinity with narratives about the dispersion of primordial wisdom. To use von Stuckrad's terms, esoteric discourse, based on the dialectic of concealing and revealing higher knowledge, is structurally similar to antiquarian discourse about the recovery of the past. Or, as Walter Stephens puts it, "the struggle of human continuity and human culture against the erasures of the elements," embodied in legends about the survival of antediluvian knowl-

---

53. *De Iside*, 9.354b; ibid., 131.
54. See *OA* III, 558.
55. On the persona of the "professor of secrets" in early modern science see Eamon, *Science and the Secrets of Nature* (1994).
56. Stuckrad, *Locations of Knowledge* (2010), 58. See also Wilding, "If You Have a Secret" (2001).

edge, is simultaneously "the perennial charm of archeology."⁵⁷ This affinity lay at the heart of Kircher's marriage of antiquarianism and esotericism.

Antiquarianism and philosophy, occult or otherwise, were by no means incompatible. Kircher admired the profound wisdom to be found in the testaments of the *prisca theologia*, and it is quite possible to extract a philosophy from *Egyptian Oedipus*. By reinterpreting Kircher's hieroglyphic studies as esoteric antiquarianism rather than Hermetic philosophy, I do not seek to downplay occult philosophy's influence, without which his project would have been inconceivable. Rather, I mean to insist that Kircher used occult philosophy in ways that were fundamentally different from those of the Renaissance authors on whom he relied, ways more historical than philosophical. He was less interested in the hieroglyphic doctrine's truth or falsehood than in its ability to illuminate the beliefs and practices of ancient peoples. At the very end of *Egyptian Oedipus*, before praising God with a hymn from the *Corpus Hermeticum*, Kircher offered a "Summation of the Entire Work," in which he set forth "several undeniable arguments and reasons that the author's exposition and interpretation of the most secret and recondite matters has been done truthfully following the beliefs of the ancient Egyptians, and that by their famous hieroglyphic signs the Egyptians understood exactly what has been said."⁵⁸ In the ultimate defense of his enterprise, Kircher staked everything on its historical accuracy.

57. Stephens, "Berosus Chaldaeus" (1979), 118. See also 108.
58. *OA*, I, 551.

CHAPTER FIVE

The Discovery of Oriental Antiquity

It will be enough ... to have demonstrated by example the extent to which knowledge of the recondite literature of the Hebrews and Arabs may facilitate the restoration of hieroglyphic literature; and that its sources are only to be sought among the Orientals. Indeed, I would dare to solemnly affirm that the neglect of such languages results in ignorance about extraordinary things—antiquities heretofore unknown, and other arts and sciences that have been condemned to perpetual darkness—while fostering them furnishes enormous advantage and benefit, not only to the Republic of Letters, but to all of the Church and the Christian Republic.
—Athanasius Kircher[1]

THE LURE OF THE ORIENT

Kircher repeatedly emphasized that the key to his hieroglyphic breakthrough was his study of previously unknown sources in Oriental languages. Esteem for Eastern wisdom had always been characteristic of Renaissance occult philosophy. But with the notable exception of the Hebrew Kabbalah, the texts traditionally attributed to Oriental sages, including the Egyptian Hermes Trismegistus and the Chaldean or Persian Zoroaster, were Greek.[2] Despite the fact that much of the learned magic espoused by writers like Ficino and Agrippa derived, via medieval Latin translations, from Ara-

---

1. *OA* I, b2^r.
2. Festugière, *La révélation d'Hermès Trismégiste* (1981); Bidez and Cumont, *Mages Hellenisés* (1938).

bic texts, Kircher's emphasis on reconstructing the *prisca theologia* through the study of original sources in Arabic and other Oriental languages besides Hebrew was unprecedented.[3] It was another way in which his hieroglyphic studies represented a novel fusion of erudition and occult philosophy.

The middle third of the seventeenth century witnessed an Orientalist moment in European intellectual history, as Near Eastern philology emerged as the leading edge of erudite scholarship. Late Renaissance humanism was an omnivorous beast, ever seeking to expand its territory. Having largely conquered Latin and Greek antiquity, by the beginning of the seventeenth century Europeans searching to expand the frontiers of knowledge turned to the East with eagerness unseen since the Arabic translation movement of the twelfth and thirteenth centuries. Unlike their medieval predecessors, however, they searched Oriental texts primarily for historical information rather than insights into philosophy.[4] The perception that Muslims possessed a more advanced science, which had fueled the medieval study of Arabic learning, was, if not entirely moribund, increasingly marginal. Early modern scholars viewed Oriental literature not so much as a living tradition, but as an untapped storehouse of knowledge of antiquity.

Kircher's vision of rediscovering lost knowledge through the study of Oriental languages was shared by other scholars. In 1635, Peiresc wrote Francesco Barberini, encouraging the cardinal to promote the translation of Ethiopian literature, for "many extremely ancient books are conserved in that language, which have been lost for centuries in every other language."[5] His interest had been piqued by a report from the Capuchin traveler Gilles de Losches, who saw in North Africa "a Library of eight thousand Volumes, many of which bore the marks of the Antonian Age. And because among other things he said he saw *Mazhapha Einock*, or the Prophecie of *Enoch*, foretelling such things as should happen at the end of the World, a Book

---

3. The pioneering Orientalist and Christian kabbalist Guillaume Postel (1510–1581) could be considered an exception to this statement. But, even though he was an early champion of Arabic studies, it is unclear how much, if anything, he took from Arabic sources in developing his philosophical and theological views. See Bouwsma, *Concordia Mundi* (1957), esp. 46–48, 51, 60–63; see also Toomer, *Eastern Wisedome* (1996), 26–28; Secret, "Postel et les études arabes" (1962).

4. While antiquarian interests were ascendant, Renaissance and seventeenth-century scholars also continued to study Arabic scientific and medical texts. See Burnett, "Second Revelation" (1999); Mercier, "English Orientalists" (1994); Wear, "English Medical Writers" (1994).

5. ". . . in quella lengua si sonno conservati libri antiqui isquitissimi et perduti da mollti secoli, in ogni altra lengua." Nicolas-Claude Fabri de Peiresc to Francesco Barberini, Aix, 31 January 1635, BAV Barb. Lat. 6503, fol. 114^r.

never seen in *Europe*, but was there written in the Character and Language of the *Aethiopians* or *Abyssines*, who had preserved the same."[6] In the same vein, Abraham Ecchellensis proclaimed the great riches that awaited discovery in Arabic literature, supposedly including the lost books of Livy and a work by King Solomon.[7] Ecchellensis's paean found receptive ears among patrons such as Cardinals Richelieu and Mazarin in Paris and Pope Urban VIII in Rome, who vied to sponsor his research. To these scholars, Oriental literature promised something like access to the long lost Library of Babylon, as dreamed of a century earlier by Annius of Viterbo, which would have preserved a perfect record of the history of the world from the age of Adam until postdiluvian times.[8]

Kircher's hieroglyphic studies were a product of this movement in seventeenth-century scholarship. They began, in the 1630s, with the study of Arabic texts about Egypt and the hieroglyphs as well as his work on Coptic, and culminated in *Egyptian Oedipus*, which Kircher explicitly presented not only as an interpretation of hieroglyphic inscriptions, but also as an example of how the study of Oriental literature could expand knowledge of antiquity. This chapter examines Kircher's study of Near Eastern texts, focusing on two cases. In his study of Arabic literature about Egypt and Hermes Trismegistus, Kircher carried out significant, original investigations of Oriental sources. In his treatise on the Kabbalah, on the other hand, he exaggerated his firsthand study of Jewish authors and concealed his reliance on Latin secondary sources. Since *Egyptian Oedipus* was to a large extent cobbled together from the texts of unacknowledged early modern authors, a careful examination of its sources is needed to appreciate the mixture of original and derivative learning that Kircher set indiscriminately before his reader. Even Kircher's genuinely pioneering work did little to make his historical conclusions sounder, however. Intoxicated by the excitement of discovering new sources in exotic languages, he failed to scrutinize his Oriental authors, as if information found in an obscure Arabic manuscript was, ipso facto, more likely to present an accurate depiction of ancient Egypt than more prosaic sources. In Kircher's scholarship, the antiquarian imperative to expand the historical data pool trumped the skeptical imperative of critical philology.

6. Gassendi, *Mirrour of True Nobility* (1657), 90.
7. Ecchellensis, *Semita sapientiae* (1646), aiiii[r].
8. Stephens, "Berosus Chaldaeus" (1979), ch. 2. On early modern "bibliomythography," see Stephens, "*Livres de haulte gresse*" (2005).

THE ARABIC SOURCES OF KIRCHER'S HERMETIC HISTORY

Kircher's history of Hermes Trismegistus and the role of ancient Egypt in the preservation of antediluvian knowledge (discussed in the previous chapter) departed in significant ways from earlier treatments of these topics. Typically, early modern European writers placed Trismegistus roughly in the time of Moses. Usually, following Lactantius and Augustine, he was said to have lived slightly later, making it possible that he had derived his wisdom from the influence of the Mosaic Scriptures. But sometimes, as with Francesco Patrizi, for example, Hermes was made slightly older than but still contemporary to the Jewish lawgiver.[9] As we saw in the previous chapter, Kircher recognized two authentic Hermes, both of whom he located further back in time. The first was identical to the Bible's Enoch and the Koran's Idris, and had reportedly built pyramids in Egypt to preserve Adamic science from the prophesied Flood. The second Hermes, the founder of gentile theology, descended from Ham's son Canaan, was an older contemporary of Abraham, and had achieved fame after emigrating to Egypt, where he advised King Mesraim and invented hieroglyphs and obelisks as a means to protect the antediluvian wisdom he had restored.

Some Renaissance proponents of occult philosophy had speculated, like Kircher, that the origins of the *prisca theologia* lay earlier than Moses, perhaps in the time of Abraham, and that its ultimate source was not the Mosaic revelation but the wisdom of the first patriarchs. But these accounts made Zoroaster, not Hermes Trismegistus, the father of ancient theology among the gentiles.[10] Traditions about the preservation of Adamic wisdom from erasure in the Flood enjoyed great popularity in early modern Europe, but they focused on the legend of the pillars inscribed by Adam's son Seth, reported by the Jewish historian Josephus.[11] The broad outlines of Kircher's history of the survival of antediluvian knowledge would have been familiar

---

9. Allen, *Synoptic Art* (1998), 30–31; Purnell, "Francesco Patrizi" (1976), 157. See also Leijenhorst, "Francesco Patrizi's Hermetic Philosophy" (1998); Ebeling, *Secret History* (2007), 68–69. Patrizi acknowledged two Hermes Trismegisti; the author of the *Hermetica* was the second. While Patrizi's first Hermes was closer in time to Kircher's and played the role of counselor to an early Egyptian king, he was subordinate to Zoroaster in Patrizi's genealogy of the *prisca theologia*.

10. Ficino, who came to favor the priority of Zoroaster over Hermes Trismegistus in his enumeration of the *prisci theologi*, was attracted to this account, which identified Zoroaster with Ham's son Canaan, supposedly still alive in the time of Abraham. See Allen, *Synoptic Art* (1998), 31–32.

11. See Stephens, "*Livres de haulte gresse*" (2005); Popper, "Abraham, Planter of Mathematics" (2006).

to seventeenth-century readers, but not its details about the central role of Hermes Trismegistus and Egyptian monuments. In defending these novel historical claims, which enabled him to use occult philosophy to interpret hieroglyphic inscriptions, Kircher found his most important evidence in Arabic literature.

Medieval Arabic literature told of three ancient figures by the name of Hermes, of whom two were associated with Egypt, the pyramids, and antediluvian knowledge. The classic source for the "legend of the three Hermes" was the ninth-century astrologer Abū Ma'shar. According to Abū Ma'shar, the first Hermes, identified with Idris and Enoch, was the first to have discussed "upper things," such as the motion of the stars, and wrote many books about terrestrial and celestial sciences. Having been instructed by his grandfather Adam, he predicted the coming of the Flood and built pyramids and temples, carving inscriptions on their walls (or, in some versions, depositing books in their interiors) in order to preserve the antediluvian sciences. The second Hermes was a Chaldean wise man who lived in Babylon and revived the sciences after the Flood; Pythagoras was his student. The third Hermes was a physician and philosopher who lived in Egypt after the Flood, wrote books, wandered through the land, and had a student named Asclepius who lived in Syria.[12]

The history of ancient Egypt was the subject of a substantial body of medieval Arabic literature. Scholars of this material have distinguished two chief traditions, the "traditionalist history" and the "Hermetic history." The traditionalist history began with the settlement of Egypt after the Flood by Noah's grandson Misraim and focused on biblical figures such as Abraham, Joseph, and Moses. The Hermetic history began in antediluvian times, recounted the story of the Flood, and continued through the period covered by the traditionalist history.[13] Modern writers on the Hermetic history, seeking to convey its strange and marvelous character, have compared it to the *Arabian Nights* and *The Magic Flute*. The affinity of its vision of ancient Egypt to Kircher's is evident in Michael Cook's description of the

---

12. Plessner, "Hermes Trismegistus" (1954), 50–57; Plessner, "Hirmis" (1960–2005); Fodor, "Origins of Arabic Legends" (1970); Ullmann, *Natur- und Geheimwissenschaften* (1972), 372–77. Abū Ma'shar's account of the three Hermes was known to a few medieval Latin authors but does not seem to have made much impact on the reception of Hermes Trismegistus in the Renaissance; see Thorndike, *History of Magic* (1923–58), vol. 1, 340–41, vol. 2, 220; Burnett, "Establishment of Medieval Hermeticism" (2001), 115.

13. Haarmann, "Medieval Muslim Perceptions" (1996), 618; Cook, "Pharaonic History" (1983), 68–71; Wüstenfeld, "Die älteste Aegyptische Geschichte" (1861); Wüstenfeld, "Histoire de l'Égypte" (1861).

Hermetic history as "teem[ing] with priests learned in astrology and magic, sage rulers, marvelous constructions, talismans, treasures, ancient wisdom, and occasional glimpses of monotheist truth."[14] Later Arabic compilations dealing with Egypt drew from both traditions, which they supplemented with information from other sources, typically providing the reader a mass of diverse and often contradictory opinions.

The construction of the pyramids, about which a number of conflicting stories circulated, was a dominant theme in medieval Arabic literature on Egypt. A recurrent controversy revolved around the question of whether the great pyramids had been built before or after the Flood. A preponderance of authorities spoke in favor of the former possibility, attributing their construction either to a king named Saurid or to Hermes.[15] In this context, the first of Abū Ma'shar's three Hermes appeared in Arabic treatises on Egypt and the pyramids. Jalal al-Din al-Suyuti, for example, quoting al-Dimishqi, wrote: "Others say that it was Hermes, he of the triple wisdom, called by the Jews Enoch, and the same person as the blessed Idris, who deduced from the position of the stars that the deluge was about to come, and gave orders for the building of the pyramids and the deposition in them of treasures, books on the sciences, and other valuables which, one might fear, would perish and disappear."[16]

Kircher knew several Arabic works that treated the history of ancient Egypt. The most important were treatises by Gelaldinus, Abenvaschia, Abenephius, and Salamas ben Kandaathi (following the forms of their names used by Kircher). All of these texts had become accessible to European scholars only recently, and Kircher was among the first Western scholars—perhaps the very first—to make use of them.

Kircher's "Gelaldinus" was none other than the aforementioned Jalal al-Din al-Suyuti, an extraordinarily prolific Egyptian author who lived in the second half of the sixteenth century. Among his works was a history of Egypt, *Kitab husn al-muhadara fi akhbar misr wā-l-qahira*, from which the passage on Hermes and the pyramids above was taken.[17] Al-Suyuti's history was brought to Europe by Abraham Ecchellensis, who acquired the manuscript in Egypt in the early 1630s and took it to Rome in 1636, where it came immediately to Kircher's attention.[18] The treatise is an encyclope-

---

14. Cook, "Pharaonic History" (1983), 71.
15. Ibid., 87; Nemoy, "Treatise on Pyramids" (1939), 21; Fodor, "Origins of Arabic Legends" (1970), 335.
16. Nemoy, "Treatise on Pyramids" (1939), 27.
17. Ibid. See also Geoffroy, "al-Suyuti" (1960–2005).
18. See above, ch. 3, "Catholic Cosmopolis."

dic work and, despite its late composition, it provided Kircher with access to excerpts from earlier Arabic authors. Sometimes Kircher cited these embedded authors directly, without mentioning al-Suyuti, thus creating the impression that he had access to a larger number of Arabic manuscripts. Among the authorities cited by al-Suyuti are Ibn Wasif Shah and al-Mas 'udi's *Akhbar al-zaman*, the most important sources of the Hermetic history of Egypt.[19]

Kircher's "Abenvaschia" was Ibn Wahshiyya, the purported translator into Arabic of several surviving texts—the originals were supposed to have been composed in Syriac—who is thought to have been active in the late ninth and early tenth centuries. The most famous of his works is the *Nabatean Agriculture* (*Kitab al-filaha al-nabatiyya*), a massive treatise ostensibly devoted to the Babylonian science of agriculture, but filled with digressions on many topics, most notably ancient magical practices.[20] It is not certain which of Ibn Wahshiyya's works Kircher knew, but likely it was some part of the *Nabatean Agriculture*. Kircher, who reported that he had acquired his manuscript in Malta, appears to have been the first European scholar to encounter Ibn Wahshiyya's work firsthand.[21] But the *Nabatean Agriculture* was already known to European readers by reputation through Moses Maimonides' *Guide for the Perplexed*, which had been published in Latin translations in 1520 and 1629.[22] In a famous section of the *Guide*, Maimonides used the *Nabatean Agriculture* as his chief source on the idolatry of the ancient Sabians, which he assimilated to the religion of ancient Egypt. Consequently, when Kircher discovered his manuscript of Ibn Wahshiyya, he believed that he had found an important and venerable source on the primordial paganism of the ancient Egyptians that had been esteemed by the great Jewish scholar.[23]

Little is known about Kircher's "Salamas" other than that he was the author of a treatise called the *Book of the Garden of Marvels of the World*

---

19. A list of authorities cited in al-Suyuti's treatise on the pyramids is given in Nemoy, "Treatise on Pyramids" (1939), 18.

20. See Hämeen-Antilla, *Last Pagans of Iraq* (2006).

21. *OP*, 113. Kircher referred to the work by various titles including *De agricultura Aegyptiorum* and *De servitute Aegyptiorum*. These appear to follow the forms given in the 1520 Latin translation of Maimonides' *Guide for the Perplexed*. As Quatremère, "Mémoire" (1835), 237, observed, apparently the translator, not recognizing the word "Nabatean," changed the title to *Egyptian Agriculture*.

22. Maimonides, *Doctor* (1969); Maimonides, *Dux* (1964). *Nabatean Agriculture* was also cited in the *Picatrix*, the medieval Latin translation of the *Ghayat Al-hakim*, but Kircher does not seem to have been aware of this connection.

23. *OP*, c1^r.

and Regions, which was printed in Arabic in Rome by the typographer Robert Granjon and the printer Domenico Basa in 1585, with Giovanni Battista Raimondi possibly acting as editor.[24] Printed before Raimondi and Granjon inaugurated the famed Medicean Arabic press, it was one of the first Arabic books printed anywhere in the world. Kircher, who received an incomplete copy from a friend in 1636, described its contents in a letter to Peiresc:

> It is divided into seven treatises, of which the first, as is clear from the introduction, treats astronomy and the theory of the celestial bodies. 2. geography and the inhabitable world and how to measure it. 3. the seas and islands, lakes, springs, rivers, and their marvels. 4. mountains and deserts and the marvels of stones. 5. the regions and all peoples. 6. animals and demons and plants and their marvels, and their use in amulets. 7. the remains of ancient monuments, such as buildings, workshops, obelisks, pyramids, amphitheaters, the magnificence of kings and the very heavy stones which they demanded, etc.

Despite making it into print, *The Garden of Marvels* did not reach many readers. As Kircher went on to relate,

> Some years after [its] publication superstitions and various errors were detected, and all copies were confiscated and condemned to eternal darkness; indeed they say that all were burned with certain Hebrew books; however that may be, I have searched for the book with the greatest care, which nevertheless could not be found anywhere, except one copy so imperfect and mutilated that of the 7 treatises only the first and last survive. I strongly fear that this book has perished, not that I approve its contents, but since it would have been very useful to me for revealing the foundations of the superstition of the ancients.[25]

---

24. *Kitab bustan fi 'ajab al-ard wa'l biladayn* (Rome: Domenico Basa, 1585). Schnurrer and de Sacy identify the author as "Salamis ibn Kündogdu as-Salihi," which somewhat resembles Kircher's "Salamas ben Kandaathi." Vervliet has Ibn al-Abbas Ahmad b. Hadjdji al-Salihi. There is no reference to either name in the *Encyclopedia of Islam*, and I have not found any information on the author beyond the following bibliographical notes on the *Garden of Marvels*: Schnurrer, *Bibliotheca Arabica* (1811), 174–76; Sacy, "Review" (1814), 192–94; Vervliet, "Robert Granjon" (1967), 221–23; Brunet, *Manuel du libraire* (1860–65), vol. 5, 68; Zenker, *Bibliotheca orientalis* (1846–61), vol. 1, 120; Assemani, *Bibliothecae Mediceae Laurentianae catalogus* (1742), 147; Assemani, *Catologo de' codici* (1787–92), vol. 1, 151–72, which provides a detailed synopsis.

25. Kircher to Peiresc, Rome, 3 December 1636, BNF FF 9538, fol. 236^v, in Stolzenberg, "Oedipus Censored" (2004), 43–46.

As a result of the actions of the censors, only three copies are known to survive today.²⁶ Already by Kircher's time, the printed work was as scarce as a rare manuscript. Fortunately, his imperfect copy included the section treating Egyptian monuments, and by the time he published *Pamphilian Obelisk*, he claimed also to have found a manuscript version as well. Kircher observed that the content of the *Garden of Marvels* overlapped with al-Suyuti's history of Egypt.²⁷

The nature of Abenephius's manuscript treatise on the religion of the Egyptians, as we saw in chapter 2, is profoundly vexed. But regardless of its provenance, and allowing that Kircher may have fabricated some quotations, much of the material that he attributed to Abenephius reflected genuine Arabic literary traditions about ancient Egypt. Indeed, one of Kircher's citations from Abenephius—concerning Misraim, son of Ham, inhabiting Egypt after the Flood—was close enough to a passage from al-Suyuti as to suggest a common source.²⁸

Of these Arabic sources, al-Suyuti's history of Egypt was by far the most important.²⁹ Quotations from "Gelaldinus" dominate Kircher's discussion of the transmission of primeval knowledge from the first patriarchs to the postdiluvian inhabitants of Egypt. Kircher frequently appealed to Abenephius for information on ancient Egyptian religion and the meaning of the hieroglyphs. But with specific regard to the history of ancient Egypt and the figure of Hermes, that author played a role secondary to that of al-Suyuti. The same is true of Salamas. Although Kircher repeatedly stressed the importance of Ibn Wahshiyya's treatise, he quoted from it only a handful of times. Collectively these Arabic sources gave Kircher access to traditions that to Europeans were unknown or forgotten, gathering a large ensemble of Oriental authorities who provided evidence locating the origin of occult philosophy in Egypt shortly after the Flood, and placing Hermes Trismegistus at the center of a narrative concerning the preservation of antediluvian knowledge.

---

26. The condemnation and destruction of the book described by Kircher has not been known to its bibliographers. The three known copies are at the Biblioteca Laurentiana in Florence, at the Naniana (now part of the Biblioteca Nazionale Marciana) in Venice, and at the Bodleian Library at Oxford. The copy at the Bodleian is apparently incomplete, raising the intriguing possibility that it may be the copy formerly possessed by Kircher. See the above-cited catalogs of the Assemanis for the first two; the Bodleian copy is referred to by Vervliet, "Robert Granjon" (1967), 223.

27. *OP*, h4ʳ, where Kircher lists among his sources, "Salamas ben Kadaathi Arab, M.S. & Im."; Kircher to Peiresc, Rome, 3 December 1636, BNF FF 9538, fol. 236ᵛ, in Stolzenberg, "Oedipus Censored" (2004), 43–46.

28. Both passages are printed by Kircher at *OA* I, 27–28.

29. On Kircher's use of Gelaldinus, see also Grafton, "Kircher's Chronology" (2004), 180–81.

KIRCHER COMPARED TO HIS ARABIC SOURCES

Nonetheless, Kircher's account was far from identical with the material he found in his Arabic texts. His narrative attempted to synthesize a wide variety of sources, including Greek, Jewish, patristic, Syriac, and Byzantine authors. The result was a composite of elements with different origins held together by an admixture of Kircher's own imagination. For example, the key source for Kircher's story of the mixing of Cainite superstitions with the pure wisdom of the children of Seth came from the fourth-century Latin Christian writer John Cassian. In his *Conferences* Cassian described how, in the generations before the Flood, the hitherto separate races of Seth and Cain intermarried, bringing about the corruption of the Adamic teachings and the rise of magic and idolatry:

> Ham, the son of Noah, who was instructed in these superstitions and sacrilegious and profane arts, knowing that he would be utterly unable to take a book about them into the ark, which he was going to enter with his righteous father and his holy brothers, engraved these wicked arts and profane commentaries on plates of various kinds of metal which could not be ruined by exposure to water, and on very hard stone. When the Flood was over he sought for them with the same curiosity with which he had concealed them and handed them on—a seedbed of sacrilege and unending wickedness—to his descendants.[30]

The story about Trismegistus's adventures in Italy, on the other hand, depended on the *Alexandrine* or *Paschal Chronicle*. According to this seventh-century Byzantine compilation, a certain Hermes, also known as Faunus, son of Picus Jupiter, ruled Italy after his father's death, but when strife broke out among his jealous brothers he migrated to Egypt. There, the tribe of Ham received him with great honor and, following the death of their King Metrem (or Mesrem, identified by Kircher as Misraim), made him their ruler.[31] As this episode indicates, the *Paschal Chronicle* combined biblical and apocryphal material, pagan dynastic histories, and Greek mythology interpreted euhemeristically as the historical deeds of primeval men.

---

30. Cassian, *Conferences* (1997), 8.21, pp. 304–7, quotation at 307. Kircher quotes the relevant passage at *OP*, 4, and *OA* II.2, 142.

31. Kircher, quoting the *Chronicle*, gives the king's name as *Mesrem*; the edition of the *Chronicle* that I consulted has *Metrem*. See *OA* I, 88; cf. Dindorf, ed., *Chronicon Paschale* (1832), 69, 75, 80.

The Arabic accounts of Hermes Trismegistus and ancient Egypt were themselves products of a similar process of synthesis. Indeed, much of their content can be traced to earlier pagan, Christian, and Jewish sources. Apocryphal stories of the transmission of antediluvian wisdom from Adam through Seth to Enoch and ultimately to the generations after the Flood were already current in late antiquity, as the example of John Cassian illustrates. The Hellenistic Jewish historian Josephus seems to have linked this narrative to Egypt when he reported that Seth's descendants, in order to preserve science from a coming catastrophe, erected two pillars in Siriad or Seiris.[32] The pseudo-Clementine *Recognitions*, which identified Ham with Zoroaster and referred to his role as progenitor of the Egyptians through his son Misraim, demonstrate that the transmission of Ham's magic was already linked to Egypt early in the Christian era.[33] Although they contained significant innovations—most notably incorporating Hermes Trismegistus and the Egyptian monuments into the story of the preservation of antediluvian knowledge—the Arabic sources should be seen as a stage in the evolution of Mediterranean traditions with a long, pre-Islamic history.[34]

Kircher's history was thus a synthesis of syntheses several times over. As an author, he took many liberties with his sources, which were more ambivalent and contradictory than the story he wove from them. The Arabic accounts, such as al-Suyuti's history, tended to be encyclopedic compilations of conflicting traditions about subjects like the construction of the pyramids. In forging his narrative Kircher chose the bits that suited him, changed parts that did not fit his purpose, filled in the many blank spaces according to his needs, and then adduced select quotations from his sources as "proof" of his claims.

The Arabic histories were distinguished by their focus on the pyramids, which played only a minor role in Kircher's story. This preoccupation was natural since Arab authors in Egypt had direct exposure to those awesome structures, whose origin called out for explanation, and it was typically in that context that they reported the Hermetic legends. Kircher, on the other hand, living in seventeenth-century Rome, understandably fixated on obelisks. His claim that Hermes Trismegistus was the inventor of obelisks is, as far as I know, unattested in Arabic or any other literature, apart from a

---

32. Fodor, "Origins of Arabic Legends" (1970), 340–41, citing Josephus *Ant.* 1.71. Reitzenstein, *Poimandres* (1904), 183–84, located Siriad in Egypt.

33. Pseudo-Clement, *Recognitions*, 4.27. See Flint, *Rise of Magic* (1991), 335.

34. Festugière, *La révélation d'Hermès Trismégiste* (1981), I, 334; Plessner, "Hermes Trismegistus" (1954).

short quotation from Abenephius whose authenticity must be regarded with suspicion.[35] Whereas the Arabic legends emphasized the role of the first Hermes, who had preserved wisdom from the destruction of the Flood by building pyramids, Kircher emphasized the role of the second Hermes, who had revived that wisdom after the Flood and preserved it from future peril by building obelisks. Although Kircher quoted the Arabic legends about the first Hermes, he undercut their import by claiming that the deluge had lain waste to his pyramids, rendering his heroic effort useless.[36] In attempting to harmonize his sources, Kircher seems to have made a composite of the second and third Arabic Hermes, describing Trismegistus as a native of southwest Asia who had traveled to Egypt, where he achieved his fame.

But Kircher's departures remained variations on the theme laid down by the Arabic authors. His narrative, centering on the transmission of knowledge before and after the Flood and depicting Hermes Trismegistus as the inventor of Egyptian monuments as a medium for preserving wisdom against hostile forces, clearly took the Arabic legends for its primary model. On the basis of Arabic sources, Kircher told a story about Hermes Trismegistus and the origins of the *prisca theologia* that differed significantly from previous European treatments of the subject. In doing so, as throughout *Egyptian Oedipus*, he flaunted his study of Arabic manuscripts as a major contribution to Oriental philology which had allowed him to solve the meaning of the hieroglyphs.

KIRCHER'S STUDY OF THE KABBALAH

Kircher made similar claims about his study of Jewish sources, especially his investigation of the Kabbalah. Like the rest of *Egyptian Oedipus*, the 150-page treatise "Kabbalah of the Hebrews" was laden with tokens of erudition.[37] In keeping with his promise to establish every claim with evidence from authorities, he loaded the treatise with quotations from Hebrew and Aramaic primary sources, which he often printed in the original, followed by Latin translations. Kircher cited more than fifty Jewish textual authorities in "Kabbalah of the Hebrews," and the bibliography for the entire vol-

---

35. *OP*, 45. For Islamic conceptions of obelisks, see Haarmann, "Misalla" (1960–2005), 140–41.

36. Annius of Viterbo did something similar to the legend of Seth's pillars, affirming their historicity but removing their significance by making Noah with his ark the guardian of antediluvian wisdom. See Stephens, "Berosus Chaldaeus" (1979), 124.

37. On Kircher's study of the Kabbalah, see Schmidt-Biggemann, "Cattolicesimo e Cabala" (2007); Stolzenberg, "Kircher Reveals the Kabbalah" (2004).

ume listed more than one hundred Jewish sources.[38] Kircher also referred to Latin writers on the Kabbalah, typically as places for the reader to find further information on specific points, but such references were fewer and the weight of authority fell heavily on Jewish testimony. Kircher's explicit declarations reinforced the impression of a profound study of primary sources. The Kabbalah, he wrote, was like

> a secret lying hidden in the darkness of inaccessible antiquity, so difficult is it . . . to arrive at its true sense and determine its meaning. Nevertheless, what it has been granted to the scrutiny of a keen mind to investigate, and what I have been able to dig up from the memorials of ancient authors by combining one piece with another, I think, rather I affirm without hesitation, to be nothing other than that Mosaic doctrine, which, passed on orally by the Judges to those who came after them, the Hebrew theologians call the Kabbalah.[39]

Kircher's rhetoric, which, following the model for *Egyptian Oedipus* as a whole, presented his study of the Kabbalah as a pathbreaking foray into uncharted scholarly territory, was on its face improbable. Although one would scarcely know it from Kircher, a vast literature on the Kabbalah had become available in Latin (and to a lesser extent in European vernaculars) during the century and-a-half prior to the publication of *Egyptian Oedipus*.[40] Christian knowledge of Jewish mysticism may have been imperfect, but in 1650 the Kabbalah was hardly the inaccessible, as yet undivulged sanctum that Kircher claimed to unveil to his readers.

Kircher deliberately presented the reader with an idealized image of his encounter with his sources that concealed a more checkered reality. If one pries below the surface of the text by tracking down its cited and, most revealingly, uncited sources, a different picture emerges. Many of the quotations of primary sources turn out to have been lifted wholesale, with ready-made Latin translations, from early modern Christian writers on the Kabbalah whom Kircher never mentioned. Furthermore, the use of clearly

---

38. *OA* II.1, Kkk1^r–Kkk2^v.
39. *OA* II.1, 210. See also *OP*, b4^v–c1^r, where Kircher discusses his Jewish sources.
40. An idea of the quantity and variety of this literature can be had from the exhibition catalog of the Biblioteca Casanatense in Rome: Cavarra, ed., *Hebraica* (2000), which largely treats Christian responses to the Kabbalah. See also Scholem, *Bibliographia Kabbalistica* (1933); Faivre and Tristan, eds., *Kabbalistes Chrétiens* (1979); Secret, *Kabbalistes Chrétiens* (1985); Secret, *Hermétisme* (1992); Dan, ed., *Christian Kabbalah* (1997); Coudert, *Impact of the Kabbalah* (1999).

marked quotations may lead the reader to assume that where such markers were absent, Kircher spoke with his own words; but in fact the body of his text also borrowed liberally from modern Latin authors, sometimes copying pages on end practically verbatim.

Not all of Kircher's erudition was a sham, however, nor all of his rhetoric empty. His study of the Kabbalah also drew on some little-known Hebrew sources, which he studied and translated himself, although perhaps with assistance. But Kircher's text supplies no means to distinguish between those places where he was passing off the scholarship of others as his own and those where he was genuinely erudite and innovative. As a result, the naive reader is likely to assign Kircher more than his due, while the skeptic may unfairly assume that he was no real scholar at all, but merely a braggart and a plagiarizer.

RANSACKING RITTANGEL AND RICIUS

A typical example of the erudite texture of Kircher's work is found in chapter 8 of "Kabbalah of the Hebrews," where Kircher explained the so-called "mystery of the chariot." Kircher wanted to demonstrate that although the modern rabbis, "blinder than moles," denied that any of the *sefirot* (the divine "numerations" that constitute the kabbalistic godhead) existed within God, the ancient Jewish sages placed the three supreme *sefirot* in God, thus indicating knowledge of the Trinity. To support his claim, on a single page he quoted Rabbi Isachor Beer's *Imre Binah*, the *Sefer Yetzirah* (attributed to Rabbi Akiba), Moses Botrel's commentary on *Yetzirah* (twice), Rabbi Saadia Gaon, and a text attributed to Maimonides.[41] Two of the quotations were printed in Hebrew with Latin translations, while the others appeared only in Latin (fig. 28).

Kircher gave his reader no reason to doubt that he had personally consulted the original Hebrew texts, and his citations, some of which included references to page numbers or chapters, seemed to support his claim to have based his study on firsthand research. But in fact, all the quotations in this passage were taken from a single Latin-Hebrew work published a decade earlier, whose author Kircher never mentioned. In 1642, Johannes Stephan Rittangel, a Lutheran convert from Judaism and professor of Oriental languages in Königsberg, published a bilingual edition of selections from *Sefer Yetzirah*, one of the foundational texts of the Kabbalah, preceded

41. *OA* II.1, 293.

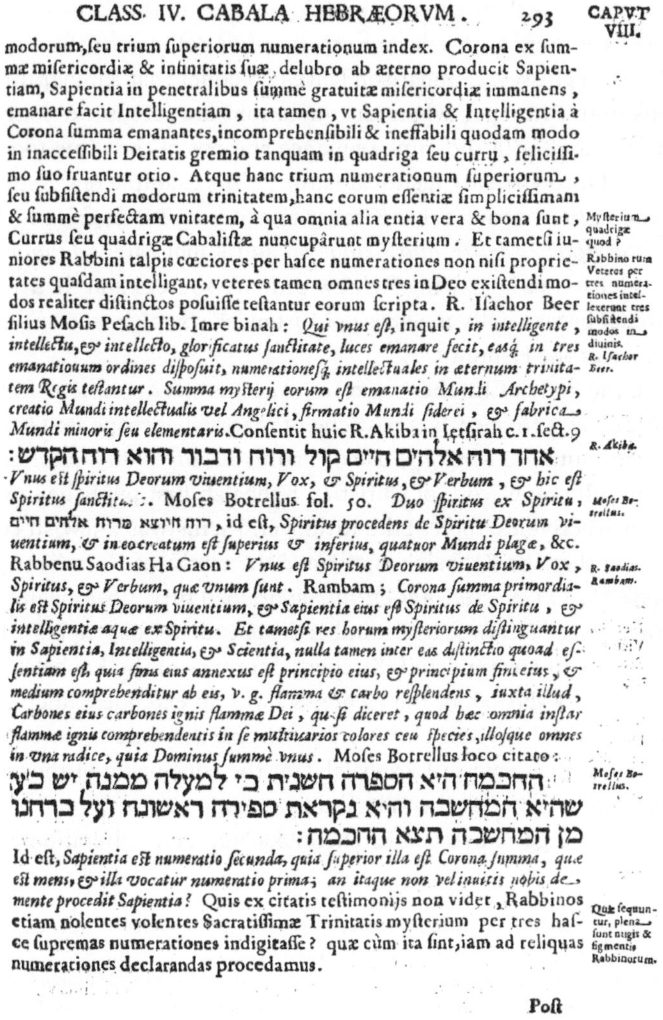

CLASS. IV. CABALA HEBRÆORVM. 293 CAPVT VIII.

modorum,seu trium superiorum numerationum index. Corona ex summæ misericordiæ & infinitatis suæ delubro ab æterno producit Sapientiam, Sapientia in penetralibus summè gratuitæ misericordiæ immanens, emanare facit Intelligentiam, ita tamen, vt Sapientia & Intelligentia à Corona summa emanantes, incomprehensibili & ineffabili quodam modo in inaccessibili Deitatis gremio tanquam in quadriga seu curru, felicissimo suo fruantur otio. Atque hanc trium numerationum superiorum, seu subsistendi modorum trinitatem, hanc eorum essentiæ simplicissimam & summè perfectam vnitatem, à qua omnia alia entia vera & bona sunt, Currus seu quadrigæ Cabalistæ nuncupârunt mysterium. Et tametsi iuniores Rabbini talpis cœciores per hasce numerationes non nisi proprietates quasdam intelligant, veteres tamen omnes tres in Deo existendi modos realiter distinctos posuisse testantur eorum scripta. R. Isachor Beer filius Mosis Pesach lib. Imre binah : *Qui vnus est*, inquit , *in intelligente , intellectu, & intellecto, glorificatus sanctitate, luces emanare fecit, easq́ in tres emanatiouum ordines disposuit, numerationesq́ intellectuales in æternum trinitatem Regis testantur*. Summa mysterij eorum est emanatio Mun.li Archetypi, creatio Mundi intellectualis vel Angelici, firmatio Mundi siderei, & fabrica Mundi minoris seu elementaris. Consentit huic R. Akiba in Ietsirah c. 1. sect. 9

Mysterium quadrigæ quod?
Rabbinorum Veteres per tres numerationes intellexerunt tres subsistendi modos in diuinis. R. Isachor Beer.

אחד רוח אלהים חיים קול ורוח ודבור הוא רוח הקדש:

Vnus est spiritus Deorum viuentium, Vox, & Spiritus, & Verbum, & hic est Spiritus sanctitu. :. Moses Botrellus. fol. 50. *Duo spiritus ex Spiritu,* רוח חיוצא מרוח אלהים חיים, id est, *Spiritus procedens de Spiritu Deorum viuentium, & in eo creatum est superius & inferius, quatuor Mundi plagæ*, &c. Rabbenu Saodias Ha Gaon : *Vnus est Spiritus Deorum viuentium, Vox, Spiritus, & Verbum, quæ vnum sunt.* Rambam ; *Corona summa primordialis est Spiritus Deorum viuentium, & Sapientia eius est Spiritus de Spiritu , & intelligentia aquæ ex Spiritu. Et tametsi res horum mysteriorum distinguantur in Sapientia, Intelligentia, & Scientia, nulla tamen inter eas distinctio quoad essentiam est, quia finis eius annexus est principio eius, & principium fini eius, & medium comprehenditur ab eis, v. g. flamma & carbo resplendens , iuxta illud , Carbones eius carbones ignis flammæ Dei , quasi diceret, quod hæc omnia instar flammæ ignis comprehendentis in se multiuarios colores ceu species, illosque omnes in vna radice, quia Dominus summè vnus*. Moses Botrellus loco citato :

R. Akiba.
Moses Botrellus.
R. Saodias. Rambam.

Moses Botrellus.

החכמה היא הספרה חשנית כי למעלה ממנה יש כ"ע
שהיא המחשבה והיא נקראת ספירה ראשונה ועל כרחנו
מן המחשבה תצא החכמה:

Id est, *Sapientia est numeratio secunda, quia superior illa est Corona summa, quæ est mens, & illa vocatur numeratio prima; an itaque non vel inuitis nobis de mente procedit Sapientia?* Quis ex citatis testimonijs non videt, Rabbinos etiam nolentes volentes Sacratissimæ Trinitatis mysterium per tres hasce supremas numerationes indigitasse? quæ cùm ita sint, iam ad reliquas numerationes declarandas procedamus.

Quæ sequuntur, plena sunt nugis & figmentis Rabbinorum.

Post

Fig. 28. A typical page from Kircher's treatise on the Kabbalah. It contains five quotations from Jewish sources, including two passages in Hebrew. Although Kircher presented himself as having studied the original texts, he copied this material from an uncited Latin-Hebrew work. Athanasius Kircher, *Oedipus Aegyptiacus* (Rome: 1652–54), vol. 2, part 1, 293. Courtesy of Stanford University Libraries.

by a Hebrew commentary.[42] To help explicate the texts—and to demonstrate that they revealed ancient Jewish knowledge of the Trinity—Rittangel interspersed them with relevant passages from other Jewish sources, as well as his own glosses. Comparison leaves no doubt that Kircher lifted his quotations from Rittangel, where the quotations are sometimes longer but never shorter, appear in the same sequence apropos of the same topic, and are translated into virtually identical Latin.[43] This example represents only a few of the Hebrew sources that Kircher mined from Rittangel, which also included the *Book of Faith and Expiation* and Rabbi Meir's *Liphne Liphnim*.[44] In addition to borrowing Hebrew primary sources and their translations, Kircher occasionally made use of Rittangel's own glosses in the body of "Kabbalah of the Hebrews."[45]

Indeed, the body of Kircher's text, as much as the quotations from primary sources, was frequently a pastiche of material appropriated from uncited modern authors. Kircher's use of Paulus Ricius's *On Celestial Agriculture*, an influential work of Christian Kabbalah first published in 1541, offers an instructive example. Like Rittangel, Ricius was a Jewish convert who believed that the Kabbalah confirmed the truths of Christianity. The fourth book of *On Celestial Agriculture* contained fifty "introductory kabbalistic theorems," followed by a number of appendices that explained some of the theorems in greater detail. Kircher's overview of the Kabbalah in the first chapter of "Kabbalah of the Hebrews" was an extended, uncited quotation of the first thirty-eight of Ricius's theorems, while his even longer introduction to the "practical Kabbalah," at the conclusion of the treatise, was lifted from the fiftieth theorem and its gloss.[46]

In appropriating Ricius's text Kircher made numerous alterations. In par-

---

42. Rittangel, *Liber Iezirah* (1642). On Rittangel, see Secret, *Kabbalistes Chrétiens* (1985), 278; Rankin, *Jewish Religious Polemic* (1970), 89–154.

43. Rittangel, *Liber Iezirah* (1642), 36–39, 55. Kircher misattributes to Maimonides a quotation from Nachmanides, perhaps due to confusion of their Hebrew acronyms, Rambam and Ramban. There are a few other minor discrepancies. In particular, Kircher partially elides or abbreviates some quotations and in one instance merges two nonadjacent passages.

44. The passage from *De fide et expiatione* is found at *OA* II.1, 290, cf. Rittangel, *Iezirah*, 50; from R. Meir, *Liphne Liphnim* at *OA* II.1, 309, cf. Rittangel, *Iezirah*, 148. In addition, Kircher took the following quotations and translations of *Sefer Yetzirah* from Rittangel: *OA* II.1, 247, cf. Rittangel, *Iezirah*, 200; *OA* II.1, 308, cf. Rittangel, *Iezirah*, 146; *OA* II.1, 330, cf. Rittangel, *Iezirah*, 153; *OA* II.1, 305 (also given at 332–33), cf. Rittangel, *Iezirah*, 197–98; *OA* II.1, 332, cf. Rittangel, *Iezirah*, 200; *OA* II.1, 334, cf. Rittangel, *Iezirah*, 204.

45. For example, Kircher's discussion of the first sentence of *Sefer Yetzirah* follows Rittangel almost word for word. *OA* II.1, 308–9, cf. Rittangel, *Iezirah*, 146–47.

46. *OA* II.1, 212–14 and 338–42, cf. Ricius, *De coelesti* (1541), 74^v–85^r Pistorius, *Artis cabalisticae* ([1587]), 120–37. Kircher used the edition of *De coelesti agricultura* printed in Pistorius,

ticular, he attenuated Ricius's unbridled enthusiasm for the "practical Kabbalah," which had an especially dim status from the standpoint of Catholic orthodoxy. He expunged several references in Ricius's theorems to the Kabbalah's excellence as a technique for "adapting the soul to the powers of heaven [regna coelorum]." These passages, whose language calls to mind Marsilio Ficino's controversial treatise "on arranging life according to the heavens [de vita coelitus comparanda]," were suggestive of what Kircher elsewhere condemned as "kabbalistic astrology." In his ninth theorem Ricius extolled this Kabbalah, despite its patina of "ignorance and dullness" attributable to modern Jews, writing that it "surpasses and excels by far the tradition of many allegorizers with respect to authority, learning, and even the facility of adapting the soul of mortals to the powers of heaven." Kircher changed the final phrase to read, "authority, learning, and sanctity." Instead of Ricius's claim that the Kabbalah "opens the way and shows the path by which he who desires a celestial life . . . may conform himself to the nature of superior things and the inseparable image of the Trinity," Kircher wrote that it "shows the paths to the entryway of eternal happiness."[47] Kircher did not borrow passively, but manipulated his uncited sources to suit his own agenda as well as the demands of ecclesiastical censorship.[48]

The examples of Ricius and Rittangel could be multiplied, but they suffice to illustrate how greatly Kircher's study of the Kabbalah depended on modern Latin authors both for quotations from Jewish primary sources and for parts of his Latin text in which he supposedly spoke in his own voice. Above all, these examples reveal the great distance that separated Kircher's idealized representation of his study of Oriental texts from his actual scholarly practice. With his citation strategies and explicit claims—in the introduction to *Pamphilian Obelisk* he declared that he had not quoted any work that he had not read firsthand—Kircher deliberately created an image of his research that differed starkly from its true nature.[49]

One should not be too shocked by Kircher's methods. Borrowing primary source quotations from secondary authors was a normal practice in Kircher's day, as it remains today. In the preface of *Pamphilian Obelisk*, Kircher described how he collected notes in something like a commonplace book. Studying countless books and manuscripts, he identified relevant pas-

---

which contains at least one passage absent from the 1541 edition. I provide citations to both editions. Kircher's use of Ricius was also noted by Secret, *Kabbalistes Chrétiens* (1985).

47. Ricius, *De coelesti* (1541) 74^v, 75^r; Pistorius, *Artis cabalisticae* ([1587]), 120; *OA* II.1, 212–13.

48. For further examples involving Ricius, see below, ch. 6.

49. *OP*, d1^r.

sages (*loca*) and, at considerable expense, had them excerpted according to designated topics (*in certas quasdam series rerum*).⁵⁰ (Apparently some of the work was done by hired hands, such as the Jewish research assistants whom he mentioned elsewhere.)⁵¹ Given Kircher's method, it is not hard to imagine that by the time he composed the treatise, he had lost track of the source of some quotations. The practice of composing a new work by pasting together passages borrowed from other authors was a venerable tradition, still practiced in the seventeenth century.⁵² It was also a common practice, and not necessarily an act of bad faith, to cite ancient sources while omitting references to modern authors. Better scholars than Kircher covered up their debts to contemporaries in order to increase the appearance of their own originality—even Joseph Scaliger, the greatest classical scholar of the preceding century, and Kircher's contemporary, the proverbially scrupulous Robert Boyle.⁵³ One should also allow other motives for suppressing references. Ricius, for example, was on the Index of Forbidden Books, and citing him might have invited trouble. Indeed, in their prepublication review of *Egyptian Oedipus*, the Jesuit censors asked Kircher not to identify too specifically the sources of certain heterodox material, in order to prevent the reader from looking them up.⁵⁴

One wants to avoid applying an anachronistic conception of plagiarism to early modern authors, but it would be equally mistaken to imagine that Kircher's failure to identify his sources was the innocent result of seventeenth-century norms. More conscientious scholars did identify modern authors and the intermediary sources from which they borrowed quotations. Kircher himself cited many modern authors in *Egyptian Oedipus*, including some, such as Reuchlin, who were prohibited by the Index. Furthermore, the rules governing the publication of books by Jesuits specifically required that authors "should not seem to be merely repeating what others have already written."⁵⁵ Perhaps Kircher was making the most of the meaning of the verb "to seem." But the Jesuit censors in charge of enforcing the rule were not fooled. They reprimanded Kircher's tendency to borrow evidence from other authors without citations, singling out one

---

50. *OP*, b4ʳ.
51. See above, ch. 3, n. 14.
52. See Moss, *Printed Commonplace-books* (1996); Moss, "Politica of Justus Lipsius" (1998); Blair, "Humanist Methods" (1992); Blair, *Theater of Nature* (1997), esp. ch. 2.
53. Grafton, *Joseph Scaliger* (1983–93), vol. 1, 109–10, 129, 150, 192, 219, 229. Newman and Principe, *Alchemy Tried in the Fire* (2002), 18–27.
54. ARSI FG 668, fol. 392ʳ, in Stolzenberg, "Oedipus Censored" (2004), 35.
55. "Regulae Revisorum Generalum," Rule 8, in *Institutum* (1892–93), vol. 3, 67.

case so egregious that Kircher was compelled to delete the section from the printed work. In his original manuscript Kircher had boasted that because some of Pico's kabbalistic conclusions were "very obscure and scarcely anyone has correctly penetrated them, I judged it my duty to play the faithful Oedipus and attack that intricate Sphinx in order by any means to conquer her." He then proceeded to lift his explanations wholesale from a book by Arcangelo da Borgonovo, as the hawk-eyed censors noted.[56] Early modern scholarly practices were different from modern ones. But when Kircher (or Scaliger, or Boyle) hid his dependence on other authors, he violated the accepted norms of his own day.

TOILING IN THE GARDEN OF POMEGRANATES

Examining Kircher's use of Rittangel and Ricius has left us with a diminished image of his prowess as a student of Hebrew literature. A final example will somewhat balance that impression. Although he borrowed much of his kabbalistic learning from uncited, modern Latin intermediaries, some of it was acquired firsthand through genuine scholarship.

The single most important source for Kircher's treatise on the Kabbalah was *Pardes Rimmonim* (Garden of Pomegranates), a Hebrew work composed a century earlier by the great Safed kabbalist Moses Cordovero (1522–1570). Kircher cited the work frequently on a variety of topics, often quoting long passages in both Hebrew and Latin translation and at other times paraphrasing. His treatment of the technical details of the *sefirot*, the production of divine names, and ritual chanting relied heavily on Cordovero's text, which he reported faithfully, though sometimes with an accompanying denunciation of its superstition. The chief virtues of "Kabbalah of the Hebrews"—its relatively comprehensive coverage of the range of kabbalistic topics and its clear, systematic explanation of complicated doctrines—can be attributed in large measure to his reliance on *Pardes Rimmonim*.

For an authentic guide to the Kabbalah, Kircher could not have done better. A kind of "Summa Kabbalistica," *Pardes Rimmonim* presented a comprehensive overview of the pre-Lurianic Kabbalah, drawing from both the theosophic tradition, represented by the *Zohar*, and the ecstatic tradition,

---

56. BNCR Ms. Ges. 1235, fol. 122^{rv}, cf. Arcangelo da Borgonovo, *Cabalistarum selectiora* (1569), 1–2; ARSI FG 668, fol. 392^r (Judgment of *OA* II.1), in Stolzenberg, "Oedipus Censored" (2004), 34–35. Borgonovo himself plagiarized many of his interpretations of Pico's conclusions from the work of his late teacher Francesco Giorgi; see Wirszubski, "Francesco Giorgi's Commentary" (1974); Secret, "Notes sur quelques kabbalistes" (1974).

based on manipulations of the letters of Holy Scripture and the chanting of divine names. Cordovero presented kabbalistic teachings in a lucid, philosophical style, supported by many quotations from earlier literature. Influenced by Italian trends in which Jewish understanding of the Kabbalah had been affected by Renaissance Neoplatonism, he communicated Jewish mysticism in an idiom that was particularly attractive to Christian scholars. In particular, he expounded the doctrine of "the three lights," which Christian kabbalists found so suggestive of the Trinity, and attempted a comprehensive inventory of kabbalistic diagrams.[57] Despite these appealing features, and despite the fact that the book circulated widely among European Jews, especially in Italy, Kircher seems to have been the first Christian writer to make use of this important work.[58]

In contrast to his recycling of predigested Hebraica from authors like Rittangel and Ricius, Kircher's use of Cordovero, like his study of al-Suyuti and other Arabic authors, required substantial, original scholarship. Although *Pardes Rimmonim* had been printed in Hebrew in Kraków in 1592, Kircher knew it only from a manuscript in the library of the College of Neophytes.[59] His study of the manuscript, which today belongs to the Vatican Library, is testified by numerous Latin marginalia in his hand, which correspond precisely to material found in *Egyptian Oedipus* (figs. 29 and 30).[60] In addition, Kircher reproduced several diagrams from the Neofiti manuscript (figs. 31 and 32).[61] While we know from Kircher's testimony that he was assisted by hired Jewish research assistants, these marginalia demonstrate his direct study of the Hebrew manuscript. As the first known exposition by a gentile of one of the major works in kabbalistic literature, Kircher's study was an important contribution to early modern Christian scholarship on the Kabbalah. Until now, however, his accomplishment has been over-

---

57. My description of *Pardes Rimmonim* (including the descriptive phrase "Summa Kabbalistica") follows Idel, *Kabbalah* (1988), 255–56; Idel, "Major Currents" (1992), 350–51 and passim; and also Busi, *Qabbala visiva* (2005), 389–94 and passim, especially with respect to diagrams. See also Ben-Schlomo, "Cordovero, Moses" (1971–72).

58. Cordovero's *Pardes Rimmonim* was referred to as early as 1613 by Buxtorf, *Bibliotheca rabbinica* (1613), 313, but this is merely a bibliographical entry; there is no indication Buxtorf saw the work.

59. BAV Ms. Neofiti 28. This is a very fine copy of 575 pages, dated 1548. See *Codici ebraici* (1893); Ben-Schlomo, "Cordovero, Moses" (1971–72). Kircher identified the manuscript as "Pardes, Hebraeus, M.S. Heb. de Cab. Heb." in *OP*, h4^r.

60. BAV Neofiti 28, fol. 101bis^r, cf. *OA* II.1, 322–23; BAV Neofiti 28, fol. 218^v, cf. *OA* II.1, 323; BAV Neofiti 28, fol. 278^r, cf. *OA* II.1, 237; BAV Neofiti 28, fol. 305^v, cf. *OA* II.1, 345.

61. BAV Neofiti 28, fols. 88^{rv}; 278^v, cf. *OA* II.1, 302; 268.

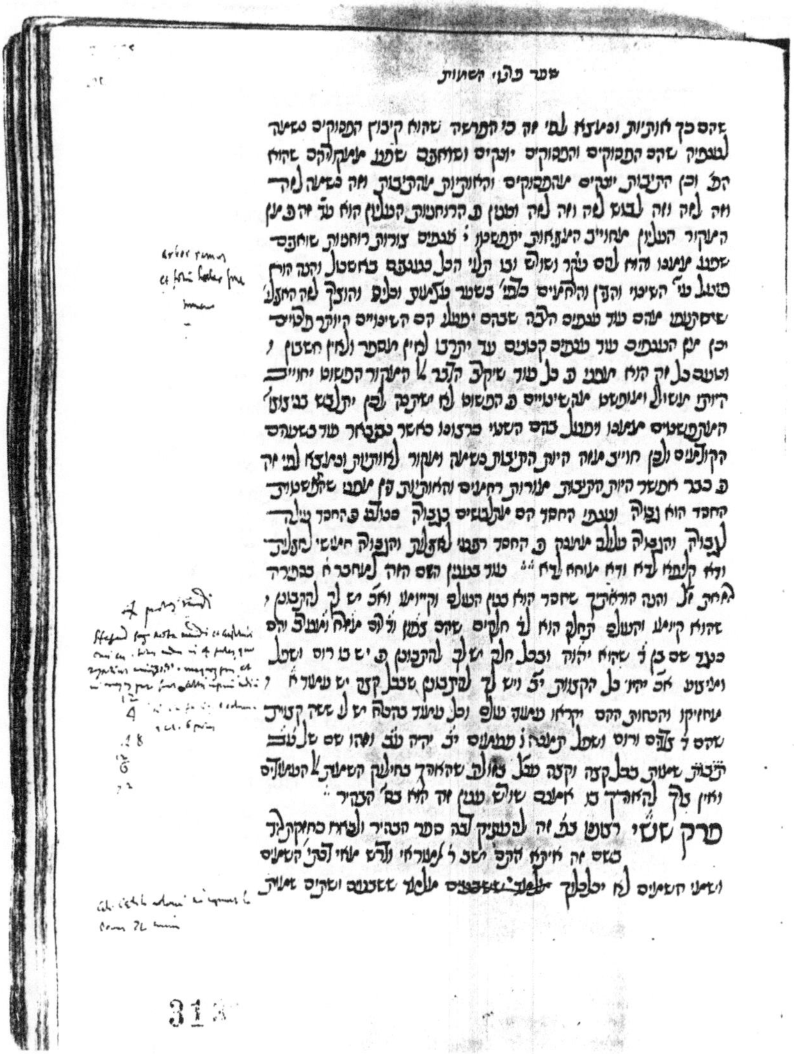

Fig. 29. The manuscript of Moses Cordovero's *Pardes Rimmonim* from the Library of the College of Neophytes, with marginal annotations by Kircher. It was the most important source for his treatise on the Kabbalah. BAV Neofiti 28, fol. 313ʳ. Biblioteca Apostolica Vaticana.

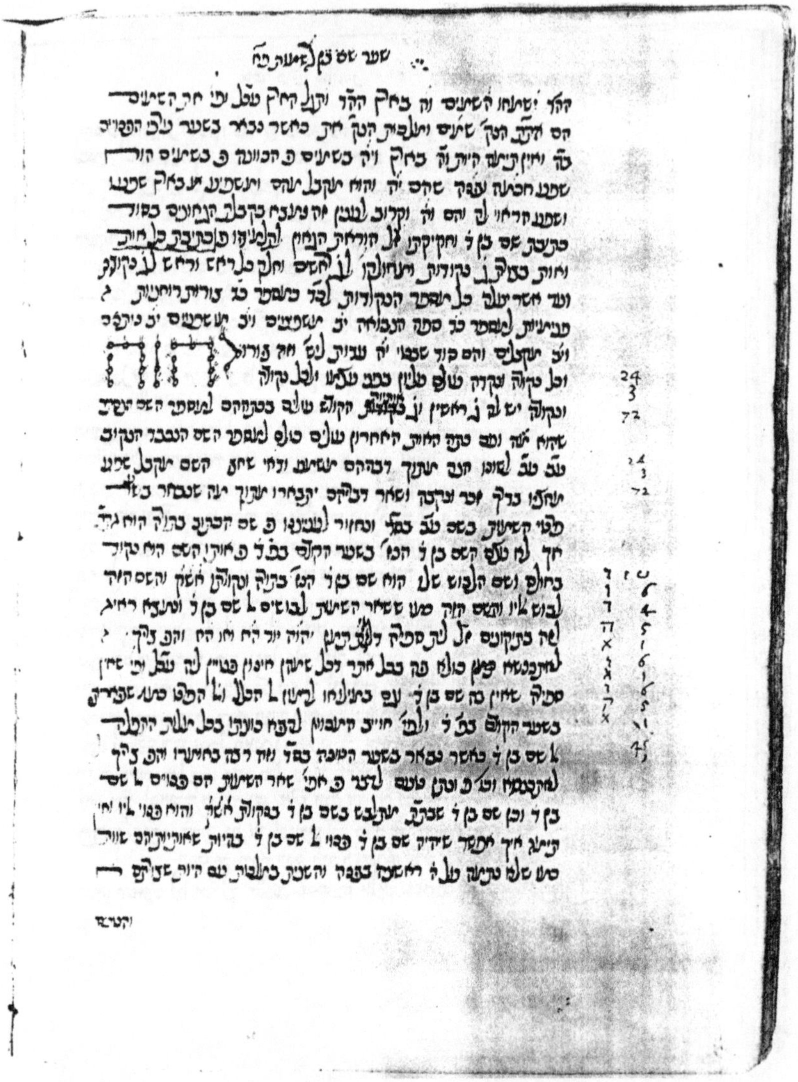

Fig. 30. Another page from *Pardes Rimmonim.* The annotations show Kircher calculating the numerical values of divine names. Corodvero's kabbalistic diagram of the Tetragrammaton, written with twenty-four circles emitting seventy-two rays, was reproduced in *Oedipus Aegyptiacus*, vol. 1, p. 268. BAV Neofiti 28, fol. 278v. Biblioteca Apostolica Vaticana.

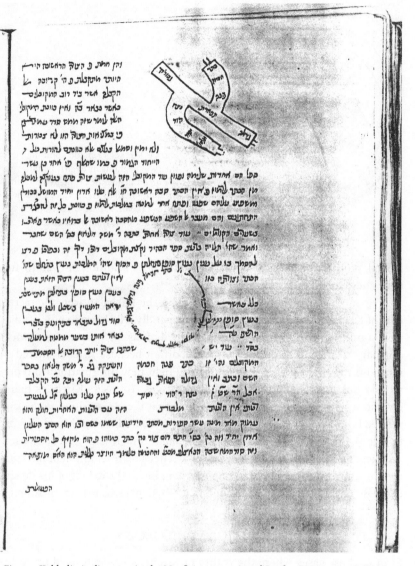

Fig. 31. Kabbalistic diagrams in the Neofiti manuscript of *Pardes Rimmonim*. BAV Ms. Neofiti 28, fol. 88ᵛ. Biblioteca Apostolica Vaticana.

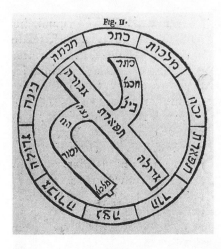

Fig. 32. Kircher's version combined the two diagrams in fig. 31, placing the *aleph* inside the circle. Athanasius Kircher, *Oedipus Aegyptiacus* (Rome: 1652–54), vol. 2, part 1, 302. Courtesy of Stanford University Libraries.

looked, probably because he never named the book's author, or even its full title, and referred to it only as "Pardes."[62]

The methods employed in the treatise on the Kabbalah were characteristic of Kircher's scholarly practice more generally. His hieroglyphic studies were a mixture of derivative research passed off as his own and genuinely original research. This conclusion is important because Kircher staked so much on the claim that he had brought to light new information from previously unknown sources in Oriental languages. To assess Kircher's hieroglyphic studies it is necessary to appreciate that these assertions were exaggerated but by no means entirely false, and that he carefully shaped his text to give the reader a distorted impression about how it had been composed. As a general rule, one should be wary of Kircher's representation of his studies, since in any given case it is impossible to know the extent of his dependence on uncredited authors without a thorough investigation of his sources.

UNCRITICAL PHILOLOGY

Even in those cases where the reality of Kircher's research lived up to his rhetoric, new Oriental sources did nothing to make his study of ancient religion, philosophy, and magic more accurate than prior European accounts.

---

62. Bartolocci, *Bibliotheca magna rabbinica* (1675–93), vol. 4, 230–35, refers to Ciantes, *De sanctissima Trinitate* (1667) as the first Christian to bring Cordovero's doctrine to light. It is unclear why Cordovero's name escaped Kircher. It appears on the first page of Neofiti 28, although it is embedded in other text.

As historical evidence his Oriental sources were problematic, requiring a more critical approach than he applied to them. To a large extent, medieval Arabic literature about Egypt and the hieroglyphs derived from the same late antique mixture of apocryphal legend and Platonic interpretations of magical rituals as did better-known Greek and Latin works. As a result, Kircher's study of these texts inevitably led him only to produce a more elaborate version of the traditional Neoplatonist imagining of ancient Egyptian mysteries. Kircher's variation on the story of Hermes Trismegistus and the *prisca theologia* is significant as a novel chapter within the history of Western esotericism; but he continued to view ancient Egypt within that fantastical framework, rather than breaking out of it. The Kabbalah likewise had late antique roots and affinities to Neoplatonism that were amplified by ongoing mutual influence during the Renaissance. The Christian Ricius described kabbalistic rituals in the vocabulary of Ficino's natural magic, explaining the correspondences between the microcosm (man) and the macrocosm (the world) in terms of their respective threefold natures: archetypal, celestial, and terrestrial. Conversely, the Jewish Cordovero was influenced by Renaissance Italian Neoplatonism. Ultimately, the similarities of Arabic and Hebrew literature to Greek and Latin traditions proved deeper than the differences.

Kircher, who has gone down in history for his credulity, methodically provided evidence to prove his every assertion. If his scholarship was unsound, it was not because he was indifferent to the importance of grounding his studies empirically. He was highly self-conscious, if not especially sophisticated, about the practice of citation and its relationship to historical proof, repeatedly addressing the subject in apologetic sections of *Pamphilian Obelisk* and *Egyptian Oedipus*, in which he sought to shore up his claims by convincing the reader of the reliability of his quotations from primary sources.[63] Such a commitment to documentary evidence, characteristic of seventeenth-century erudition, was a recent development. Beginning in the preceding century, antiquaries and ecclesiastical historians championed a historiography rooted in the consultation of primary sources and pioneered the use of modern citation techniques. Although citing authoritative authors was hardly new in itself, the full-blown emphasis on proof through citation that is visible in Kircher's work, encompassing not only precise references to sources but also extended quotations from them, emerged only in the seventeenth century.[64]

---

63. See chapter 7.
64. Grafton, *Footnote* (1999). See also, Momigliano, "Ancient History" (1950); Cropper, Perini, and Solinas, eds., *Documentary Culture* (1992).

Kircher could have imbibed these methods from many sources, but no doubt the most important was his apprenticeship with the guru of seventeenth-century erudition, Peiresc. When Kircher sent Peiresc an early version of his work in 1635, the latter responded with constructive but harsh criticism regarding the use of evidence. Kircher wrote back in the humble voice of a disciple:

> You admonish me most wisely concerning the careful use of citations and concerning the accurate demonstration of authorities, and know that it shall be the chief concern of my studies to demonstrate everything accurately by means of classical authors; indeed, consider that I am so formed by nature that I would prefer to confess my ignorance a hundred times rather than rashly assert something groundless by pursuing vague and unfounded conjectures.[65]

These words may elicit a guffaw, since Kircher's method often seems to consist precisely in pursuing vague and groundless conjectures. But the concern about documentation expressed in this early letter is evident on virtually every page of *Egyptian Oedipus*. From this angle, Kircher's work may be seen as the tragic misapplication of Peiresc's teaching. Kircher took to heart the lesson of the antiquaries about proving assertions with detailed citations of evidence, but not the concomitant lesson of the Renaissance philologists regarding how to assess that evidence critically. Without the latter the former was severely, perhaps fatally, undermined. The best early modern antiquaries were, unlike Kircher, also keen textual critics. Ideally, the antiquarian imperative to accumulate evidence and the critical imperative to evaluate evidence skeptically were complementary. But they pushed in different directions. It will not do to see Kircher as an anachronism who failed to keep up with the scholarly developments of his day. The essential tension that defined his hieroglyphic studies was not between old-fashioned occult philosophy and modern scholarly rigor. His most conspicuous shortcomings were, rather, the result of his attempt to practice state-of-the-art antiquarian research without a critical approach to his sources.

Although Kircher's Arabic authors proved to be unreliable guides to the

---

65. "Et primùm quidem quod me admonet de cautelà in citationibus adhibendis, et de fideli auctoritatum exhibitione, omnino prudenter monet, sciatque me in hoc principalem studiorum meorum curam hactenus ponere, ut omnia fideliter per authores classicos ostendantur; imò ita meum natura comparatum esse, existimet, ut ignorantiam meam centies fateri malim, quàm vagis coniecturis aut leviusculis persuasionibus insistendo, aliquid temerè ac sine fundamento asserere." Kircher to Peiresc, Rome, 8 February 1635, BNF NAF 5173, fol. 25^{rv}; cf. FF 9362, fol. 13^{rv} (copy).

mysteries of Pharaonic Egypt, modern scholars value such literature as evidence of medieval Islamic culture and for its preservation of late antique traditions, if not earlier ones. (Ibn Wahshiyya's *Nabatean Agriculture* is particularly important, not only for its preservation of pre-Islamic lore, but because of the significant role it has played, via Maimonides' discussion of the Sabians and the doctrine of divine accommodation, in European intellectual history.[66]) Already in the seventeenth century, a scholar like Peiresc—who prodded Kircher, in vain, to approach his Oriental authors more critically—understood that something valuable could be learned from a text such as Abenephius even if it did not reflect genuine ancient Egyptian traditions. Even though the historical context in which Kircher placed these textual traditions was incorrect, by treating them primarily as evidence of ancient beliefs and practices rather than timeless guides to be judged as true or false, pious or superstitious, he took a significant step toward historicizing them, which would eventually acquire critical foundations. As a pioneer in the study of this material, he anticipated scholars like Richard Reitzenstein, Julius Ruska, and Walter Scott, who at the beginning of the twentieth century expanded the study of ancient Hermetic literature by adding Arabic sources to the classical Greek and Latin ones.[67]

To understand why early modern scholars took Kircher as seriously as they did despite his flaws, it is essential to understand that his work was conceived and received as a contribution to Oriental studies. Thus, defending Arabic studies in 1659, the Spanish Orientalist Tomas de Léon singled out the historical works cited by Kircher in *Egyptian Oedipus*, alongside Averroes and Avicenna, as examples of the value of Arabic learning.[68] In the seventeenth century the study of Oriental literature was at a much more rudimentary stage than classical philology. Consequently, textual criticism was relatively less important than the preliminary tasks of discovering and disseminating new materials. As chapter 8 documents, even though *Egyptian Oedipus* was widely seen as a failed solution to the hieroglyphs, it was appreciated by many scholars who treated it as a compendium of rare notices about antiquity, extracted from unpublished Oriental sources.

---

66. See; Benin, "Cunning of God" (1984); Funkenstein, *Perceptions* (1993), 131–55; Assmann, *Moses the Egyptian* (1997); Elukin, "Maimonides" (2002); Stroumsa, "Sabéens de Harran" (2004).

67. Reitzenstein, *Poimandres* (1904); Ruska, *Tabula Smagdarina* (1926); Scott and Ferguson, eds., *Hermetica* (1926–36), vol. 4, "Testimonia." Ruska was impressed by Kircher's precocious use of Arabic Hermetic sources (59–60); see also 216ff. for Ruska's comments on Kircher's treatment of the Emerald Tablet.

68. García-Arenal and Rodríguez Mediano, *Un Oriente español* (2010), 385.

ARCHIVE FEVER

Even as an uncritical accumulator of evidence, Kircher was far from exemplary, as his deceptive citation methods make clear. Having chosen to practice a type of scholarship, erudition, whose essence was a preference for substance over style, he repeatedly sacrificed content to form, lavishing more care on the outward appearance of his imposing scholarly apparatus than on the solidity of its internal supports. More revealing than the gap between appearance and reality, however, is the kind of appearance that Kircher labored so hard to achieve. He staked a great deal on the originality of his work, especially his use of primary sources, which had required "Herculean labor" to collect and interpret. His emphasis on the central role of Oriental languages and newly discovered manuscripts in his hieroglyphic studies points to the new context in which he made use of the doctrine of occult philosophy. *Egyptian Oedipus* was not offered to the Republic of Letters as a restoration of timeless philosophy, in the manner of Ficino's *Platonic Theology* or Patrizi's *New Philosophy of Everything*, nor as a defense of practical magic, like Ficino's *Three Books on Life* or Agrippa's *On Occult Philosophy*. It was meant to be a different kind of restoration: an antiquarian reconstruction of ancient beliefs and practices, both pious and superstitious, which were preserved in inscriptions and artifacts. It was supposed to demonstrate how much the study of antiquity would benefit from the study of Oriental literature.

Regardless of how well Kircher realized that goal, he was as eloquent as any early modern scholar in championing the search for new sources, especially from the Near East. His genuine studies of Oriental manuscripts, as well as his misrepresentation of derivative research to fit the same mold, were products of that ideal. Did anyone before Leopold von Ranke express the dusty pleasures of archival research with so much visceral pleasure? Kircher wrote in the introduction of *Egyptian Oedipus*:

> I consider there to be . . . no art so thoroughly extinct or lost, of which no vestige glimmers in some corner of the world unknown to us, or in some foreign library buried under the covering of a foreign language, as if under ashes and cinders, among the dusty cadavers of half-eaten codices; or which, broken up and dispersed in many pieces by the neglect of time, may not be discovered scattered among the ancient authors. Now the fact that they lie buried happens in part through carelessness and neglect, and in part through the jealousy and ignorance of those who possess the treasures of such books. Oh, how numerous, how excellent,

and incomparably valuable are the memorials of books which on this account waste away in famous libraries, struggling against worms and cockroaches, amid squalor and decay; which, if they were brought to light by men expert in languages—O immortal God!—in a brief time how exceptional an increase we would see in the instruments of learning, so long neglected and unpolished in many regards, above all from the treasures of the Hebrews and Arabs.[69]

Kircher's mission to recover lost knowledge certainly owed a great deal to Renaissance occult philosophy and its search for primordial wisdom. But his encomium to archival research also vividly expressed a vision informed by the latest developments in erudite scholarship—a vision of the heroic scholar, in search of historical knowledge, recovering forgotten, worm-eaten manuscripts in exotic languages and publishing them for the collective benefit of the Republic of Letters.

69. *OA* I, b1^v.

CHAPTER SIX

Erudition and Censorship

It is the nature of these books that they can nowhere serve me better than where they treat the worst things. For in books of this sort I do not seek Catholic truths, which infinitely many other authors may provide, but the errors of the ancients and whatever teachings or opinions are expressed in the hieroglyphic signs.
—Athanasius Kircher[1]

Like all early modern authors, Kircher was constrained by censorship. For a member of a religious order living in Rome—the seat of the Catholic Church, including its chief instruments of intellectual discipline, the Congregations of the Holy Office of the Inquisition and of the Index of Prohibited Books—this pressure was especially tangible.[2] Not long after arriving in the Holy City, Kircher wrote Peiresc about the difficulty of publishing Arabic books dealing with magic. "Surely they will never be permitted and cannot be permitted," he opined, "especially here in Rome, where the censorship of all books is so strict that not even the least straw of error or false opinion is tolerated."[3] Given Kircher's subject matter, this was no idle worry. In his hieroglyphic studies he often treated non-Christian beliefs and rituals more positively than Catholic doctrine allowed, and long sections of *Egyptian Oedipus* described illicit magical practices with the detail of an instruction manual. Works containing similar material, such as Agrippa's

---

1. Kircher to Peiresc, Rome, 3 December 1636, BNF FF 9538, fols. 236ᵛ–37ʳ, in Stolzenberg, "Oedipus Censored" (2004), 46.

2. See Prosperi, *L'inquisizione romana* (2003); Godman, *Saint as Censor* (2000); Infelise, *I libri proibiti da Gutenberg all' Encyclopédie* (1999).

3. Kircher to Peiresc, Rome, 3 December 1636, BNF FF 9538, fol. 236ʳ, in Stolzenberg, "Oedipus Censored" (2004), 44.

On Occult Philosophy, had been placed on the Index of Prohibited Books. As Kircher recounted to Peiresc, an Arabic book of the kind he would use as source material had been found, following its publication in Rome in 1585, to contain "superstitions and errors," on which account all copies had been confiscated and burned.[4]

Kircher's hieroglyphic studies never ran afoul of the Holy Office or the Index. On the contrary, they went to press with the imprimaturs of his Jesuit superiors and the Master of the Sacred Palace. Nonetheless, censorship left its mark on them. This chapter looks at how the Jesuit censors responded to Kircher's exposition of magical practices and esoteric wisdom, and how Kircher responded to them. It offers a different kind of case study than previous investigations of Jesuit censorship, which have concentrated on the Society's efforts to enforce orthodoxy in the realm of natural philosophy, including Kircher's controversial astronomical work.[5] Because of constraints on free expression, interpreting early modern texts about heterodox subject matter is inherently difficult. In the case of *Egyptian Oedipus* we are lucky to have an uncommon trove of archival material—Jesuit censorship reports, the surviving manuscript of half of *Egyptian Oedipus*, and unpublished correspondence—which allows a detailed reconstruction of how censorship shaped Kircher's text, providing insight into how to interpret the published work.[6] Kircher's strategic response to the censors sheds light on his scholarly priorities, showing that the explanation of ancient artifacts and the publication of information about ancient magic and religion, especially from rare Oriental texts, were the core of his hieroglyphic studies.

DISCUSSING THE "ERRORS OF THE ANCIENTS"

As a Jesuit author, Kircher was required to submit his works to superiors for internal review prior to publication—a task entrusted to the College of Revisers, a panel of five theologians that made recommendations to the general of the Society about whether a book could be printed. Although

---

4. See above, ch. 5, "The Arabic Sources of Kircher's Hermetic History," for more details.

5. On Jesuit censorship, see: Baldini, *Legem impone subactis* (1992), ch. 2; Gorman, "A Matter of Faith?" (1996); Hellyer, "Because the Authority of My Superiors Commands" (1996): 319–54; Hellyer, "Construction of the *Ordinatio*" (2003); Hellyer, *Catholic Physics* (2005). On the censorship of Kircher's *Itinerarium exstaticum*, see above, introduction, n. 10.

6. The complete texts of the censorship documents pertaining to *OP* and *OA*, including all passages quoted in this chapter, are printed in Stolzenberg, "Oedipus Censored" (2004). Parts of this chapter appeared in Stolzenberg, "Utility, Edification, and Superstition" (2006), where I also discuss how Jesuit censorship functioned as a form of peer review.

nondoctrinal concerns such as scholarly quality and authorial decorum constituted an essential dimension of the revisers' work, the primary purpose of Jesuit censorship was to maintain the "soundness and uniformity of doctrine" required by the Society's constitutions—in other words, to police doctrinal orthodoxy, represented by Aquinas in theology and Aristotle in philosophy. Nothing was to be allowed that did not agree with the common sense of the schools and doctors, was "not entirely congruent with Christian faith and piety, or which might rightly offend others, or which seems inappropriate for the reputation of the Society." Thus, in theory, no "new opinions disagreeing with common doctrine" or anything that "overturns the common reasons which theologians confirm about Christian dogma" were permitted. In particular, theological matters were not to diverge from Saint Thomas. Furthermore, Jesuit books were required to be "edifying and useful."[7]

In both *Egyptian Oedipus* and *Pamphilian Obelisk*, the revisers found clear-cut, but easily remedied, conflicts with Catholic theology. For example, the manuscript of *Egyptian Oedipus* called attention to the imperfect understanding of the Trinity in the Chaldean Oracles, which failed to acknowledge the primacy of the Father, and declared that, "according to orthodox theology there is no origin of divine emanations except the Father."[8] The revisers censured this statement, commenting, "As if the Holy Spirit didn't proceed from the Son!"[9] Elsewhere, the revisers asked Kircher to clarify his exposition of an Egyptian teaching about the reception of divine gifts, lest it be read as supporting the heresy of Semipelagianism, an accusation then being levied at the Jesuits by critics of the Molinist doctrine of divine grace. In other passages they asked him "to correct that way of speaking which seems to make God the author of sin," as well as statements seeming to suggest that souls are mortal or material.[10]

A much larger part of Kircher's text fell into a grayer area. The heterodox, non-Christian beliefs and practices that constituted the subject matter of *Egyptian Oedipus* and *Pamphilian Obelisk* could be discussed, but the revisers required that such topics be presented in a particular manner and within certain limits, which varied for different kinds of material. The principal distinction visible in the judgments of these books was between er-

---

7. "Regulae Revisorum Generalum," *Institutum* (1892–93), vol. 3, 65–68. For further details, see Stolzenberg, "Oedipus Censored" (2004), 6–12.

8. BNCR Ms. Ges. 1235, fol. 87^r; cf. *OA* II.1, 133.

9. ARSI FG 668, fol. 391^r. Kircher's theological gaffe might reflect his dependence on the Byzantine Psellus.

10. ARSI FG 668, fol. 391^v.

roneous *beliefs* and magical *practices*, with the latter perceived as a greater danger and thus in need of more aggressive censorship. Although the Tridentine Church considered right belief as important as right practice, there was a venerable tradition of opposing heretical doctrines by exposing and refuting them. Kircher cannily invoked this tradition, defending his discussions of unorthodox doctrines by comparison to the writings of Irenaeus, father of the church, against the Gnostics.[11] No comparable tradition supported the pious exposition of illicit magical practices. These matters were not merely theoretical. In this period the Holy Office and other authorities actively prosecuted illicit magic, both popular and learned.[12]

In the realm of descriptions of pagan theology, the revisers were troubled by such things as Kircher suggesting that the Egyptians had worshiped the Nile "not without reason," and depicting Hermes Trismegistus as a holy man despite his having introduced superstitious rites and having believed in perceptible gods. Where Kircher discussed Psellus's opinion that souls are produced from seed, he was instructed to indicate that this was an error *in fide*. The treatment in *Egyptian Oedipus* of the Chaldean Oracles, the Orphic verses, and the sayings of Pythagoras, which contained unorthodox doctrines about such theologically sensitive matters as the creation of souls and the process of divine emanation, received special attention from the revisers. Here as elsewhere the revisers were primarily concerned that Kircher make clear that erroneous and superstitious beliefs were such and that he not appear to endorse them. Thus he was not to say that Zoroaster was divinely inspired, or that Orpheus took many things from the books of Moses, or that the followers of Pythagoras lived a "heavenly life on earth."[13]

Kircher's dependence on Jewish and Muslim sources was especially problematic, since these texts described the beliefs and practices not of extinct pagans, but of living infidels whose presence was closely felt, whether in the form of Europe's barely tolerated Jewish minority or the Ottoman empire that threatened Christendom from the east. The reviser Honoré Fabri complained that Kircher gave too much credence to Hebrew and Arabic authors, even seeming to prefer them to familiar Christian ones. This criticism ignored or at least called into question the whole rationale of Kircher's study, which was explicitly to bring to light knowledge hidden in "Oriental

---

11. *OA* II.1, 3.
12. For examples in Rome, see Cohen, *Words and Deeds* (1993); Fiorani, "Astrologi, superstiziosi e devoti" (1978); Ernst, "Astrology, Religion and Politics" (1991); Dooley, *Morandi's Last Prophecy* (2002).
13. ARSI FG 668, fols. 398^r, 390^r–91^{rv}; cf. 394^v.

memorials," especially those of the Arabs and Jews.[14] When they reviewed the second part of *Egyptian Oedipus*, the revisers commanded Kircher to completely rewrite the treatise on the Kabbalah, as well as the following treatise devoted to Arabic magic.[15]

The status of Jewish literature was vexed due to the 1592 bull of Clement VIII, included in the 1596 edition of the Index of Prohibited Books, which removed previous exceptions to the ban on the Talmud and forbade outright all books by "Talmudists, Kabbalists, and other impious Jews."[16] But despite its sweeping language, the ban was generally interpreted in more nuanced terms. Catholic theologians widely acknowledged that there was a good Kabbalah and a bad one, and that the ban applied only to the latter. Within the Society of Jesus there was a literary tradition that sought to clarify this question, which found its most authoritative expression in the work of Jacques Bonfrère, to whom the revisers appealed as a guide.[17] Thus the revisers were not opposed in principle to Kircher's positive assessment of parts of the Kabbalah, but they were dissatisfied with his definition of good and bad Kabbalah. They called on Kircher "to show clearly which Kabbalah is legitimate and which prohibited; so that he may not seem to contravene the Bull of Clement VIII . . . and also the Index of Prohibited Books itself; which includes the Kabbalah of Johannes Reuchlin . . . let the author distinguish, with Bonfrère . . . and with others, between a good kabbalah, an indifferent kabbalah, and a bad kabbalah."[18]

The more substantive problem, however, was not that Kircher failed to distinguish adequately between the good and bad Kabbalah, but that he discussed many beliefs and practices that, by his own assessment, belonged to the illicit variety. The revisers demanded that Kircher omit substantial sections of the treatise whose subject matter fell into this category. With regard to beliefs, even erroneous ones, there was room for discussion so long as the material was presented in the correct way. In their judgment of the first volume of *Egyptian Oedipus*, the revisers reprimanded the author for

---

14. Fabri excluded from this censure the Arabic "philosophers, physicians, and mathematicians" who had long formed part of the Scholastic canon. ARSI FG 668, fol. 394^r.

15. ARSI FG 668, fol. 391^v.

16. Bujunda, ed., *Index* (1993), vol. 9, 930.

17. See Bonfrère, *Pentateuchus Moysis Commentario* (1625), ch. 21. Other influential Jesuit works dealing with the Kabbalah were Possevino, *Apparatus sacer* (1606); Serarius, *Rabbini et Herodes* (1607); Serarius, *Iosue* (1609–1610); and the biblical commentaries of Cornelius à Lapide. See Secret, "Jésuites et le kabbalisme" (1958): 543–55.

18. ARSI FG 668, fol. 391^v. See also ARSI FG 668, fol. 390^v.

"repeatedly . . . cit[ing] too respectfully the Talmud and other Jews, which are despised names or persons, as is to be seen in the Index of the Council of Trent."[19] The problem, then, was not that Kircher cited the Talmud and other Jewish authors condemned by the Index, but that he did so "too respectfully." The revisers advised Kircher to treat these authors in the circumspect manner that the *Ratio Studiorum* (the Jesuit educational code) prescribed for teaching Averroes and other theologically problematic interpreters of Aristotle.[20]

LANGUAGE AND SUPERSTITION

In other words, a Jesuit author had to obey linguistic rules if he wanted to discuss certain topics. In his original manuscript, Kircher had already accompanied his discussions of pagan, Jewish, and Moslem beliefs and practices with refutations of their superstitious errors and stern warnings to the Christian reader "not to try this at home." The revisers accepted the principle that some heterodox ideas could be presented if they were framed with such language, but they found Kircher negligent for not distinguishing adequately between his own convictions and the false opinions of his sources, for not condemning superstitions forcefully enough, and for sometimes even seeming to endorse them. They called on him to make more frequent use of expressions such as *"inquiunt"* ("they say") to better distinguish the opinions of his sources from his own, to augment his denunciations of falsehood, and to abbreviate his discussions of superstitious material.[21]

There was a tradition of early modern writers employing such expressions to inoculate themselves against the charge of unorthodoxy. A classic example was Agrippa's preface to *On Occult Philosophy*, in which the author claimed that the views expressed in the book were written "more narratively than affirmatively."[22] Savvy early modern readers understood the tactics of "honest dissimulation" and knew to take such expressions with a grain of salt. Agrippa's disclaimers did not save his work from the Index. Kircher's use of similar expressions often rings hollow and may seem like

---

19. ARSI FG 668, fol. 398^r.
20. See "Regulae Professoris Philosophiae," 3–5, in *Institutum* (1892–93), vol. 3, 189–90.
21. See especially, ARSI FG 668, fol. 396^v. These linguistic strategies parallel those used by Jesuit authors to discuss unorthodox natural-philosophical theories, as discussed by Hellyer, "Because the Authority of My Superiors Commands" (1996), 336ff., who calls attention to the importance of the judicious use of words such as "true," "false," "probable," and "hypothesis."
22. Agrippa von Nettesheim, *De occulta philosophia* (1992), 66.

a linguistic trick to smuggle illicit material past the gate. But the Jesuit revisers did not view it that way. In the case of heterodox beliefs, they were willing to take such language at face value and accepted its efficacy, even offering Kircher advice on the kinds of phrases he needed to insert in order to receive their stamp of approval.

Nicolaus Wysing, among the most scrupulous of the revisers, accepted the inevitability that such a work would contain "some things, not only contrary to the opinion of the author, but also contrary to truth, which nevertheless should be tolerated since they are necessary to prove things." To mitigate the danger posed by such material, Wysing gave instructions for a series of disclaimers to be placed at the beginning of the work. Among these was to be a declaration defining precisely what authority should be granted to the Talmud, the Kabbalah, and other Jewish authors. Another disclaimer was to warn the reader that authors cited in the course of the work were to be granted authority "only to the extent that they demonstrate that on account of which they are cited, although they may happen to mix in something inconsistent." Furthermore, the reader was to be told that

> In that way of speaking used sometimes by the ancient authors, it is not to be thought from this that in matters so obvious (in the light of faith . . .) the author believes otherwise than is received in the Church. But he uses the very words of the authors, even those far from the truth, both so that it may appear how far they attained in such secret matters, and because it would be too difficult and irksome to reduce those ways of speaking, repeated so many times, to the words customary today. Examples include where there is discussion with the ancients about the soul of the world . . . about the shape of the trinity, about the triform God . . . . For although they may sometimes have a sound meaning, they nevertheless sometimes seem to be wrongly understood by these authors, and can draw into similar suspicion those who may read them without an attached warning.

Wysing was doubtless sincere in his effort to maintain orthodoxy and protect readers. But the careful use of language that he prescribed risked sending an ambiguous message to the reader, who might perceive the author to be speaking from both sides of his mouth. Kircher's treatment of chronology, for example, which assembled evidence that implied a heretical view of antediluvian history, was allowed to stand unchanged so long as the introduction of the work informed the reader that statements about chronology and genealogy "should be understood as things said only in passing," unless they agreed with opinions that Kircher explicitly endorsed as his own.[23]

There was other material, however, that posed so much danger to "curious souls" that no kind of language could render it safe for publication. The revisers demanded that Kircher entirely omit his discussion of the "practical Kabbalah," since it was "filled with dangerous superstition." Similarly, they demanded that he drastically cut the section on Arabic magic, removing all specific details of those practices. Thereby, they said, the reader might have a taste of the "Kabbalah of the Saracens" without being drawn into its depths.[24] A like verdict was pronounced regarding the lengthy treatment of Arabic planetary magic seals in the section on "hieroglyphic astrology."[25]

USELESS KNOWLEDGE

The linguistic strategies that the revisers sanctioned in the case of superstitious beliefs were deemed insufficient in the case of magical practices, especially those attributed to the Jews and Arabs. In the manuscript submitted to the revisers, Kircher had already wrapped his discussions of magical practices, such as the practical Kabbalah, in all manner of disclaimers, reprobation, and invective. But the revisers were not appeased. They frequently invoked the eighth rule of the Revisers General, justifying their decisions by weighing the potential danger of the material in question against its utility. "In the correction of [the treatises on the Hebrew and Saracenic Kabbalah]," they wrote, "it will be sufficient (for the author's purpose) if the whole thing contains much less, seeing as it is undoubtedly for the most part useless, if not also dangerous." Kircher's discussion of the practical Kabbalah was to be deleted in its entirety, while his discussions of the divine names and the *sefirot* were to be abbreviated by removing sections that were, according to Kircher, false, and thus, according to the revisers, useless. For example, Kircher had described the kabbalistic techniques for manipulating the alphabet, known as *gematria*, *notarikon*, and *temurah*, as the foundation of the bad kind of Kabbalah. On this basis, the revisers instructed him to remove his discussions of the forty-two- and seventy-two-letter names of God that were based on them. Even if such techniques were merely indifferent rather than bad, the revisers argued that

23. ARSI FG 668, fol. 390^{rv}; Grafton, "Kircher's Chronology" (2004).
24. ARSI FG 668, fol. 392^v.
25. ARSI FG 668, fol. 396^r.

they are hardly of any genius or significance, and thus nothing of substance can be concluded [from them] . . . . Hence it is clear that the work spent by the author in explaining this kind of kabbalah so exhaustively is useless; it will be even more useless to print it, unless the whole thing is shortened within much narrower limits.

Likewise, the revisers claimed that Kircher's discussions of the inner workings of the *sefirot* "have no great utility and in addition contain a great deal of obscurity and fantastical rubbish." They made similar calculations concerning Kircher's treatment of other non-Christian traditions. The judgment of the second volume of *Egyptian Oedipus* concluded with the warning:

> Finally, the author must carefully beware, lest in this work, whatever he has drawn from the Platonists, Pythagoreans, Kabbalists, Talmudists, and other authors of similar character, he recklessly spills it all onto the paper and thrusts on the world things long buried, and which should always remain buried. Therefore let him adduce these authors in such a way, where it is necessary, that he may explain doubtful things clearly, condemn blameworthy things, not assert magical or superstitious matters in detail, establish safe limits to his claims, and let him be convinced that the usefulness for the general public of treating such things can hardly be compared to the danger that must be feared on the part of some curious individuals.[26]

For his part, Kircher tried to establish the pious utility of his descriptions of certain superstitious practices, such as kabbalistic amulets, by claiming that they would teach the Christian reader, who might happen on such an object, to recognize its dangerous character. He even asserted that the Holy Office had requested him to publish such a description.[27] Kircher hoped that invoking the authority of the Inquisition would lend his discussion of magical seals a sheen of propriety. But the revisers were not impressed. "It is not enough," they insisted, that after describing minutely how such things were used in practice, Kircher "reproves them as superstitious and to be avoided, since overly curious and insufficiently pious individuals might esteem them and put them into use."[28]

There was a scholarly culture clash between Kircher and the revisers

---

26. ARSI FG 668, fols. 391^v–92^v.
27. *OA* II.2, 474. See also ibid., 211.
28. ARSI FG 668, fol. 396^r. Cf. ARSI FG 668, fol. 392^r.

about the utility and edification to be had from studying ancient superstition. Like other erudite scholars, Kircher was dedicated to a total recovery of antiquity: no part of the past should remain buried. The revisers, on the other hand, thought that much of this kind of material, because false and superstitious, was ipso facto useless and unnecessary. Where Kircher judged this material by its usefulness for historical explanation, the revisers evaluated its utility in terms of its truth and falsehood. At best, they allowed that some superstitious material might have value to scholars, but argued that this was outweighed by the danger it posed to "curious souls."[29]

THE LIMITS OF CENSORSHIP

It was one thing for the revisers to request changes to Kircher's manuscript and quite another to make him obey. Comparison of their judgments with the printed book and the surviving sections of the manuscript reveals that Kircher did not comply with many of their most serious demands.[30] For the most part he dutifully carried out specific requests for minor changes. He added, for example, more of the kind of language (already present in the original draft) requested by the censors to distinguish his own opinions from those of his impious sources, inserting phrases such as "so say the Rabbis," "as the Platonists say," or "according to the opinion of the Arabs."[31] In many cases, particularly in the treatise on the Kabbalah, he merely inserted an asterisk next to an opinion that had been singled out for correction and placed a note in the margins labeling it a "ridiculous Rabbinic superstition" or something similar. He added numerous passages refuting specific errors, and more general disclaimers such as: "If anything heterodox occurs in these things, I wish the reader to know that it is not asserted by my judgment, but from the opinion of the ancients."[32]

---

29. For a case study of conflict between erudition and censorship outside the Society of Jesus, see McCuaig, *Carlo Sigonio* (1989), 251–90.

30. BNCR Ms. Ges. 1235. This consists of the fair copies of parts 1 and 2 of *OA* II that were submitted to the revisers and subsequently corrected and sent to the printer. (It also includes a few loose, unidentified sheets from the otherwise lost manuscript of *OA* III.) Some parts are missing, but the majority of the manuscript is intact. The text is in the hand of an amanuensis, with corrections added by Kircher. For further details, see Stolzenberg, "Utility, Edification, and Superstition" (2006), 14.

31. When I say that Kircher changed the text, I mean that in the manuscript, the text in question appears in Kircher's hand as an addition to the original text. Such additions need not have been made after the revisers' judgments; but when the changes deal with material the revisers asked to be changed and address their criticisms, I make this assumption.

32. BNCR Ms. Ges. 1235, fol. 85^r; cf. *OA* II.1, 130.

While Kircher generally complied with requests for such small alterations, the revisers' most urgent demands—most notably their order to delete large portions of the treatises on the Kabbalah and Arabic magic—went unheeded. The "useless" sections on the forty-two- and seventy-two-letter names of God remained essentially unaltered, as did the problematic parts of his discussion of the doctrine of the *sefirot*. Kircher responded to the demand that he completely delete descriptions of magical practices by doing more of exactly what the revisers had declared inadequate. That is, he left these sections intact but added more disclaimers and invective (fig. 33). The chapter on the practical Kabbalah remained as long and detailed in the printed book as in the original draft, but Kircher added new statements indicating that he spoke "not according to my own opinion, but according to the Jewish theologians." He left his discussion of the forty-two-letter name of God unabridged, but inserted new phrases emphasizing that these doctrines were vain Jewish superstitions.[33]

All four volumes of *Egyptian Oedipus*, as well as *Pamphilian Obelisk*, were sent to press in a form much closer to Kircher's original drafts than to that commanded by the revisers in their judgments. Evidently, the constraints placed on Kircher by ecclesiastical censorship were not so great in practice as the letter of the law would suggest. The censorship process did not take place entirely on paper—at least not for an author like Kircher, who was a colleague of the revisers at the Collegio Romano. There were likely conversations and negotiations involving Kircher, the revisers, and the general of the Society. A glimpse of the face-to-face dimension of the process, which by its nature has left few traces in the archive, can be caught in Nicolaus Wysing's judgment of *Obeliscus Pamphilius*, which refers to the changes requested in his report as being "in addition to those which I have discussed personally with the author."[34] Kircher's ability to disregard the revisers almost certainly depended on the complicity of the general.[35] At the time of the review of the first volume of *Egyptian Oedipus*, the reviser Wysing addressed a poignant plea to General Alessandro Gottifreddi:

> I fear that the work expended by the Father Revisers in the judgment of this book may be of little use with Father Athanasius; for even recently

---

33. BNCR Ms. Ges. 1235, fols. 335^r, 187^r.
34. ARSI FG 668, fol. 390.
35. For similar outcomes in Jesuit censorship of natural philosophy, see Hellyer, *Catholic Physics* (2005), 35–52, and Hellyer, "Because the Authority of My Superiors Commands" (1996), 339–41, on cases of authors appealing to the general to override the revisers.

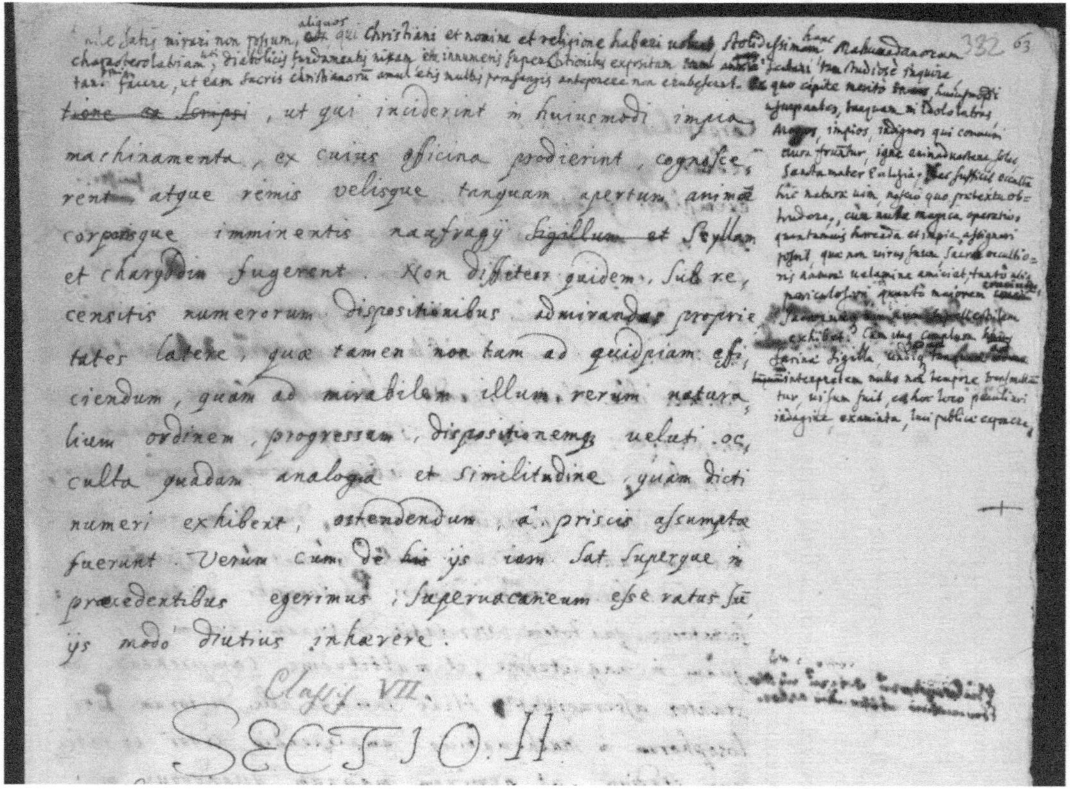

Fig. 33. The Jesuit censors ordered Kircher to delete this discussion of "hieroglyphic arithmetic" in the manuscript of *Egyptian Oedipus* since it described the fabrication of Arabic magic seals, which, on Kircher's own account, involved a demonic pact. Instead of removing the offending material, Kircher added the text in the margins, denouncing the practices as "most foolish Muhammadan characterolatry," whose practitioners the Church rightly punished by death as "impious, shameful, magicians and idolaters," and reaffirming that "it seemed right to publish these things... so that anyone who should encounter an impious device of the this sort would know from whose workshop they came, and flee by oar and sail as if from an evident Scylla and Charybdis." BNVE Ms. Ges. 1235, fol. 332. Biblioteca Nazionale Centrale di Roma.

in that Synopsis, he followed the judgment of these fathers only as much and in the manner that he wished. Then, he once said to me that he had significantly enlarged the book on the *Pamphilian Obelisk* after the judgment; and I even hear elsewhere that he boasted, on account of experience in these matters, that he can safely [engage in] practices of this sort. Further, I know from personal experience that things in a work to be printed (i.e. at the time of printing) are sometimes changed by Father Athanasius, at least with respect to sequence, so that it cannot easily be determined whether he has followed the judgment or rather has neglected it.[36] Indeed, as these things seem to me capable of highly prejudicing our judgments, I have decided they should be deferred to the providence of your paternity. . . . I entrust myself most humbly to your Father's paternal benevolence.[37]

It was the duty of the College of Revisers to apply rigorously the rules governing Jesuit publications and pass their recommendations on to the general. In making a final decision the general was at liberty to consider extenuating circumstances and to make exceptions.[38] All evidence suggests that the three Jesuit generals who approved the successive volumes of *Pamphilian Obelisk* and *Egyptian Oedipus* declined to enforce the revisers' demands. With his high profile in the Republic of Letters and the sponsorship of Innocent X and Ferdinand III, Kircher had more leverage than most Jesuit authors.[39]

HOW TO WRITE A JESUIT TREATISE ON MAGIC AND SUPERSTITION

Kircher's victory over the father revisers should not be taken as evidence that the threat of magic and superstition was not taken seriously in seventeenth-century Rome, or that censorship did not leave its mark on Kircher's text.

---

36. In his denouncement of Kircher's *Iter Extaticum* II in 1655, the reviser François Duneau stated that "it is known from experience" that Kircher often had not corrected his books as instructed. ARSI FG 661, fols. 30^r, 34^r.

37. ARSI FG 668, fol. 399^r.

38. Cf. Baldini, *Legem impone subactis* (1992), 87, who remarks that the generals often possessed a greater "political-cultural" sensitivity than the revisers, and thus were relatively more tolerant of innovative propositions, sometimes mediating between authors and revisers and occasionally even permitting works that the revisers had condemned.

39. I discuss the likely reasons for the indulgence granted Kircher at more length in Stolzenberg, "Utility, Edification, and Superstition" (2006), 347–48.

Kircher made concessions to the revisers, though fewer, as we have seen, than they had ordered. More significantly, the threat of censorship played an important role in the genesis of *Egyptian Oedipus*.

In 1635, Kircher wrote to Peiresc about the difficulty of publishing Arabic manuscripts relating to Egypt and the hieroglyphs. "It is sure," he declared with regard to the Barachias manuscript, "that the whole [text] cannot be published, since it treats many magical things, and in many places entirely concerns incantations, which neither the Holy Office nor our Society will permit, because of the scandal that they could cause to souls."[40] Peiresc pressed him on the matter, causing Kircher to explain the situation at greater length:

> Since books of this sort are full of superstitious magical seals and other opinions condemned by the church . . . [and] the magic arts are so mixed up with the hieroglyphic works . . . that [the authors] seem not to recite but to teach, indeed they seem to pave the way to revive the necromancy of the ancients. . . . surely these things will never be permitted and cannot be permitted . . . .

It would not do, he continued, to publish an expurgated version that removed the controversial parts, since "it is the nature of these books that they can nowhere serve me better than where they treat the worst things. For in books of this sort I do not seek Catholic truths, which infinitely many other authors may provide, but the errors of the ancients and whatever teachings or opinions are expressed in the hieroglyphic signs, and their use." Kircher explains that to overcome these obstacles he devised *Egyptian Oedipus*, in which

> all the said authors are introduced in such an order that the authors' words may be sincerely and faithfully alleged, but, the scandalous parts having been prudently and discreetly refuted, the errors of the ancients may be like witnesses, and nothing which may be of use for emending antiquity may be omitted from the cited authors. . . . I have shared this plan of mine with excellent friends and most learned men. . . . And all judiciously proclaim this to be the only way that these authors may be delivered from eternal darkness, perhaps even the flames. . . . For in this

---

40. Kircher to Peiresc, Rome, 8 February 1635, BNF NAF 5173, fol. 25^v; cf. BNF FF 9362, fol. 13^v (copy), in Stolzenberg, "Oedipus Censored" (2004), 42–43.

way hidden truth shall be made known, the dangerous rocks of scandal shall be avoided and, what I wish most of all, greater benefit will be rendered to the Republic of Letters.[41]

As Kircher tells it in this letter, *Egyptian Oedipus* was conceived as a means to publish Oriental texts about pagan religion and magic in a manner acceptable to ecclesiastical censors. Since these superstitious texts could not be published on their own, he would instead present passages from them within a larger work, in which he would interpret them and refute their errors. In other words he would employ the kind of linguistic strategies discussed above. When Kircher finally executed this plan years later, the censors intimidated him rather less. And when the revisers found Kircher's "refutations of the scandalous parts" insufficient to render certain parts of his books innocuous, he ignored them, confident that he could prevail over their judgments.

HOW TO READ A JESUIT TREATISE ON MAGIC AND SUPERSTITION

Although Kircher was unwilling to excise his expositions of magical practices and "useless" heterodox doctrines, he readily accompanied them with the language of reprobation that Catholic censorship, at a minimum, required. There is thus good reason to be skeptical of the negative value judgments Kircher attached to various heterodox teachings, including many traditions associated with occult philosophy. Nevertheless, while it is evident that he was not much troubled by the danger that such material posed to readers and that he used such language because censorship required him to do so, we cannot simply assume that because Kircher had to say certain things, he did not believe them. In places where Kircher appears inconsistent, wavering between positive and negative opinions of certain beliefs or practices, we may reasonably give less weight to the opinion that orthodoxy required. Likewise, his willingness to publish detailed descriptions of illicit magical practices suggests that he did not consider them to be as dangerous as he felt compelled to state. But what Kircher really thought is beyond our grasp. For example, was he sincere when he stated that planetary magic seals, whose construction he described in detail, were dependent on demonic powers—apparently caring less about weak-willed readers' souls than about his own need for evidence of ancient practices? Or did he

---

41. Kircher to Peiresc, Rome, 3 December 1636, BNF FF 9538, fols. 236^r–37^r, in ibid., 43–46.

believe, despite his declarations to the contrary, that the seals were not demonic but naturally efficacious? Or did he simply consider them inefficacious and harmless? The evidence does not allow us to say.

Similarly, what was his real opinion about the kabbalistic techniques for achieving union with God (*adhaesione in Deum*) and receiving divine gifts in this life and the next? Modifying Paulus Ricius's descriptions of these methods (see chapter 5), Kircher, independently of the revisers, inserted language distancing himself from such superstitions and excised references to the New Testament that Ricius had used to defend them. "As long as they [the Jews] refuse to hear the true sense [of the Gospels]," Kircher added, "enveloped in darkness they deservedly fall into the deepest pit of superstitions."[42] But there are also indications that Kircher esteemed some aspects of these practices. Ricius, after describing the intermediary role played by angels in achieving *adhaesione in Deum*, made the terse but pregnant comment: "whence the cult and invocation of saints (if you will reflect) traces its true origin, about which [I will speak] at greater length elsewhere."[43] Inspired by this suggestive remark, Kircher launched into a discussion of the cult of saints and the use of external rituals in the Catholic Church. Although he refrained from stating with Ricius that the cult of saints had its origin in kabbalistic invocations of angels, he lent support to those kabbalistic practices by situating them in the context of intercessory prayer and the sacraments, rather than illicit theurgy. He even added a comparison with the Eucharist to Ricius's discussion of Rabbi Ismael's mystical matzoh, which, when eaten while reciting certain invocations, caused the kabbalist to be filled with the light of supernal wisdom.[44] Given Protestant arguments that some Catholic rituals derived from Jewish superstitions, this was a dangerous theme to broach, and Kircher revised his original draft to soften these comparisons.[45]

The anti-Jewish invective that occurs frequently in Kircher's discussion of the Kabbalah may seem like a sop to the revisers, and some of it was demonstrably added in response to their judgments. But even in the absence of any external pressure, Kircher partook of his share of Christian bigotry, as we know from his encounter with the Rabbi Salomon Azubi in Provence.

---

42. *OA* II.1, 342; cf. Ricius, *De coelesti* (1541), 84^r; Pistorius, *Artis cabalisticae* ([1587]), 136.
43. Ricius, *De coelesti* (1541), 84^v; Pistorius, *Artis cabalisticae* ([1587]), 136.
44. *OA* II.1, 341–42. Apropos of this and similar practices, Ricius expressed astonishment that the modern Jews nonetheless mocked baptism and the Eucharist, but he did not make so explicit an analogy as Kircher. Ricius, *De coelesti* (1541), 84^v–85^r; Pistorius, *Artis cabalisticae* ([1587]), 137.
45. *OA* II.1, 340; cf. BNCR Ms. Ges. 1235, fol. 222^r.

Kircher and Azubi, who had been introduced by Peiresc in 1633, had a brief correspondence, which deteriorated when Kircher took offence at the way Azubi responded to what appears to have been Kircher's attempt to convert him.[46] Afterward Kircher wrote to Peiresc:

> Regarding the letters that relate to Rabbi Salomon, I would find in them incivility, obstinacy, palpable arrogance, and hard-headedness, in a word, the mark of the Jewish character. But I could not fail to send them to you, translated from Hebrew into Latin, so that your lordship may see for himself the perversity of the character of the Jews, who, when we labor for their benefit that they may be converted and live well, due to a certain innate hostility to our holy law, would rather remain perverted.[47]

When Kircher described the modern Jews as "blinder than moles"—however conveniently that opinion may seem to have satisfied the censors' concern to respect the spirit of Clement VIII's bull—there is no reason to doubt that he was expressing his genuine opinion.[48]

In sum, there is no sure way to know to what extent Kircher believed in the perfidious, superstitious character of the beliefs and practices that he so condemned. In the absence of evidence from unpublished sources, we must be skeptical, but ultimately agnostic, in judging the relationship between Kircher's published statements and true opinions about heterodox material. Even so, appreciating Kircher's encounter with the censors can help us to understand some inconsistencies in his work. For example, the great difference in tone between the preface and conclusion of the treatise on the Kabbalah, both of which express almost blanket condemnations, and the body of the text, which is much more ambiguous, can be understood as an effort by Kircher to mitigate the reaction to controversial parts of his exposition by surrounding them with firm expressions of Catholic orthodoxy. A similar strategy can be seen at work in other parts of *Egyptian Oedipus*—notably the section on "hieroglyphic astrology," whose denunciatory preface hardly prepares the reader for the mostly neutral, descriptive treatise that follows.

---

46. See Miller, "Mechanics of Christian-Jewish Intellectual Collaboration" (2004), 86–88.

47. "Ad literas porro Rabbi Salomonis quod attinet, nescio quid in iis inhumanitati, contumaciae, & manifestae superbiae, uno verbo, signa Iudaïci ingenii, & durae cervicis deprehenderim. Verùm omittere non potui quin eas ex hebraeo in latinam translatas vobis transmitterem; ut D.V. aspiciat coram perversitatem ingenii Iudaïci, ut dum ad eorum bonum laboramus, ut convertantur et vivent, illi magis, ex insita quadam virulentia in sanctum legem nostram, pervertantur." Kircher to Peiresc, Rome, 9 August 1633, BNF FF 9538, fols. 227^v–28^r.

48. *OA* II.1, 293.

Most significantly, Kircher's response to censorship offers insight into the overall nature of his hieroglyphic studies. He was quite willing to accommodate the censors when it came to the value judgments that accompanied his discussions of unorthodox beliefs and practices. But he refused to compromise on the details of his descriptions. To an extent, this may reflect the limits of what even an author as powerful as Kircher could get away with. But the battles that Kircher chose to pick also reflect his fundamental priorities. Unlike many Christian writers on subjects such as the Kabbalah, learned magic, and the *prisca theologia*, Kircher was not primarily concerned with rehabilitating suspect traditions and convincing readers of their probity. His chief interest in this material lay, rather, in finding support for his interpretation of the hieroglyphic inscriptions and, more generally, in displaying his erudite mastery of "recondite antiquity." While Kircher's treatment of the practical Kabbalah, for example, was ambiguous, there is no evidence that he was interested in practicing such theurgical techniques himself. For Kircher, knowledge of the Kabbalah was "useful" because it enabled him to explain the Jewish amulets sent to him for interpretation by perplexed collectors, not because he wanted to fabricate and use them.[49] The same may be said with regard to his discussion of Arabic magic. For all his attention to the minute details of the construction of planetary talismans, there is no evidence, and no reason to suspect, that he employed them. It is not surprising, then, that Kircher was unperturbed by the requirement that he accompany his discussions of such subjects with statements condemning their moral and theological probity. But expunging such discussion altogether, or even eliminating the technical details, would have deprived him of the tools to carry out the esoteric antiquarianism that lay at the heart of his enterprise.

49. See Stolzenberg, "Kircher Reveals the Kabbalah" (2004).

CHAPTER SEVEN

Symbolic Wisdom in an Age of Criticism

Our most learned friend, Reverend Father Athanasius Kircher of the Society of Jesus, whom we greeted twice on this journey, superbly explained with an entire massive volume as much as could be accomplished by human conjecture in so ancient a matter.
—Peter Lambeck, imperial librarian, on Kircher's interpretation of the obelisk in Piazza Navona.[1]

How did Kircher translate Egyptian hieroglyphs? The basic principles that informed his understanding of the hieroglyphs are well known and have been the subject of numerous studies.[2] Most fundamentally, these were the ideas that Egyptian priests had used hieroglyphic inscriptions to encode profound wisdom, and that hieroglyphs functioned as ideal symbols which had an essential, nonconventional relationship to their referents. The first of these principles was the dominant theory about the hieroglyphs before the eighteenth century; the second was shared by the many scholars who subscribed to what in chapter 1 I defined as the strong theory of symbolism, as opposed to the soft theory that considered hieroglyphs as products of human convention rather than a reflection of ultimate reality. While these under-

---

1. Lambeck, *Commentariorum de Augustissima Bibliotheca liber* (1766–82), vol. 1, 25.
2. Inter alia: Beinlich, "Athanasius Kircher" (2002); Marrone, *I geroglifici* (2002); Donadoni, "I geroglifici" (2001); Strasser, "Das Sprachdenken Athanasius Kirchers" (1999); Strasser, "La contribution d'Athanase Kircher" (1988); Eco, *Search for the Perfect Language* (1995), 154–65; Cipriani, *Gli obelischi* (1993); Iversen, *Myth of Egypt* (1993) 89–100; Marquet, "La quête" (1987); Pfeiffer, "Il concetto di simbolo" (1986); Rivosecchi, *Esotismo in Roma barocca* (1982), ch. 1; Pastine, *Nascita dell'idolatria* (1978), ch. 3; Allen, *Mysteriously Meant* (1970), 120–33; David, *Le débat sur les écritures et l'hiéroglyphe* (1965), ch. 3.

lying principles had broad currency, Kircher's translations of hieroglyphic inscriptions were unique. Earlier interpreters, from Horapollo to Valeriano and Caussin, wrote what were essentially dictionaries of symbols proffering allegorical explanations of isolated "hieroglyphs," a few of which were genuinely Egyptian, many of which were not. It is not surprising that these scholars restricted their efforts to this level, since the symbolic conception of the hieroglyphs, under which they, like Kircher, labored, did not offer a ready scheme for translating entire inscriptions.[3] As Daniel Russell observes, the Renaissance concept of hieroglyphs as "unmediated" ideograms "could never produce anything other than a jumble of unconnected signs that could never function as a real language."[4] Kircher, undaunted, went further: he took complete hieroglyphic inscriptions and converted them into coherent works of Latin prose. Under his treatment, obelisks yielded metaphysical treatises. Lacking a Rosetta Stone or any understanding of what we now know to be the nature of hieroglyphic writing, how did he proceed?

In contrast to the many studies of Kircher's theory of hieroglyphic symbolism, little attention has been paid to the mechanics by which he produced the translations printed in *Egyptian Oedipus* and *Obeliscus Pamphilius*.[5] These were based on the idea that symbolic images could only be understood in relationship to texts (a notion inspired by the European emblem tradition), combined with the Neoplatonic theory that the hieroglyphs were a "nondiscursive" language, which communicated without grammar or syntax. The hermeneutic practice that Kircher derived from these principles allowed him to produce translations that were not unmethodical, but which were inherently liable to charges of arbitrariness. Kircher was keenly aware of this vulnerability. His characteristic expressions of certainty were not at all the product of serene self-confidence, as is often assumed. On the contrary, they reflected Kircher's anxiety about the tenuousness of his claims, which was rooted in his relationship to seventeenth-century erudite scholarship. This chapter first examines Kircher's method of translating hieroglyphs and then turns to his efforts to defend their reliability against philological criticism.

---

3. I would apply this generalization even to Pignoria and Herwart. Although they published interpretations of the Bembine Table, they did not attempt the kind of prose translation that Kircher did. See ch. 1, "Renaissance Egyptology."
4. Russell, "Illustration, Hieroglyph, Icon" (2002), 83.
5. Of the studies mentioned above, Marrone, *I geroglifici* (2002), especially ch. 5, offers the richest consideration of Kircher's semiotics, emphasizing its systematic nature, but does not attempt to describe the process by which he produced his translations.

IDEAL READINGS IN THE PIAZZA NAVONA

For the sake of example, let us consider one inscription, which Kircher singled out as a showcase for his method.[6] The so-called Pamphilian obelisk, which now stands atop Bernini's aquatic pedestal in the Piazza Navona, was originally commissioned by the emperor Domitian (81–96 CE), probably to form part of the complex around the temple of Isis in Rome (fig. 34). The style of the figures indicates that the obelisk was carved by Roman craftsmen, although the inscriptions are intelligible and demonstrate an adequate understanding of hieroglyphic writing. Modern Egyptology informs us that the figures at the top, on the pyramidion, are purely iconic while the body of the obelisk bears a short text referring to the god Horus and the emperor Domitian. The pyramidion's southern face depicts a king presenting an offering to the gods Maat and Atum or Amun, beneath a winged sun disk depicted with a double uraeus (fig. 35).[7] The hieroglyphs below refer to Domitian, using epithets borrowed from inscriptions of Ptolemy III Euergetes (256–221 BCE) and originally associated with Ramesses II (1292–1225 BCE). Those visible in the figure read: "Horus, strong bull, beloved of Maat."[8] Kircher rendered the same signs as follows:

> To the Triform Divinity Hemphta—first Mind, motor of all things; second Mind, craftsman; pantamorphic spirit—Triune Divinity, eternal, having no beginning nor end, Origin of the Secondary Gods, which, diffused out of the Monad as from a certain apex into the breadth of the mundane pyramid, confers its goodness first to the intellectual world of the Genies, who, under the Guardian Ruler of the Southern Choir and through swift, effective, and resolute follower Genies who partake in no simple or material substance, communicate their participated virtue and power to the lower World. Their likeness is presented to the priests so that they may worship and propitiate them in sacrifices by an analogous rite. From these [Genies] derives the power communicated from the supreme Divinity to the sidereal World, where powerful Osiris, sensible Divinity of the Sun, with all his power fructifies and bestows essence, life, and motion on things; true soul of the sensible World, ruler

---

6. *OA* III, 561–62.

7. Although Kircher identifies this side as the southern face, it corresponds to the western face as the obelisk now stands in Piazza Navona. I follow Kircher's compass. I thank Antonio Loprieno for explaining the images and hieroglyphs.

8. Marucchi, *Gli obelischi egizi di Roma* (1898), 125–31; D'Onofrio, *Gli obelischi* (1992), 291–92.

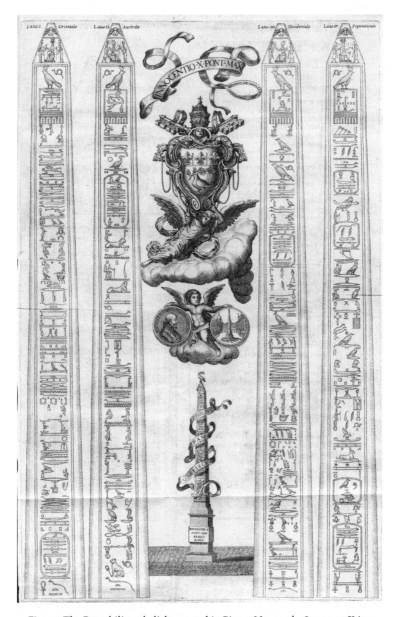

Fig. 34. The Pamphilian obelisk, erected in Piazza Navona by Innocent X in 1649. (Bernini's ornamental fountain was completed in 1651.) Unbeknownst to Kircher, it had originally been commissioned by the emperor Domitian, whom the inscriptions honor borrowing epithets from Ptolemy III Euergetes. Athanasius Kircher, *Obeliscus Pamphilius* (Rome: 1650), fp. 1. f GC6 K6323 6500, Houghton Library, Harvard University.

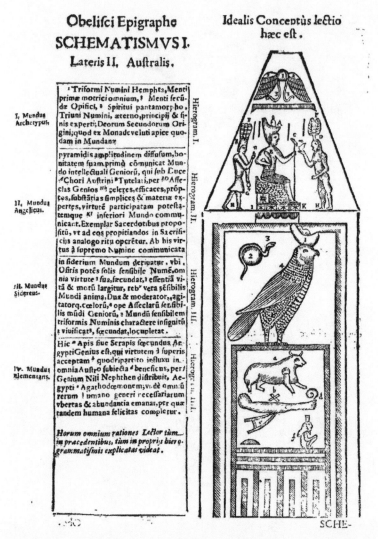

Fig. 35. The top of the southern face of the Pamphilian obelisk. Due to a typographical error, the headings have been reversed: the Latin translation on the left constitutes the "ideal reading." Kircher considered this translation to be a showcase of his method. He divides the inscription into four "hierograms" and associates each with a level of reality: (I) the archetypal world, (II) the angelic world, (III) the celestial world, and (IV) the elementary world. The Latin translation is keyed to the inscription with numbers and letters, allowing Kircher to explain his interpretation symbol by symbol. Athanasius Kircher, *Obeliscus Pamphilius* (Rome: 1650), 444. Courtesy of Stanford University Libraries.

and moderator and charioteer of the Heavens, with the support of the follower Genies of the sensible world, marked by the character of the triform Divinity, he fructifies and enriches the sensible World. Here is generous Apis or Serapis, fertile Genie of Egypt, who distributes the power received from above by a fourfold influence to all things subject to the South through Nephthe, Genius of the Nile, Agathodemon of Egypt; whence emanates the fruitfulness and abundance of all things necessary to the human race, through which at last human happiness is fulfilled.[9]

There was a method to Kircher's translation, mad though it may seem. "The reader may see the reasons for all this explained," he wrote below his translation, "both in the preceding pages and in the discussions of the individual hierograms."[10] (By "the discussions of the individual hierograms," Kircher meant the relevant entries in the hieroglyphic dictionary in book 4 of *Obeliscus Pamphilius*.)

He divided the inscription into four groups of hieroglyphs, which he called "hierograms," each of which he took to correspond to a different level in the hierarchy of emanation. (Although Kircher had textual evidence that the hieroglyphs described such a world, his imposition of this scheme onto the physical structure of the obelisk was his own a priori assumption.)[11] He labeled the different phrases in his Latin translation, which he called an "ideal reading," with numbers and letters, which corresponded to the elements of the hierograms, and referred to the sections of the book that explained their symbolism.

For example, the first hierogram, at the apex of the pyramidion, consisted of a winged circle from which dangled a snake. Kircher explained it as a symbol of God, whom the Egyptian sages understood in Trinitarian terms as the triform divinity. To defend this interpretation, he first offered evidence from the Coptic lexicon that he had translated in 1643. (Although Kircher denied that the hieroglyphs recorded spoken language, he believed that the ancient Egyptians had spoken a language closely related to Coptic, whose vocabulary thus preserved valuable information about ancient Egyptian beliefs).[12] This allowed him to state that the Egyptians called the su-

---

9. *OP*, 444.
10. *OP*, 444.
11. See the discussion above, ch. 4, "Reading the Bembine Table," for an analogous example.
12. Kircher also acknowledged a phonetic dimension to the hieroglyphs, but he considered this alphabetic function a late, vulgar development, unrelated to their symbolic use in the obelisk inscriptions. See *OP*, 126–30, 149; *OA* III, 42ff.

preme deity *Hemphta*, or *Deus maximus*, and to offer other Egypto-Coptic epithets for God, which he translated as "expressing all things by the word of its power," "eternal vivifier of all things," and "cause of causes" or "universal soul of the world."[13] In a very loose way, these ideas inform the description of the "triform divinity" at the beginning of Kircher's translation.

Next, Kircher examined the circle, wings, and snake individually. The circle, numbered "1" in the diagram, corresponded to the part of the translation that read, "first mind, motor of all things." In support of this translation Kircher adduced various ancient testimonies to prove that circles symbolized "the eternal, immense, infinite, pure and simple essence of God, which Trismegistus sometimes calls the Father, sometimes first Mind, and sometimes supreme Intellect." First he cited the famous Hermetic maxim, "God is an intellectual circle whose center is everywhere and circumference nowhere." Then he quoted from the pseudo-Aristotelian "Egyptian" *Theology*: "Intellect (sc. the first intellect) is the center of a circle, which contains angles, sides, lines, surfaces . . . but is itself indivisible, immeasurable." Then he quoted another Hermetic text, the *Pimander*, which stated, "God is driven in circles," and "Mind and reason fulfilling itself is free from every bodily mass, impassible, intangible, assisting itself, purging itself, and serving everything, its rays are good, principal light, and the first form of souls." Kircher followed up with a little Plato, "God always turns circles," before concluding with Eusebius, who stated that God and all divine things were represented by a circle. For further information he remanded the reader to the entry on circles and spheres in the hieroglyphic dictionary in book 4 of *Pamphilian Obelisk*. In this manner Kircher demonstrated his interpretation of the circle as "first mind, motor of all things." He claimed to have shown not only that the ancients understood circles in this way, but also that this symbolism was rooted in the nature of things, and not mere convention. "By a circle or sphere," he wrote, "appropriately, sensibly, and most properly . . . is expressed the eternal essence of God, unconfined by any boundaries, the first form of the divinity, to use the Egyptian expression."[14]

Kircher then considered the serpent and wings in a similar fashion. In the first case he cited Horapollo and the ancient authors Epeeis and Pherecydes (both preserved by Eusebius) as well as the Bible to support his interpretation of the serpent (numbered 3 in the diagram) as representing the second part of the Egyptian Trinity, translated as "second Mind, craftsman."

13. *OP*, 398.
14. *OP*, 399–401.

Then he established that the wings (numbered 2) represented the third part of the divinity, "pantamorphic spirit," which the Platonists called the universal spirit or world soul, by citing an Arabic text, Eusebius, Clement and Plutarch, as well as another Coptic etymology.[15] Having dispatched the first hierogram, he subjected the rest of the inscription to similar treatment.

In each case, Kircher's method was to pile up quotations from authorities about the symbolic significance of hieroglyphic figures which supposedly supported his interpretation. By his own reckoning, this proved that the first hierogram represented the triune God, the second represented "genies" or intelligences, the third represented the heavenly bodies, and the fourth represented the terrestrial world. Kircher's translation of the obelisk thus described a Neoplatonic cosmos, in which the power of the "triform divinity" manifested itself in the three realms of creation.

Kircher's evidence was, to be generous, uneven. For example, his quotations made a plausible case for the idea that the circle was associated with divinity among the ancients in general and the Egyptians in particular (inasmuch as the *Pimander*—i.e., the *Corpus Hermeticum*—and the pseudo-Aristotelian *Theology* were taken as documents of Egyptian wisdom). But they offered little to support Kircher's interpretation of the circle as specifically representing the first person of the Egyptian Trinity. Although two of his quotations associated the circle with intellect, this was the result of fudging: He added a gloss, inserting the word "first" before "intellect" in the passage from pseudo-Aristotle, and he rendered the Hermetic maxim that commonly reads "God is an intelligible sphere" as "God is an intellectual circle." Two quotations almost seemed to clinch his interpretation, but these were highly suspect. First he adduced the Arabic words of Abenephius:

> When they wish to indicate the three divine virtues or properties, they draw a winged circle from which a serpent goes out; through the figure of the circle they indicate the incomprehensible, inseparable, eternal nature of God, which has no part in anything with beginning or end; through the figure of the serpent, the universal creative power of God; through the figure of two wings, the power of God in motion, which gives life to everything that is in the world.[16]

---

15. As these examples suggest, Eusebius was one of Kircher's most important sources of symbolic lore. Eusebius's *Preparation for the Gospel* was one of the foundational texts for the Christian interpretation of the *prisca theologia*. See Walker, *Ancient Theology* (1972).

16. *OP*, 403.

Then he quoted the ancient Phoenician author Sanchuniathon:

> Jupiter is a winged sphere, from it a serpent is brought forth; the circle proclaims the divine nature without beginning and end, the serpent proclaims its word, which animates and fructifies the world, its wings the spirit of God, which vivifies the world by its motion.[17]

As discussed in chapter 2, the existence of Abenephius cannot be confirmed independently of Kircher's testimony, and the passage in question may well have been altered or fabricated by him. Sanchuniathon was a genuine ancient author, whose work survives only in fragments. But he is not known to have said what Kircher said he did.[18] As Ralph Cudworth wryly commented about these two citations, "How far credit is to be given to this, we leave others to judge."[19]

In general, Kircher's evidence that the ancient Egyptians held the metaphysical and cosmological views that he read in the inscriptions was stronger than his evidence that the symbols before him held such meanings. But, for the sake of argument, let us grant that he made a convincing case about the meaning of each symbol in the inscription. This left him, in effect, with a series of nouns: first mind, guardian genies, fiery nature, and so forth. Up to this point, Kircher was following standard Renaissance operating procedure for interpreting symbols. Kircher's audacity was to go to the next level by translating large groups of hieroglyphs into paragraphs of Latin prose. What warrant, if any, did he have to string together his jumble of abstract concepts by imposing grammar, playing with syntax, and inserting linking words in order to produce the translations that he called "ideal readings?"

---

17. *OP*, 403.

18. The passage in the surviving fragments of Sanchuniathon that most resembles Kircher's quotation reads: "The Egyptians still portray the cosmos according to this same notion. They draw an encompassing sphere, misty and fiery, and a hawk-shaped snake dividing the middle.—The entire device is rather like our letter Theta.—They indicate that the circle is the cosmos, and they signify that the snake in the middle holding it together is the Good Demon. [¶] Also the magus Zoroaster, in his sacred collection of Persian lore, says just this: 'The one who has the head of a hawk is god. He is the first, imperishable, everlasting, unbegotten, undivided, incomparable, the director of everything beautiful, the one who cannot be bribed, the best of the good, the wisest of the wise. He is also the father of order and justice, self-taught, and without artifice and perfect and wise and he alone discovered the sacred nature.'" Philo of Byblos, *Phoenician History* (1981), 67. Kircher claimed to possess a fragment of Sanchuniathon from a manuscript passed on to him by Peiresc, but it is unlikely that it contained information that has not come down to Sanchuniathon's modern editors. See *OP*, 111–12.

19. Cudworth, *True Intellectual System* (1743), 317.

EXPRESSING THE INEXPRESSIBLE

To ancient Neoplatonist philosophers, Egyptian hieroglyphs seemed to offer an instantiation of the notion of a "nondiscursive," ideal language that represented things by symbols corresponding to their essential natures.[20] The authoritative source for this idea was Plotinus, who wrote,

> The wise men of Egypt ... when they wished to signify something wisely, did not use the forms of letters which follow the order of words and propositions and imitate sounds and the enunciations of philosophical statements, but by drawing images and inscribing in their temples one particular image of each particular thing, they manifested the non-discursiveness of the intelligible world, that is, that every image is a kind of knowledge and wisdom and is a subject of statements, all together in one, and not discourse or deliberations.[21]

According to this influential Greek misunderstanding, the hieroglyphs were not phonetic and in no way reproduced normal language, which obeyed rules of grammar and syntax and was restricted by the limits of discursive reason. Instead, as signs directly connected to the ideal realm, they required a symbolic reading. As Marsilio Ficino, glossing Plotinus, wrote:

> When the Egyptian priests wished to signify divine mysteries, they did not use the small characters of script, but the whole images of plants, trees or animals; for God has knowledge of things not by way of multiple thought but like the pure and firm shape of the thing itself.
>
> Your thoughts about time are multiple and shifting, when you say that time is swift or that, by a kind of turning movement, it links the beginning again to the end, that it teaches prudence and that it brings things and carries them away again. But the Egyptian can comprehend the whole of this discourse in one firm image when he paints a winged serpent with its tail in its mouth, and so with the other images which Horus [i.e., Horapollo] described.[22]

From this commonplace about the hieroglyphs' nondiscursive nature,

---

20. Rappe, *Reading Neoplatonism* (2000).
21. Plotinus, *Enneads* (1966–1988), 5.8.6; vol. 5, 256–57.
22. Ficino, *Opera* (1962), vol. 2, 1768; translation from Gombrich, *"Icones Symbolicae"* (1985), vol. 2, 1768. See also Giehlow, "Hieroglyphenkunde" (1915), 23.

Kircher drew an original conclusion with major methodological consequences. When we discuss the interpretation of the obelisk, he wrote in *Pamphilian Obelisk*,

> let no one imagine that this hieroglyphic writing is accomplished in the way in which alphabetic writing is accomplished, produced in the manner employed by us with syllables and words. Rather, since this writing is symbolic and pertains to ideal concepts, hence it also needs to be presented through an ideal reading (*lectio idealis*). In reading them, to quote Iamblichus, "put aside words and receive the meaning."[23]

As we have seen, however, far from putting aside words, his "ideal readings" consisted of dense paragraphs of scholastic Latin mumbo jumbo. Kircher had an explanation: The fact that the hieroglyphs were nondiscursive meant, paradoxically, that they could only be apprehended by the mediation of normal language. The hieroglyphs, he wrote in *Coptic Forerunner*, are a kind of writing

> much more excellent, sublime, and nearer to abstract thoughts, by which the whole reasoning and conception of the highest things, or some remarkable mystery hiding in the bosom of nature or divinity, is presented to the wise man in a single view with an appropriate, skillful connection of symbols. Therefore in writings of this kind, the attributes of speech . . . having been abandoned, it is necessary to be led from the external visible image to the hidden forms of things, and from the sensible object to the idea of the intelligible, in the manner of that common saying of the Kabbalists, "When I found a pomegranate, I ate the seeds and threw away the rind."
>
> Since, nevertheless, the hieroglyphic mental concept of the things depicted can hardly be grasped without conceiving of things indicated by names, words, and other parts of speech, on account of the dependence of the formal concept on the object or sensible things, a certain [kind of] reading was established appointing names and words for things signified through symbols.[24]

Here, Kircher justified his method by arguing that because the hieroglyphs lacked grammar and syntax (i.e., because they were nondiscursive),

23. *OP*, 398.
24. *PC*, 260–61.

and because they could only be apprehended if they were expressed with grammar and syntax (i.e., discursively), they were *supposed* to be translated in the manner of his "ideal readings"—that is, by imposing grammar and syntax. While other early modern thinkers shared the notion that hieroglyphs were nondiscursive symbols, no one else seems to have drawn Kircher's conclusion or produced similar interpretations of hieroglyphic inscriptions. The original impetus behind the Neoplatonic symbol was precisely to express that which could *not* be mediated by ordinary language (since it could not be apprehended by discursive reason) but which might be understood by a different, intuitive faculty. The contradiction involved in translating an ideal symbol into ordinary language might be minimized either by restricting oneself to interpreting a single symbol—so that the sense of apprehending "all at once" was less compromised—or by describing the translation as an imperfect reflection of ideal reality. But those were not limitations that Kircher would accept.

EGYPTOLOGY BY THE COMBINATORY METHOD

To Kircher, then, the fact that the hieroglyphs were "nondiscursive" provided the license to link the inscribed symbols with maximum flexibility in order to produce translations that said what he knew they should. But how did he know what they should say? According to Kircher, the "Egyptian Oedipus" needed to be familiar with three things in order to decipher the hieroglyphs: the doctrine of the ancient Egyptians, the inscriptions on the surviving obelisks, and the method of combining one with the other.[25] We have seen in previous chapters that Kircher was working within well-trodden territory in believing that esoteric texts like the *Corpus Hermeticum*, the Chaldean Oracles, and the Kabbalah preserved the teachings of the ancient Egyptians and other *prisci theologi*. On this basis he felt warranted to assume a priori that the Pamphilian obelisk recorded something like the doctrines of Neoplatonist occult philosophy. But Kircher assumed an even closer connection between esoteric literature and Egyptian inscriptions, on many occasions treating hieroglyphic inscriptions as paraphrases of specific texts. In chapter 4, for example, we saw how he believed that the Bembine Table paraphrased the Chaldean Oracles.

Kircher's belief that he could explain Egyptian inscriptions by associating them with extant texts in Greek, Hebrew, or Arabic followed the example of the early modern emblem tradition, which had taught him that

25. *OA* III, 552.

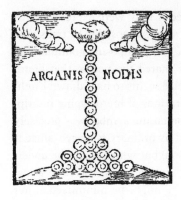

Fig. 36. The *impresa* of the Parthenian Academy of the Collegio Romano, depicting a chain of metal rings hanging from a lodestone, accompanied by the motto, "By secret knots." Athanasius Kircher, *Oedipus Aegyptiacus* (Rome: 1652–54), vol. 3, 560. Courtesy of Stanford University Libraries.

symbols could only be understood in conjunction with words. The type of emblem called an *impresa*, or device, had the most influence on Kircher's thinking, serving in effect as a template for his hieroglyphic interpretations. An *impresa* consists of an enigmatic picture and an accompanying motto, and its meaning is a function of their relationship.[26] Kircher discussed *imprese* in the section of *Egyptian Oedipus* devoted to "*symbolica*," where he wrote: "An *impresa* necessarily must be composed of a figure and motto, since the author's meaning cannot by any means be properly expressed by one without the other."[27]

For Kircher, the difficulty posed by the hieroglyphic inscriptions was that the "mottoes," so to speak, had been lost. But he believed that he had found them in various esoteric texts and especially in Oriental literature, which he interpreted in the framework of occult philosophy. He made this point explicitly in the final section of *Egyptian Oedipus*, where he sought to defend his method. After presenting arguments intended to show that the hieroglyphs were symbolic and not alphabetic, he compared them to *imprese*, whose meaning resided in the conjunction of image and text. By way of example, he first described the *impresa* of the Parthenian Academy of the Collegio Romano, which consisted of an image of a chain of rings hanging from a lodestone and the motto "*arcanis nodis*," or "by secret knots" (fig. 36). After showing how its meaning could be teased out by considering the symbol and motto in tandem, he declared: "Behold how much meaning and

---

26. Strictly speaking, *imprese* and emblems are distinct forms. The principal distinction relevant here is that an *impresa* or device has two elements, while an emblem has three (in addition to the motto and image, an emblem has an accompanying explanatory text, often in poetic form).

27. *OA* II.1, 11.

contemplation hide under only one device. It happens in exactly the same way, I say, in the hieroglyphic doctrine, in which the subject and argument of an entire philosophical observation is present through abstract concepts in but a single symbol."[28]

Kircher followed this remark with two Egyptian examples. The first was a representation of Osiris, which he analyzed symbolically in a similar manner. The second was the portion of the Pamphilian obelisk that we have been considering as an example of Kircher's method. He again reproduced the inscription of the top of the obelisk's southern face, with the four sections now labeled, A, B, C, and D, alongside quotations from the pseudo-Aristotelian *Theology*, which he considered its verbal counterpart (fig. 37). After citing the *Theology* on the symbolic nature of hieroglyphic writing, he quoted its description of Egyptian wisdom:

> "For that reason the Sages indicate with these signs how God differs from other things, since the world and time exist under him; all the more so since God is the author of causes, and since he created all things for their own unique reasons, and since in them he illuminates intelligences; through these he illuminates the middle intellect, but through the intellect he illuminates the middle universal soul and heaven, and through the middle soul he illuminates the nature of things subject to generation and corruption."

[Having quoted pseudo-Aristotle, Kircher continues:] Who does not see that this perfectly agrees with the present diagram? In it the globe represents the symbol of the divine mind and the triform Divinity, which, being near the vertex of the pyramid, declares the diffusion of the divine mind, as if from a kind of center, into each of the Worlds in order: first into the Intellectual or Angelic World, as figure B teaches; then into the Sidereal World, C, represented by the Hawk, symbol of the Solar Divinity; and finally into the Material World, D, of which the Cow is the figure. In other words, the Archetypal and Supramundane Osiris by a marvelous agreement links together the divine influences communicated to Osiris *Pantamorphos*, chorus leader of the Intellectual world, and then to solar Osiris, soul of the Sidereal World, and then to Material Osiris, protector of the Elementary World.[29]

Just as in an *impresa*, Kircher declared, the hieroglyphic doctrine con-

---

28. *OA* III, 560–61.
29. *OA* III, 561–62.

CAP. I.   562   OEDIPI ÆGYPTIACI THEAT. HIEROGL.

Ægyptiorum sapientia à sacra Scriptura commendatur.

cant itaque hisce Sapientes, quomodo Deus differt ab alijs, quòd Mundus
A & tempus existant sub eo; ampliùs quòd Deus est author causarum, quòdque creauit singula pro ratione singulorum : & quòd in ijs illuminat Intelligentias, per eas autem medium intellectum, per intellectum verò mediam animam vniuersalem, cæle-
B stemque, & per animam mediam illustrat naturam generabilium corruptibiliumque. Quis non videt, hæc aptissimè huic præsenti Iconismo congruere? Vbi globus diuinæ mentis ac triformis Numinis symbolum exhibet, quod pyramidij vertici proximum,
C ostendit diuinæ Mentis veluti ex centro quodam in omnes Mundorum series ordine consequentes effusionem, primò in Intellectualem seu Angelicum, vti figuræ B docent; deinde in Sidereum C, quem Accipiter Solaris Numinis symbolum
D exprimit ; ac tandem in Hylæum D, cuius Bos typus est, videlicet, Osiris Archetypus & Supramundanus, in Osirin Pantamorphum, Intellectualis Mundi choragum, & hic in Osirin Solarem Siderei Mundi animam, & hic in Hylæum Osirin, Elementaris Mundi præsidem, mirâ quâdam analogiâ diuinorum influxuum communicationem continuat.

Ex quibus, ni fallor, luculentissimè patet, nihil aliud hac interpretatione, quàm quod dixi, innui posse & debere. Quod adeo ipse Plato, & Aristoteles, eorumq; sectatores non tantùm asseruerunt, sed tanquam dogma sacrosanctû amplexati sunt. Authoritates circa hoc negotium propositas, vide in huius Tomi vestibulo folio secundo, & in Classe Sexta Systematica Mundorum circa finem allatas, ex quibus

Fig. 37. The hieroglyphs from the top of the Pamphilian obelisk (see fig. 35) reprinted beside a passage from pseudo-Aristotle's *Theology*, which Kircher treated as its textual counterpart. The letters A, B, C, and D indicate how the sections of the inscription and text correspond. Athanasius Kircher, *Oedipus Aegyptiacus*, (Rome: 1652–54), vol. 3, 562. Courtesy of Stanford University Libraries.

sisted of *symbola* and *sententiae*, which could not be understood in isolation from one another. But the two parts had been transmitted separately, which was why no one before him could solve their mystery. Based on the model of *imprese*, Kircher's "combinatory method" involved matching hieroglyphic inscriptions with the texts he had collected from different linguistic traditions—a "Herculean labor," as he put it, to which he devoted decades.[30]

The link between hieroglyphs, emblems, and allegory was identified a century ago by Karl Giehlow, who showed the seminal role played by Horapollo's *Hieroglyphica* in the development of the emblem and related symbolic arts.[31] But, as we saw in chapter 1, that tradition had little influence on the translation of genuine Egyptian inscriptions. With Kircher, the influence of hieroglyphs on emblems, in effect, turned full circle. Informed by the Renaissance way of thinking about emblems, Kircher studied the hieroglyphs on surviving Egyptian antiquities and declared that to be understood they had to be collated with texts. The fruits of this method, for better or worse, were his unprecedented translations, rendering entire hieroglyphic inscriptions as coherent Latin texts.

IDEAL READINGS AND MORAL DEMONSTRATIONS

Method does not amount to rigor, however. Kircher was acutely aware that his interpretations were not entirely persuasive. Consequently, he expended vast sums of rhetoric to convince readers of the reliability of his translations. At the outset of *Pamphilian Obelisk* he wrote:

> I foresee that there will be some who, incited by spite, are going to object that the interpretation that we have proposed in the fifth book is the product of our own cleverness, above all because it is established with symbolic figures in such a way that they may easily be forced and twisted to mean whatever anyone wishes; for this is to be believed about the hieroglyphs as it is about all other symbols. And consequently no one can be sure about the truth of any interpretation that is made of this sort.[32]

Unfortunately for Kircher, a symbolic language without grammar or syntax was extremely equivocal. The ambiguity of the hieroglyphic sym-

30. *OA* III, 552–53.
31. Giehlow, "Hieroglyphenkunde" (1915).
32. *OP*, b2^r.

bols was essential to Kircher's "success"—without it, he could not have produced his translations—but at the same time it rendered his interpretations insecure. Their vulnerability was not simply due to the fact that they were incorrect and based on dubious evidence. It was the inevitable outcome of the idea that the hieroglyphs were nondiscursive mystical symbols. Suppose that the hieroglyphs really were a symbolic language of the kind that many ancient Greeks and early modern Europeans imagined. A translation from such a language could never attain the precision of a translation from an alphabetic language, and would remain open to criticism. In the absence of a living tradition and community of readers in consensus about such a language's meaning, doubts could never be easily dispelled. That is why most scholars avoided making the kind of detailed assertions about hieroglyphic inscriptions that Kircher flaunted in *Egyptian Oedipus* and *Pamphilian Obelisk*. Even a sympathetic reader such as Peter Lambeck (quoted at the beginning of this chapter) granted Kircher's translation of the Pamphilian obelisk only a high degree of plausibility, and not the certainty on which Kircher insisted.

Acknowledging that some critics would find his translations unconvincing because they believed that symbols were by nature susceptible to arbitrary interpretations, Kircher set out to show that this complaint was unjust. He would demonstrate, he declared, that his explanation of the hieroglyphs "so agrees with the intention of the ancients that . . . it is evident that through the said symbols they wished to indicate nothing other than what is said [in my interpretation]." To this end, he invoked a traditional classification of knowledge according to degrees of certainty, identifying his conclusions with the standard of proof called "moral demonstration," as distinguished from mathematical and physical proofs. Only mathematical demonstration, he explained, produced knowledge that was absolutely certain, since it investigated abstract quantities and proceeded through principles that were always true and known by the light of nature. Physical demonstration, which arrived at knowledge of the hidden causes of things through experience, did not achieve the certainty of mathematics, since the senses can sometimes be deceived, but it nonetheless counted as genuine knowledge (*scientia*), as Aristotle testified. "Finally," Kircher wrote, "moral demonstration, since it depends upon experience of human actions, indeed produces knowledge [*scientia*], but of the sort which the condition of moral things allows, which is also called human trust [*fides humana*]." Kircher explained that this kind of knowledge depended on the authority of verbal testimony (*authoritas dicentis*), and described it as the human analogue of the authority of God's revelation (*authoritas revelantis Dei*) that produced faith. This human authority, he warned, "must be relied upon unless we wish to provoke and destroy the histories of all past things." In other words, reliable knowledge of past human affairs could be established by appealing to credible textual authorities. Moral demonstration, Kircher declared,

> has its certainty that is established by the unanimous reckoning and testimony of all teachers [*doctores*] and is accepted up to this time; and it includes all secular and ecclesiastical histories, as well as the memorials of ancient authors accepted in different places that have been left to posterity. These [histories and memorials], since they have been accepted by everyone, from so many centuries ago until these times, are worthy of compulsory trust [*fides*]. Without this the acts of all human affairs, as I have said, would dissolve, nothing certain could be written or said, and all would be murky and obscured by doubt. I do not speak here of authors and histories of suspect credibility [*fides*], but about those with an excellent reputation who have been granted authority for many centuries.[33]

Kircher here entered into the contested territory of *fides historica*.[34] Although in the realm of human events one could not expect the kind of certainty associated with mathematics, Kircher argued, knowledge of the past was nonetheless possible and it possessed a degree of certainty sufficient to its subject matter. Invoking the standard of moral demonstration, he claimed that his interpretation of ancient Egyptian wisdom was sound because it was based on credible textual authorities. As we have seen in previous chapters, Kircher was relentless, if not scrupulous, in providing textual evidence for his claims. In order to translate the hieroglyphic symbols, he drew on a wide range of ancient and modern authors, and he supported his translations by showing their congruence with texts that supposedly preserved the beliefs of the ancient Egyptians.

But could Kircher rely on his audience to assent to the credibility of his sources? Two of his most important authorities—the Hermetic Corpus (the *Asclepius* and the *Corpus Hermeticum*) and the pseudo-Aristotlean *Theology*—would seem to have been, at the time he wrote, weak foundations to support bold and controversial claims. Decades earlier, in 1614, the Protestant scholar Isaac Casaubon had published his devastating critique showing the *Corpus Hermeticum* to be a pseudonymous production of the early

33. *OP*, b2^{rv}.
34. See Völkel, *Pyrrhonismus historicus* (1987).

Christian era.³⁵ The work that circulated as Aristotle's *Theology* is now known to be a medieval Arabic redaction of parts of Plotinus's *Enneads*. It became known to the West only in the sixteenth century, after a copy found in Damascus was translated into Latin and published in Rome in 1519 with the pregnant subtitle *The Mystical Philosophy of the Egyptians*.³⁶ The pseudo-Aristotelian *Theology* never received a single blow as devastating as Casaubon's against the Hermetica. But from the moment of its publication the work's authenticity was subjected to mounting criticism, and, despite the apologetic efforts of authors like Patrizi, by the seventeenth century most European scholars had come to consider it spurious.³⁷

Kircher acknowledged these doubts. In the introduction to *Pamphilian Obelisk*, after cataloguing the many authorities on which he based his interpretation—so that "no one can doubt the truth of the interpretation unless they are unbelieving and obstinate"—he admitted that there were two counts on which critics might challenge him: the Hermetic Corpus and the *Theology*.³⁸ In responding to the criticisms against two of his chief sources on ancient Egyptian theology, Kircher adopted three strategies.

First, Kircher attempted to define the debate over the authenticity of the Hermetic Corpus as essentially a question of theology rather than philology. As he saw it, the controversy pivoted on the fact that many things in the Hermetic texts "taste of doctrines of a later age," and that "they too openly discuss the Trinity and the divine Word." According to Kircher, the Hermetica's critics (although he didn't call Casaubon by name, he addressed his arguments throughout this section) thought that the texts must be post-Christian forgeries because they spoke too clearly about the mysteries of the faith to have been written by an ancient pagan.³⁹ In other words, the controversy centered on the question of how much knowledge it was possible for a pre-Christian pagan sage to have achieved. Kircher's analysis was accurate enough as an assessment of the motivation of the Hermetic Corpus's most important critic. As Anthony Grafton has shown, it was precisely Casau-

---

35. Casaubon published his analysis of the Hermetica in a section of his refutation of Cesare Baronio's *Annales ecclesiastici*: Casaubon, *De rebus sacris* (1614). See Grafton, *Defenders of the Text* (1994), chs. 5 and 6; Mulsow, ed., *Ende des Hermetismus* (2002).

36. Kraye, Ryan, and Schmitt, eds., *Pseudo-Aristotle* (1986), especially Kraye, "Pseudo-Aristotelian Theology" (1986).

37. Nonetheless, some editions of Aristotle's complete works continued to include the *Theology*—sometimes identified as spurious—into the seventeenth century, including the 1605 Geneva reprint of Casaubon's edition. See Kraye, "Pseudo-Aristotelian Theology" (1986), esp. 274–75.

38. *OP*, b4ᵛ.

39. *OP*, 36.

bon's discomfort with the notion that an ancient pagan could have been privy to the deepest religious mysteries that inspired his critique.[40] Accordingly, Kircher sought to counter the critics by refuting the notion that a pagan living before the time of Moses could not have possessed even imperfect knowledge of Christian mysteries. Having shown that Hermes Trismegistus lived in the time of Abraham and received many teachings that came from the first patriarchs, Kircher declared, "It is not surprising that he so nearly arrived at the mysteries of orthodox theology."[41]

If accepted, Kircher's argument would have disarmed some of the charges of plagiarism and anachronism that had been levied against the Hermetic Corpus on the basis of parallels to the Old and New Testaments. But other anomalies remained, such as the mention in *Corpus Hermeticum* 18 of the Greek sculptor Phidias, who lived in the fifth century BCE, long after the supposed time of Hermes Trismegistus. Kircher's second defensive strategy was to challenge the significance of inconsistencies and anachronisms that could not be removed by a more generous assessment of the limits of pagan philosophizing. Regarding the anachronistic Greek philosophical language of texts supposedly composed in the Egyptian language long before the development of such terminology, Kircher repeated the old rationale, from Iamblichus, that the texts had been translated from Egyptian into Greek by a later author familiar with Greek philosophy.[42] Kircher did not accept the claim that the Hermetic Corpus had been edited in the Christian era by Gnostics. But even if it had been, he argued, it would not logically follow that it did not contain genuine wisdom that originated with Hermes Trismegistus.[43]

Kircher's last point was valid. Casaubon was right that the Hermetic Corpus was composed in Greek in the early Christian era; but, as recent scholarship has argued, this does not mean that it does not preserve Egyptian as well as Greek traditions.[44] Casaubon's philology was constrained by the conviction, not shared by Kircher, that the vector of influence between Judeo-Christian and pagan wisdom pointed only in one direction. Thus, he considered the resemblance of Hermetic teachings to ones in the New Testament as evidence of inauthenticity, not only because those passages used language characteristic of the early Christian era but also because a pagan text that spoke accurately about such matters must, ipso facto, derive from

---

40. Grafton, *Defenders of the Text* (1994), chs. 5 and 6, esp. 149–51, 167–71.
41. *OP*, 36.
42. *OP*, c2^r. See Iamblichus, *On the Mysteries* (2003), 8.4; pp. 314–16.
43. *OP*, c2^r.
44. See Mahé, *Hermès* (1978); Fowden, *Egyptian Hermes* (1986).

Jewish or Christian sources. On this basis, Casaubon considered correspondences between the Hermetica and Genesis as evidence of plagiarism, although in this case he could not support the point with the argument from linguistic anachronism.

Kircher used similar arguments about the ambiguous implications of anachronism to defend the authenticity of the pseudo-Aristotelian *Theology*. Following Francesco Patrizi, he considered the book to represent the secret oral teaching of Plato, committed to ink by his prodigal disciple, the elderly Aristotle, and reflecting ancient Egyptian wisdom. Regarding the anachronisms in the text that appeared to contradict such an attribution, Kircher wrote: "Granting but not conceding that some anachronisms occur in the said book; nevertheless, I say that it can by no means be concluded from this that it is not a book by Aristotle. For anachronisms of this sort are constantly added by later translators . . . either in order to remove obscurity in the text or due to the inspiration of [the translator's] own imagination."

As an example of the complications presented by ancient texts that did not survive in their original form, Kircher referred to the fact that ancient treatises were originally written in continuous text; only much later did editors divide them into books and chapters. But, having momentarily donned the cloak of the skeptic, Kircher reverted to his more characteristic stance. "Therefore," he wrote, "if anachronisms appear in ancient memorials, they are to be ascribed not to the original author but to his annotators and translators."[45] Kircher's comment about the entanglement of an author's words with those of later commentators suggests a textual critic's sensitivity to the problematic accretions that formed around texts as they were copied and passed on over the centuries. But the indiscriminate and unwarranted conclusion that he drew from this observation undermined the very possibility of textual criticism. If every anachronism was to be attributed to later editors, how could any work ever be recognized as spurious by textual analysis?

This brings us to Kircher's third and dominant strategy: the appeal to canonical authorities. His attempt to cast the controversy over the Hermetica as a disagreement about the limits of pagan wisdom evaded the philological component of Casaubon's argument. His caveats about the conclusions that might be drawn from the presence of anachronistic elements in ancient texts were an attempt to turn the philologist's skepticism against him. But Kircher's doctrine of "moral demonstration" was an outright rejection of the entire enterprise of critical philology. Ruling out the possibility of de-

---

45. *OP*, c2^v.

termining the authenticity of evidence through textual criticism, Kircher endorsed an alternative criterion: the compulsive force of the testimony of established authorities. Ultimately, fine points concerning textual inconsistencies and arguments about the limits of pagan wisdom were irrelevant: the Hermetic treatises were proven to be genuine works by Hermes Trismegistus because the testimony of the doctors of the church said so. Since Justin Martyr cited the *Asclepius* and the *Corpus Hermeticum* and called Hermes Trismegistus "most ancient" (*vetustissimus*) in the middle of the second century, Kircher argued, these works could by no means have been written in the period after Christ. He then quoted the corroborating testimony of Lactantius and cited the further confirmation of Origen, Tertullian, Clement, and Augustine. "Thus," he affirmed,

> it is clear that the two books of Hermes, *Pimander* and *Asclepius*, were always held in the highest esteem not only by profane authors, but also by the holy fathers, who, bolstered by the authority of those time-honored [books], strove to establish their opinion about God being one and three . . . but they would not have done this with such zeal if they did not know that they were very ancient and the true and genuine issue of Hermes . . . .[46]

Instead of dealing with Casaubon on his own terms, Kircher identified his method of textual criticism as a form of corrosive skepticism that threatened to do away with all received knowledge hitherto held dear:

> If any obstinate individuals persist in attacking [Hermes Trismegistus's authorship of the Hermetic Corpus], I can counter them with nothing else but that, with the same rashness with which they reject the Hermetic books, regardless of right or wrong, I could attack and reject the Pythagoreans, Platonists, and the writings of all the ancients, indeed Holy Scripture itself, and the books of all good authors. This presumption and hateful audacity is intolerable to both God and men.[47]

THE DIVORCE OF MERCURY AND PHILOLOGY

Kircher's insistence on obedience to ancient authors no doubt seems like weak ground on which to challenge Casaubon's philological claims. But his

46. *OP*, 36–38.
47. *OP*, 44.

position should be given its due. The notions of tradition and authority on which he based his definition of moral demonstration lay at the very heart of the ideology of the Catholic Church, which claimed spiritual authority on the basis of a continuous transmission of doctrine, vouchsafed by the testimony of canonical authors. Kircher attempted to apply this standard to a pagan rather than a Christian corpus, but the basic principle was the same, and in both cases the most reliable witnesses were identical: the doctors of the church. Similar claims made by subsequent Catholic authors circulated into the eighteenth century. In his *Demonstration of the Gospel*, Pierre Daniel Huet (1630–1721) asserted, "Every book is authentic which has been believed authentic by all people around that time and continuously since then.... If anyone will have called this axiom into question he will have nothing certain in letters."[48] What is remarkable is that whereas Huet resorted to such arguments in order, as he saw it, to defend Holy Scripture from skepticism in the wake of Spinoza, Kircher did so merely to defend his interpretation of the hieroglyphs. The comparison speaks volumes about his vanity and priorities. Kircher opposed the critics of his antiquarian theses with the fervor one is accustomed to see in pious authors who believed that the very foundation of religion was in jeopardy.

Invoking the standard of "moral demonstration" to defend the reliability of his interpretations was superficially plausible. The distinction between knowledge of matters relating to human nature (so-called "moral" affairs), of which it was only possible to obtain probable knowledge, and the certain knowledge produced by mathematical demonstration went back to Aristotle and was a feature of Thomist epistemology.[49] The threefold epistemology referred to by Kircher enjoyed wide currency in the seventeenth century, as scientific and philosophical reformers promulgated new definitions of "moral certainty" at the same time that Neo-Scholastic philosophers continued to hone the traditional Thomist one.[50]

When Kircher used this vocabulary to defend his translation of the hieroglyphs, he was attempting to harness a classic defense of knowledge based on human testimony against skeptical attacks. He would have his

---

48. Huet, *Demonstratio Evangelica* (1722; first edition, 1679); quotation from Shelford, "Thinking Geometrically" (2002), 615.

49. Aristotle, *Nicomachaean Ethics* 1094 b13; Thomas Aquinas, *Summa Theologica* 1a 2ae. 96, 1. See *Oxford English Dictionary*, 2nd ed., s.v. "moral, a.," 7.

50. Shapiro, *Probability and Certainty* (1983), esp. 31–32. Kircher's terminology likely follows that of the Spanish Jesuit Francisco Suàrez, who distinguished metaphysical, physical, and moral truths and also spoke of *fides humana* in this context. See Suárez, *Disputationes metaphysicae* (1965), vol. 2, p. 60 (XXIX, 3, 36–37).

reader believe that the doubts that might be raised about his conclusions were those that could be raised about any historical claims. Thus, to reject Kircher's interpretation of the hieroglyphs because it lacked mathematical certainty would be tantamount to rejecting all knowledge that depended on human testimony. But, despite Kircher's invocation of the language of a well-established defense of historical knowledge and probabilistic truth, his idiosyncratic definition of moral demonstration rode roughshod over the distinctions and criteria that normally constituted the substance of that defense. In reality, the weaknesses that beset Kircher's sources had nothing to do with the distinction between mathematical and moral demonstration. By commonly accepted standards of moral demonstration and *fides historica* they were wanting.

The rhetorical-philological tradition of historical proof—rooted in Aristotle and Quintillian and developed to new levels of sophistication by early modern pioneers of textual criticism like Lorenzo Valla—was based on methods for discriminating authentic and spurious texts, above all, by means of internal criteria such as verisimilitude, self-contradiction, and anachronism.[51] Antiquaries, too, did their part to develop a more solid science of the past by emphasizing the importance of primary sources and by adding nontextual evidence to the historical data pool. These hard-won developments provided the study of the past with critical foundations intended to withstand the skeptical challenge that had become so fearsome a feature of the intellectual environment by the seventeenth century.[52] Kircher's effort to disarm one of the key weapons of textual criticism, anachronism, did not altogether lack insight. But it ignored the cumulative persuasive power that comes from the convergence of multiple arguments. An isolated example of anachronism could be explained away, but the combined force of numerous circumstantial items allowed erudite scholars to establish the authenticity of historical evidence to a high degree of probability without recourse to unquestionable authorities.

In order to defend texts like the Hermetic Corpus, Kircher, though steeped in antiquarian veneration for documentary evidence, did away with such standards, reverting to a resolutely anticritical obedience to authorities. Normally, notions of moral demonstration, including the traditional scholastic version, presupposed the possibility of discriminating reliable and unreliable claims within the sphere of human events by rational criteria. But Kircher's definition of moral demonstration implied a rejection

---

51. See Ginzburg, *History, Rhetoric, and Proof* (1999), esp. chs. 1, 2.
52. Popkin, *History of Scepticism* (1979).

of that possibility. Dismissing the "mitigated skepticism" that characterized the erudite traditions of antiquarianism and critical philology, he espoused what might be called a "fideist" approach to historical evidence.[53] Just as he saw the *authoritas dicentis* that produces moral demonstration as analogous to the *authoritas revelantis* that produces religious faith, his historical fideism implied a kind of suprarational trust in privileged human authorities. In effect, he endorsed the most devastating conclusions of the Pyrrhonists regarding the critical evaluation of truth and declared that the only way to escape their nihilistic implications was to place faith in the authority of tradition.

Or at least this is the position that Kircher staked out in key passages of *Pamphilian Obelisk* and *Egyptian Oedipus*, where it answered his need to defend texts like the Hermetic Corpus and the pseudo-Aristotelian *Theology*. But Kircher was not a thinker who proceeded from first principles, and he repeatedly demonstrated his readiness to adopt whatever argument was expedient in a given situation, even if it contradicted a position he took elsewhere. Kircher did not set out impartially to assess the authenticity of these texts. Rather, he set out to defend them as best he could, by whatever means available, because he wanted their support for his interpretation of the hieroglyphs. Kircher not only rejected the criteria of critical philology in favor of a standard based on necessary obedience to authorities; he did not even apply his own criteria consistently. When it served his purpose, he would reject as spurious a treatise enshrined with the mystique of ancient Hermetic wisdom. Thus, in the section of *Egyptian Oedipus* devoted to "hieroglyphic alchemy," he attacked the attribution of the famed Emerald Tablet to Hermes Trismegistus, using arguments similar to those whose validity he rejected when they were deployed by others against the Hermetic Corpus and pseudo-Aristotle's *Theology*. Although attributed to Trismegistus, the Emerald Tablet was not especially useful to Kircher for reconstructing the hieroglyphic doctrine. On the contrary, he associated it with the quest for the philosopher's stone among a school of alchemists whose beliefs he considered delusional and absurd.[54] Consequently, he argued that the failure of any ancient Greek philosophers to mention the Tablet spoke

---

53. "Mitigated skepticism" refers to the position that accepts skeptical arguments against certain knowledge, but allows for reliable knowledge based on probable truths about appearances. "Fideism" acquiesces more completely to the skeptical critique, especially concerning religion, and responds by declaring the need for a supernatural foundation to knowledge: faith in God's revelation and its traditional interpreters. See ibid.

54. The *Emerald Tablet* was the key text for a tradition of Hermeticism associated with alchemy whose history was distinct from the Ficinian Neoplatonist tradition. See Leinkauf, "In-

powerfully against its composition in antiquity. The story of the miraculous discovery of the Emerald Tablet by Alexander the Great in Hermes' tomb, reported in a work by Albertus Magnus that Kircher also rejected as spurious, was nothing but a figment of the *"chimiasters."* Instead, Kircher declared the Tablet's text to be a pastiche forged from genuine Hermetic and Neoplatonist texts, but also containing some "barbarous words" that would never be found in the work of the real Hermes.[55]

On many occasions Kircher demonstrated freedom of thought and sought to identify himself with the *novatores*, even using the antitraditional rhetoric associated with the new science to defend his hieroglyphic studies. In the introduction to *Egyptian Oedipus*, he compared his unprecedented decipherment to the discovery of America, the identification of new heavenly bodies by "the lynx-eyed astronomers of the day," and other examples that showed that "every day many things are discovered that once seemed to surpass the subtlety of human understanding."[56] The mention of lynx-eyed astronomers was a blatant and approving reference to the telescopic observations of Galileo, the most famous member of the *Accademia dei Lincei*. But the language with which Kircher attacked the critics of the Hermetica could have come from the mouths of Galileo's persecutors:

> Certain minds are so formed by nature that they devote themselves to the labor of expunging, weakening, and completely abolishing those very things that have been prized and esteemed by all learned authors in the long course of centuries and which have preserved their authority by most solid teaching without violence until today. In this they seem to have no other goal than completely to abolish a teaching produced by the judgment of so many eminent and serious men, and they proffer to the world, with a truly arrogant and insufferable display, that the writers of all past ages were blind and that they themselves are the only judges [*critici*]. . . . These exertions go beyond the limits of modesty, and prove their arrogant and reckless character. It is astounding how they inflict grave harm on the Republic of Letters, since in this way you may find no sacred or profane memorial that cannot be called into doubt, causing the greatest confusion of minds and inextricable perplexity, and that may not be wrenched up and torn to pieces. Wherefore, I have believed

---

terpretation und Analogie" (2001); Mulsow, "Das schnelle und langsame Ende" (2002); Ebeling, *Secret History* (2007), 89–90.

55. *OA* II.2, 427–28.
56. *OA* I, br^r.

the safest path to be the one well trodden by the course of time, the daily study of the ancients, the persistent labor of the most learned men, and the painstaking investigation of the wisest Fathers.[57]

Kircher was not by nature a reactionary, and although no great textual critic, he could play the game of critical philology when he wanted. But in crucial sections of his hieroglyphic studies where he elaborated an explicit defense of his method—the introduction to *Pamphilian Obelisk* and the conclusion of *Egyptian Oedipus*—he challenged the foundations of critical historical scholarship with the argument from authority, taking refuge in an extreme form of ipse-dixitism. It is worth asking: Why?

Kircher was in a difficult position. He needed to convince his readers of the reliability of evidence that had been convincingly shown to be unreliable. He did the best he could by attempting to frighten his readers with the specter of the Pyrrhonist dragon—asking them to choose between blind obedience to textual authorities and the destruction of all historical knowledge—and by wrapping his claims in the sanctimonious authority of church fathers. Had he been content simply to demonstrate the existence of an ancient tradition of occult wisdom without claiming a definitive solution to the hieroglyphs, he could have avoided many of the assertions that, as he well knew, rendered him vulnerable. (For example, he could have made arguments like those of Ralph Cudworth, discussed in the next chapter.) But Kircher was determined to decipher the Egyptian inscriptions, no matter what. If critical philology stood in the way, critical philology had to go.

Kircher's acceptance of dubious sources, especially the *Corpus Hermeticum*, has received more attention than the translation methods examined in the first part of this chapter.[58] But his need to defend the credibility of such sources was in large measure the product of those methods and their underlying assumptions about symbolic communication. Kircher needed the support of external textual authorities so badly because his translations did not defend themselves as would, for example, the correct solution of an alphabetic cipher or Champollion's eventual translation of the hieroglyphs—by virtue of internal rigor and univocality. A translation from a symbolic, nondiscursive language could not have been otherwise. Lacking a Rosetta Stone, had Kircher not believed that the hieroglyphs were ideal

---

57. *OP*, 35–36.
58. See, e.g., Rowland, "Kircher Trismegisto" (2001), 113–21; Yates, *Giordano Bruno* (1964), 416–23; Pastine, *Nascita dell' idolatria* (1978), 14. See also Grafton, *Defenders of the Text* (1994), 159; Assmann, *Moses the Egyptian* (1997), 20.

symbols without grammar or syntax, he would never have gotten anywhere. But, believing that they were, his results were necessarily uncertain, despite his claims to the contrary, because such a language was necessarily multivalent and ambiguous.

At the center of Kircher's project were translations of hieroglyphic inscriptions. These translations were not a secondary matter—no mere "vehicle" to promote a revival of Hermetic philosophy. They were the culmination and raison d'être of *Egyptian Oedipus* and *Pamphilian Obelisk*. The pronounced apologetic strain that runs through both works laid bare Kircher's priorities. Unlike earlier writers on Renaissance occult philosophy whose work Kircher has been seen as having continued, he expended relatively little energy defending the *truth and piety* of ancient pagan (or esoteric Jewish) beliefs and practices. Instead, he devoted his apologetic efforts to defending the *historical accuracy* of his description of those beliefs and practices, which would suffice to establish the credibility of his interpretations.

Kircher's apologetic efforts point to the intellectual context to which his work belonged: seventeenth-century erudition. *Egyptian Oedipus* was an antiquarian and philological enterprise—the recovery of a lost ancient literature and the interpretation of ancient artifacts and inscriptions based on reconstructing ancient religious beliefs and customs. But the highly selective manner in which Kircher implemented erudite methods generated tensions and, ultimately, fissures in his work. It is frequently claimed that Kircher "ignored" Casaubon's critique of the Hermetic Corpus.[59] This chapter has shown that this claim, which encourages the idea that Kircher was out of touch with the scholarship of his time, is badly mistaken: Kircher labored very hard to mount a defense against the arguments of Casaubon and like-minded critics. He did so because he conceived of his own work as a contribution to the field in which those arguments held sway. And yet, by both temperament and necessity, Kircher rejected the rigorous methods of the leading practitioners of his chosen discipline. He did so not out of ignorance or a failure to understand their significance, but as a conscious response to the threat they posed to the success of his undertaking. Again, we are confronted by the central irony of Kircher's enterprise: in order to become the antiquary-hero who would solve the great hieroglyphic enigma, he turned his back on the critical methods that provided the study of the past with solid foundations.

59. For example: Ebeling, *Secret History* (2007), 97; Pastine, *Nascita dell' idolatria* (1978), 95.

CHAPTER EIGHT

Oedipus at Large

> Kircher, bewildered as he was, had yet some ground for his rambles. He fairly followed Antiquity: unluckily indeed, for him, it proved the *ignis fatuus* of Antiquity; so he was ridiculously misled. However he had enough of that fantastic light to secure his credit as a fair writer.
> —Bishop Warburton[1]

"DAVUS SUM, NON OEDIPUS"

"I am Davus, not Oedipus," declared Jean-François Champollion of his initial efforts at hieroglyphic interpretation in 1818.[2] The phrase, taken from a play by Terence, in which the simple slave Davus declared his inability to solve profound riddles, was a commonplace for indicating that one was out of one's depth. Issuing from Champollion, however, it was no confession of debility. Rather, the young Egyptologist was announcing that he was not Kircher, the self-declared Oedipus of Egyptian mysteries, who, as far as Champollion was concerned, had been most assuredly out of his depth. Champollion's disdain is unsurprising. It is noteworthy, however, that in the second decade of the nineteenth century the man who would finally decipher the enigmatic writing of ancient Egypt studied Kircher's century-and-a-half-old volumes and used them as a foil.

The deficiencies of Kircher's hieroglyphic studies had never gone unnoticed. In particular, many early modern critics lambasted the two defects that most stand out to the modern reader: Kircher's reliance on discredited

---

1. Warburton, *Divine Legation of Moses* (1846), vol. 2, 244.
2. From the unpublished *Premier essai d'un Dictionnaire des hiéroglyphes égyptiens* (1818, 1819), quoted by the editor in Champollion, *Dictionnaire égyptien* (2000), i.

sources like the *Corpus Hermeticum* and the fantastic quality of his interpretations of hieroglyphic inscriptions. (As we saw in the previous chapter, these were precisely the areas that Kircher himself realized to be most vulnerable.) And yet one cannot say that Kircher's work was unsuccessful. It was read, cited, and commented on by other scholars with considerable frequency for more than a century.[3] As the *Yverdon Encyclopedia* observed, despite Kircher's reputation as a bit of a crank, copies of *Egyptian Oedipus* were still "expensive and sought-after" more than a hundred years after its publication.[4] In examining the reception of Kircher's hieroglyphic studies, this chapter aims to explain their surprising endurance among scholars who deemed them fundamentally flawed.

To make sense of the *fortuna* of Kircher's hieroglyphic studies, it is necessary to distinguish carefully between those ways in which Kircher followed or flouted the scholarly principles of his time, and to determine when and how those principles changed over the course of the seventeenth and eighteenth centuries. Above all, the prevailing chronology of the fate of occult philosophy, which is based on an overestimation of the effect of Casaubon's dating of the Hermetic Corpus, must be revised. It is easy to laugh along with readers like Hermann Wits, who in 1683 chastised Kircher for "hugging and kissing" the *Corpus Hermeticum* "with faith and love," and dismissed his Trinitarian interpretation of Egyptian hieroglyphs as "beyond belief."[5] But most seventeenth-century critics agreed with Kircher about many other things that, from a post-Enlightenment perspective, seem equally risible. The significant shortcomings that were widely criticized by contemporaries have tended to obscure how much of *Egyptian Oedipus* fit the paradigm of normal scholarship at the time Kircher wrote.

OEDIPAL CONFLICTS AMONG THE COPTICISTS

The reception of Kircher's work on Coptic is instructive because it isolates the technical aspect of his scholarship from the larger issues about pagan wisdom and sacred history that surrounded *Egyptian Oedipus*. In this small domain, the dichotomy of denigration and appreciation was particularly stark. Unquestionably, Kircher's writings on Coptic were deeply flawed, de-

---

3. This is true not only of his hieroglyphic studies but of other works as well, especially Kircher, *Magnes* (1641); Kircher, *Ars magna lucis* (1646); Kircher, *Musurgia universalis* (1650); Kircher, *Mundus subterraneus* (1665); Kircher, *China illustrata* (1667).

4. See above, introduction, n. 35.

5. Witsius, *Aegyptiaca* (1739), 8, 72–76.

spite the assistance he had received from native Arabic speakers and Coptic priests. This was due both to the challenges that he faced as a pioneer and to his limitations as a philologist. But his Coptic-Arabic-Latin lexicon and grammar, *Egyptian Language Restored* (1643), had the supreme virtue of existing. Kircher was not only the first European author to write at length on the language; for more than a century his was the only Coptic dictionary in print. Until the posthumous publication of Marthurin Veyssière de la Croze's vastly superior lexicon in 1775, would-be students of the language relied on Kircher, often to their consternation.[6] Many hours spent annotating Kircher's dictionary so frustrated Louis Picques that he declared the Jesuit ignorant of all languages other than German and a smattering of Latin. Other seventeenth- and eighteenth-century Coptic scholars rendered similar verdicts. Even so, Kircher remained within the spectrum of early modern scholarly practice. David Wilkins was among those who disparaged Kircher's lexicon, but after reading Wilkins's edition of the Coptic New Testament, the estimable La Croze ranked Wilkins's ability below Kircher's. La Croze himself, for all his acumen, once considered Coptic a key to understanding Chinese, a reminder of the relatively rudimentary state of non-European philology at that time.[7] As late as the 1820s, Champollion, despite complaining about Kircher's "habitual charlatanry," nonetheless consulted *Egyptian Language Restored* while working out his epoch-making solution of the hieroglyphs.[8]

In a thorough recounting of the progress of Coptic studies, published in 1808, Étienne Marc Quatremère offered a more balanced view of Kircher. Of *Egyptian Language Restored*, he wrote:

> This useful work, which has greatly contributed to making the Coptic language flourish, was bound to earn its author the recognition of scholars; so it was initially received with universal applause. M. Woide, excellent judge of this subject, speaks very favorably of it. The numerous mistakes that the author committed doubtlessly deserve indulgence, since, traveling an uncleared path, plagued with obstacles, it was impossible

---

6. Grammars played a less important role than dictionaries in the progress of Coptic. The next grammar after Kircher's was published by Blumberg in 1716 and represented only a modest improvement. During these years some scholars consulted the very superior unpublished grammar of Bonjour, as well as La Croze's lexicon, in manuscript. Hamilton, *Copts and the West* (2006), ch. 13.

7. Quatremère, *Recherches* (1808), 50–81; Hamilton, *Copts and the West* (2006), ch. 12.

8. Champollion, *Grammaire égyptienne* (1836), ix; Champollion, *Lettre à M. Dacier* (1822), 36.

for him not to make some errors; but one cannot pardon his arrogance, and the bad faith with which he inserted in his work some words that are not found in the original ... The enthusiasm that Kircher had excited soon cooled down and he justifiedly endured bitter reproaches on the part of scholars who followed in his path.[9]

In a recent study, Alastair Hamilton arrives at a similar verdict. After discussing Kircher's many gross errors, he cautions that they "should not blind us to the immense influence which Kircher had. For many years one scholar after another appealed exclusively to Kircher when discussing Coptic."[10]

Egyptology after Kircher

In his summa of pre-Champollion Egyptology, published in 1797, the Danish scholar Georg Zoega concisely summarized the reception of Kircher's work on the hieroglyphs. "There were learned men who applauded Kircher," he wrote, "and believed that he had explained the signs on the obelisks according to the ideas of the ancient Egyptians. But most dismissed his interpretations as fantasies."[11] Both parts of Zoega's assessment deserve consideration. The Roman antiquary G. B. Casali was among the first to applaud, publishing a treatise on ancient Egyptian religion in 1644 which followed Kircher's interpretation of the hieroglyphs, based on the preview of his method in *Coptic Forerunner*.[12] In *Of Idolatry* (1678) the future archbishop of Canterbury, Thomas Tenison, accepted Kircher's translation of the Flaminian obelisk without expressing reservations, and relied on *Egyptian Oedipus* for insight into Egyptian theology.[13] In the middle of the eighteenth century, the archeologist William Stukeley still took Kircher as a reliable guide to ancient symbolic wisdom. A confidant of Newton (he was the source for the story about the apple) and Fellow of the Royal Society, Stukeley believed that "learned Kircher" had "unlocked the Springs of this kind of Learning," and based his interpretation of ancient Druid architecture on Kircher's Trinitarian explanation of the hieroglyph of the winged globe and serpent (fig. 38).[14]

9. Quatremère, *Recherches* (1808), 52–53.
10. Hamilton, *Copts and the West* (2006), 217.
11. Zoega, *De origine et usu obeliscorum* (1797), 178.
12. Casali, *De veteribus Aegyptiorum ritibus* (1644), 35 and passim.
13. Tenison, *Of Idolatry* (1678), 76 and passim.
14. Stukeley, *Abury* (1743), 9, 92, and Tab. XI; Haycock, *William Stukeley* (2002), 208–11 (quotation at 210).

More commonly, however, Kircher's hieroglyphic interpretations were rejected as "visionary," a pejorative that indicated they were the products of an overactive imagination more than sound scholarship.[15] Those who rejected Kircher's translations in these terms did not necessarily disagree with his premises, however. The idea that hieroglyphs were symbols encoding sacred mysteries remained the dominant theory at least until the early eighteenth century. It was the scope as much as the content of what Kircher claimed to know that met disapproval in a scholarly culture that increasingly valorized intellectual modesty and candid admissions of nescience. Eighteenth-century writers on the hieroglyphs often praised Pignoria, in explicit contrast to Kircher, for acknowledging uncertainty and limiting himself to conjectures.[16] This self-consciously modest attitude found concrete form in the remarkable restoration of the Solarium obelisk carried out by Pius VI in 1792. The project's architect, Giovanni Antinori, forewent the customary practice of replacing missing inscriptions, instead filling gaps in the monument with smooth granite (fig. 39). Antinori explained his purpose as follows: "All that is antique will remain with its surviving hieroglyphs; let that which is modern appear so with bare stone and flat surfaces, without summoning imposture to carve there those symbols it does not understand."[17]

The center of gravity of erudite Egyptology moved away from the meaning of the hieroglyphs in favor of things that could be measured, counted, and otherwise described precisely. For example, more attention was paid to topics such as the engineering aspects of Egyptian monuments and the materials from which they were wrought. While Rome remained the main center of research, in situ descriptions of Egyptian ruins, communicated in travel literature, became increasingly important.[18] None of these dimensions was absent from Kircher's work, but they were subordinate to his reconstruction of the "hieroglyphic doctrine." Following the explosion of archeological research that took place during the eighteenth century, Zoega built on the work of Montfaucon and Winckelmann, who had placed Egyptian remains in the larger context of ancient art, framing his fin-de-siècle

---

15. E.g., Warburton, *Divine Legation of Moses* (1846), 196; "Kircher, Athanase" (1770–75). Cf. Lafitau, *Moeurs des sauvages ameriquains* (1724), vol. 2, 144–45, who wrote that Kircher's explanations "peuvent avoir quelque chose d'Idéal, que les Sçavans pourroient lui disputer."

16. E.g., Montfaucon, *L'Antiquité expliquée* (1719), vol. 4 (tom. 2, part 2), 332.

17. Quoted by Collins, "Obelisks as Artifacts" (2000), 63. See also Curran et al., *Obelisk* (2009), 199–201.

18. Curran et al., *Obelisk* (2009), 205–27.

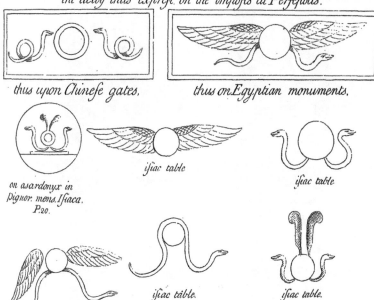

Fig. 38. In his interpretation of Stonehenge and other ancient Druid monuments, the Newtonian antiquary William Stukeley followed Kircher's Trinitarian interpretation of the winged circle and serpent. William Stukeley, *Abury, a Temple of the British Druids* (London: 1743), plate, p. 78. f Arc 855.214*, Houghton Library, Harvard University.

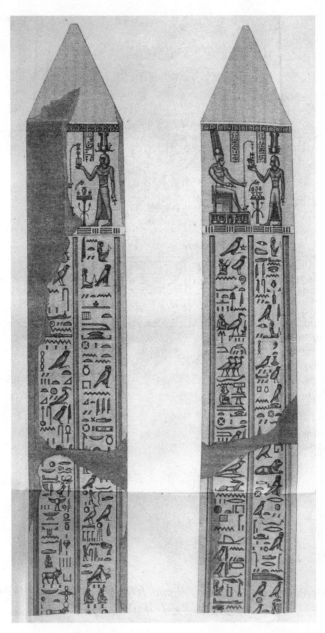

Fig. 39. The Solarium obelisk after restoration by Pius VI in 1792. Instead of replacing missing pieces with conjectural inscriptions, the architect Giovanni Antinori filled the gaps with bare granite, reflecting the eighteenth century's more circumspect approach to hieroglyphs. Georg Zoega, *De origine et usu obeliscorum* (Rome: 1797). Courtesy of Stanford University Libraries.

opus on obelisks in the comparative framework of ancient architectural practices, especially funerary monuments.[19]

The upshot of this change in scholarly attitudes was that no significant attempt to decipher the hieroglyphs took place between Kircher and the efforts of Champollion and Thomas Young at the beginning of the nineteenth century. Investigators of Egyptian antiquities continued to address the hieroglyphic inscriptions, of course, but they mostly confined themselves to general discussions of their probable nature and meaning, rather than hazarding character-by-character interpretations. As Montfaucon put it, "Most sensible persons concur that it is hardly possible to undertake the explanation of these enigmas, without running the risk of running wild."[20] Many eighteenth-century Egyptologists decided that the hieroglyphs would never be deciphered, a pessimistic conclusion that Kircher's optimism no doubt helped to consolidate. Turning their backs on the Oriental literature Kircher had introduced, they declared ancient Greek and Roman authors the only reliable sources about the hieroglyphs. Moreover, among classical authorities they privileged those who spoke most plainly about Egyptian inscriptions over the many others who described them as mystical symbols. Without categorically rejecting the well-attested opinion that ancient Egyptian priests had possessed symbolic wisdom, they avoided imputing such content to most surviving hieroglyphic inscriptions. Zoega, who described Kircher as "more prone to making confident assertions than to examining and judging things diligently," rejected the view that all hieroglyphic inscriptions concerned "theosophy, occult sciences, religious and magical rites." Instead, he praised Angelo Maria Bandini's 1750 treatise, which followed those ancient writers who described obelisk inscriptions as praises for kings, mixed with honors to the gods.[21] As reasonable as such judgments may have been (Bandini turned out to be exactly right), they did not reflect new information about Egypt so much as neoclassical aesthetic preferences and, above all, new assumptions about what was historically plausible that originated outside of Egyptology.[22]

Through all these changes, however, Kircher remained an inevitable reference point for writers on Egypt and the hieroglyphs. Zoega, the leading Egyptologist of the generation before Champollion, who had trained

19. Zoega, *De origine et usu obeliscorum* (1797).
20. Quoted in Iversen, *Myth of Egypt* (1993), 101.
21. Zoega, *De origine et usu obeliscorum* (1797), viii, 178–82; Bandini, *Dell' obelisco* (1750); Curran et al., *Obelisk* (2009), 215–18.
22. See Griggs, "Antiquaries and the Nile Mosaic" (2000), for a similar comparison of Kircher to eighteenth-century antiquaries.

in Heyne's cutting-edge philology seminar at Göttingen, treated Kircher as a significant interlocutor. While his interpretations of the hieroglyphs were usually rejected, they were rarely ignored. Montfaucon, for example, deemed Kircher's interpretation of the Bembine Table as obscure as the table itself, but felt obliged nonetheless to summarize it.[23] Other dimensions of Kircher's studies received praise. Antoine Court de Gébelin, Thomas Maurice, and Daniel Wyttenbach hailed Kircher's diligent collection of physical evidence, which laid the groundwork for future scholars, even as they rejected his conclusions.[24] Montfaucon recycled a number of Kircher's illustrations in his authoritative *Antiquity Explained through its Images* (1719), and even Champollion and Young relied on his "tolerably faithful, though inelegant, representations of Egyptian art."[25]

THE PERSISTENCE OF OCCULT PHILOSOPHY

If Kircher's interpretations of the hieroglyphs were widely if not unanimously found unconvincing, what about the larger historical claims that he set forth in *Egyptian Oedipus*? Here the picture is more fluid and complex. The basic framework of sacred history, which set the ultimate parameters of Kircher's research, went virtually unchallenged before the second half of the

---

23. Montfaucon, *L'Antiquité expliquée* (1719), vol. 4 (tom. 2, part 2), 332, 340–41.
24. Court de Gébelin, *Monde primitif* (1773–81), vol. 3, 395; Maurice, *Indian Antiquities* (1800); 36; Wyttenbach, *Animadversiones* (1821), 160–62.
25. On Montfaucon, see Curran et al., *Obelisk* (2009), 209. Champollion, *Lettre à M. Dacier* (1822), 29. The quotation is from Young, *Account of Some Recent Discoveries* (1823), 2.

Fig. 40. The Jesuit missionary Joachim Bouvet repurposed Kircher's theories by substituting China for Egypt as the fountainhead of pious gentile theology. In this drawing Bouvet has meticulously copied a diagram from *Egyptian Oedipus* (see fig. 41), to which he has added phrases from early Chinese classics, which Figurist missionaries read as evidence of primordial knowledge of the God of the Bible. At the top, surrounding the Hebrew Tetragrammaton, Bouvet's insertions read: "*Huang tian shang di*" (Lord of august heaven above); "*Wei huang shang di*" (Sole august lord above), "*Tian huang da di*" (Great lord, august one of heaven); "*Huang huang tian di*" (August, august, lord of heaven), "*Hao tian shang di*" (Lord of the highest heaven above). Below the central image and name of Jesus, Bouvet has written Yao, the name of an ancient sage, which can be read as three crosses—evoking the Holy Trinity—above the word for origin or mighty. It simultaneously evokes ιαω, the Greek form of Yahweh. (I thank Haun Saussy for explaining the Chinese text.) "Mémoires sur le rapport des anciennes croyances des Chinois avec les traditions bibliques et chrétiennes," BNP NAL ms. 1173. Bibliothèque nationale de France.

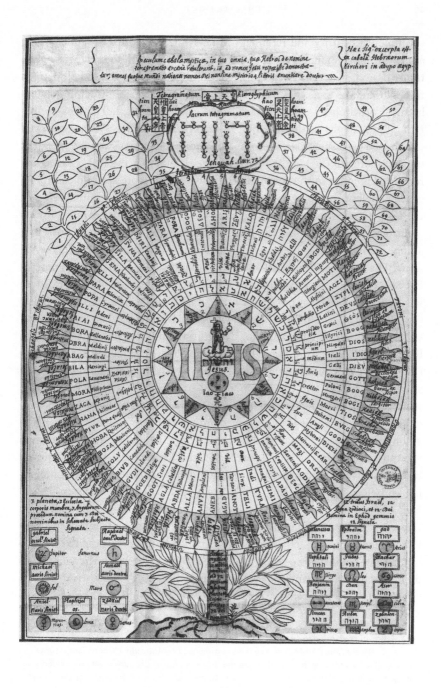

seventeenth century and retained favor with the majority of scholars long afterward. That an inquiry into ancient Egyptian wisdom should focus on the transmission of antediluvian knowledge and genealogies linking all nations to Noah and his sons was fully in accord with historical principles shared by most of Kircher's contemporaries, his critics included. The idea of "perennial philosophy"—that the first human beings, such as Adam, Seth, and Enoch, had been scientists, philosophers and theologians—was the prevalent understanding of the history of knowledge throughout the Renaissance and remained popular in the seventeenth century. The more specific claim of a gentile ancient theology, or *prisca theologia*, had been the dominant view of the sixteenth century, but became more controversial in the seventeenth.[26]

In the Catholic world the ancient theology enjoyed a late flowering among missionary scholars, especially French members of the Society of Jesus. The so-called "Figurists" in the China mission, such as Jean-François Foucquet and Joseph Henri Marie de Prémare, interpreted ancient Chinese wisdom as a manifestation of the *prisca theologia*, which prefigured many Catholic truths. Kircher's influence was especially profound on Joachim Bouvet (1656–1730). A late exemplar of the Jesuit missionary-cum-mandarin, Bouvet applied his impressive linguistic mastery to a radical reinterpretation of ancient Chinese literature. Following Kircher, he claimed that the ancient patriarchs had used numerology to encode their perfect wisdom, which was transmitted to different cultures by the *prisci theologi*. This antediluvian science was best preserved among the Jews in the Kabbalah and among the Chinese in classics like the *I Ching*, which Bouvet claimed was the oldest book in the world, predating the Pentateuch. Bouvet believed he had recovered the true meaning of ancient Chinese literature, including knowledge of Christ and the Trinity, which had been lost for millennia beneath "the shadow of hieroglyphs" (figs. 40 and 41).[27] Figurism may seem like an exotic and isolated florescence at the margins of European intellectual developments; but Bouvet corresponded for ten years with no less a figure than Leibniz, who recognized a deep affinity between the *I Ching*, as described by Bouvet, and his own theories concerning binary arithmetic.[28]

Kircher's influence was also visible in the work of Joseph-François

---

26. Bottin et al., *Models of the History of Philosophy* (1993).
27. Collani, *P. Joachim Bouvet* (1985), quotation at 127; Mungello, *Curious Land* (1985), 300–328. Bouvet's ideas were too radical for publication—especially after the papal condemnation of the Jesuits' synthesis of Christianity and traditional Chinese ceremonies in 1715—and circulated only in manuscript. See also Rowbotham, "Jesuit Figurists" (1956); Walker, *Ancient Theology* (1972), 194–230.
28. Porter, *Ideographia* (2001); Perkins, *Leibniz and China* (2004).

Fig. 41. "The Mystical Tree Planted in the Middle of Paradise for the Salvation of the Seventy-two Nations, Whose Fruit Are the Seventy-two Names of God," from Kircher's treatise "Kabbalah of the Hebrews." It represents Kircher's theory that all the world's peoples possess a divinely inspired four-letter name of God. Athanasius Kircher, *Oedipus Aegyptiacus*, vol. 2, part 1, plate, p. 287. Courtesy of Stanford University Libraries.

Lafitau, whose *The Customs of the American Indians Compared with the Customs of Primitive Times* (1724) was an innovative study of comparative ethnography. Drawing on his firsthand experience as a Jesuit missionary among the Iroquois, Lafitau argued that all pagan religions, ancient and modern, were corruptions of the revealed religion of the first biblical patriarchs. He sought to prove his thesis by demonstrating the vestiges of true religion in paganism, claiming, for example, that traces of the doctrine of the Trinity were visible in the religions of China, Japan, and Egypt. As opposed to Pierre Daniel Huet (1630–1721), who derived all pagan religion and philosophy from Moses, Lafitau followed Kircher in seeking common origins in the first generations of mankind. He interpreted pagan mythology as "symbolic theology" and relied on *Egyptian Oedipus* for insights into the culture of Egypt and other Oriental nations (fig. 42).[29]

Another influential outpost of ancient theologizing in the late seventeenth century was found in Protestant England among the Cambridge Platonists, especially Henry More, who believed in an esoteric "cabbala" passed on from Moses to Greek philosophers like Pythagoras and Plato, and Ralph Cudworth.[30] In *The True Intellectual System of the Universe* (1678), Cudworth relied on the traditional narrative of the *prisca theologia* as he sought to refute "atheists" and demonstrate the consonance of reason and religion by arguing that ancient pagan philosophers had believed in one supreme creator God. Like Kircher, Cudworth believed that Hermes Trismegistus was the founder of Egyptian theology, and he sought to salvage the reputation of the Hermetic treatises in the wake of Casaubon's critique. But Cudworth realized that Casaubon's criticisms could not credibly be dispatched, as Kircher had tried to do, by rejecting the principles of textual criticism on which they rested. On the contrary, he accepted the validity of Casaubon's arguments and, with rhetorical flourish, even offered additional evidence of forgery in *Corpus Hermeticum* 13, which Casaubon had overlooked, although it "seems to be more rankly Christian than any other." Instead of challenging the textual critic's evidence, Cudworth met him on his own ground and called him to task for considering the various books of the *Corpus Hermeticum* as a single work, rather than as a collection of discrete treatises. Cudworth willingly conceded that Casaubon had demonstrated that two of the treatises were bogus, but no more. As for the rest, he acknowledged that various considerations, including linguistic anachronisms, made it unlikely that Hermes Trismegistus himself was their author; but

---

29. Lafitau, *Moeurs des sauvages ameriquains* (1724); Huet, *Demonstratio Evangelica* (1722).
30. On More, see Coudert, *Impact of the Kabbalah* (1999).

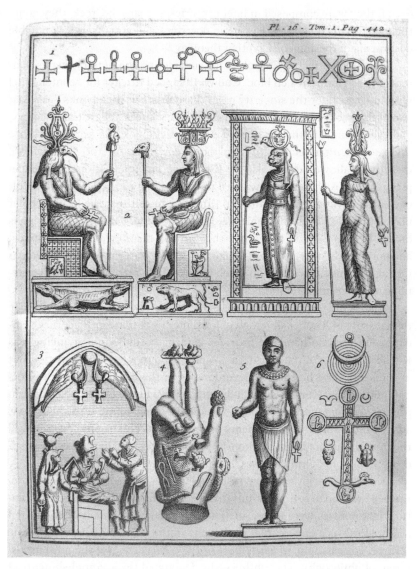

Fig. 42. Evidence of the Holy Cross among ancient pagans, according to Joseph-François Lafitau. Following Kircher, Lafitau sought vestiges of the primeval religion of the first patriarchs among pagan nations. He took some of his textual and visual evidence from *Egyptian Oedipus*, including the Hermetic Cross depicted in the bottom right corner, which Kircher attributed to Hermes Trismegistus. In fact it derived from the "hieroglyphic monad" invented by John Dee in the sixteenth century. Joseph-François Lafitau, *Moeurs des sauvages amériquains comparées aux moeurs des premiers temps* (Paris: 1724), 442. Courtesy of Stanford University Libraries.

even if they were composed in later centuries, perhaps in Greek, they nonetheless contained genuine Hermetic doctrines and could serve as reliable documents of the ancient Egyptian theology.[31]

Just as he corrected Casaubon for considering the heterogeneous contents of the Hermetica en bloc, Cudworth chided Kircher for "contend[ing] with much zeal for the sincerity of all these Trismegistick books." At least two of the books had to be rejected as reliable sources of Egyptian theology, and even more so "that pretended Aristotelick book, *De secretiore parte divinae sapientiae secundùm Aegyptios* [i.e., pseudo-Aristotle's *Theology*], greedily swallowed down also by Kircherus, but unquestionably pseudepigraphous." Although Cudworth's understanding of the Hermetic theology of the Egyptians was similar to Kircher's, he felt that the Jesuit scholar played too loose with the evidence. He did not doubt that the hieroglyphs on the Bembine Table concerned the "*arcane theology* of the Egyptians, and not mere history, as some imagine." But, he wrote, "the late confident *Oedipus* seems to arrogate too much to himself, in pretending to such a certain and exact interpretation of it."[32]

Because the two thinkers shared so much, Cudworth enables us to take the measure of Kircher's hieroglyphic studies by seventeenth-century standards. On the one hand, he provides evidence of how unconvincing was Kircher's defense of the Hermetic Corpus, even to a fellow Hermetist. But at the same time he shows the endurance of the larger framework of occult philosophy. Not only did Cudworth believe in the existence of a pagan *prisca theologia* (understood in traditionally Neoplatonic terms, but now, significantly, fused with elements of mechanical philosophy); he believed that much of its traditional canon still deserved respect, including the Chaldean Oracles, the Orphic verses, and parts of the Hermetic Corpus.[33]

Occult philosophy and the historical claims of the *prisca theologia* remained credible to serious minds in the second half of the seventeenth century. Its premises were increasingly under attack, but even its critics often accepted the idea of perennial philosophy in the form of Adamic or Mosaic science. For example, in *The Court of the Gentiles* (1676) the puritan historian of philosophy, Theophilus Gale, denigrated the accomplishments of

---

31. Cudworth, *True Intellectual System* (1743), vol. 2, 308–55; quotation at 319.
32. Ibid., 317, 320.
33. Although Cudworth referred to them as "monuments of pagan antiquity, as are altogether unsuspected and indubiate," his discussion of the Chaldean Oracles and Orphic verses was nuanced. He did not insist on the authorship of Zoroaster and Orpheus, but he was certain that they were very ancient, and that they offered excellent evidence "of the theology of the ancient pagans." See ibid., vol. 1, 281, 298.

pagan philosophers, arguing that all their genuine knowledge of divine matters derived from Moses. He was not overly impressed by the wisdom of the hieroglyphs, which he described as a mode of communication characteristic of "the infancie of the world." Gale had no doubt, however, that the history of philosophy began with the revealed wisdom of Adam and his progeny. As "the first created divine Institutor of all Philosophie," he wrote, Adam knew how to "anatomise, and exactly pry into the very natures of things, and there contemplate those glorious Ideas, and Characters of created Light and Order, which the increated Light and Divine Wisdome had impressed thereon; and thence he could by the quicknesse of his apprehension, immediately collect, and forme the same into a complete systeme or bodie of Philosophie; as also most methodically branch forth the same into particular Sciences, &c."[34] Gale's historical framework was similar to that of Cudworth and More—not to mention Kircher, on whom he relied in discussing matters Egyptian—but his attitude toward pagan tradition was much more negative. He, too, believed in the existence of a divine perennial philosophy that Noah and his sons passed on to the entire world, including, for example, Plato. But for Gale, pagans had not significantly contributed to the preservation of truth; instead he emphasized how they invariably misunderstood and corrupted the Adamic wisdom they had received from Moses.

As long as philosophical truth was equated with divine wisdom and the legend of Adamic science endured, the debate over occult philosophy would remain primarily a debate about how much knowledge pagans could be imagined to possess.[35] Despite the rapid diffusion of Casaubon's critique, its impact was more limited than is often imagined. Even among those who accepted that the *Corpus Hermeticum* was a post-Christian forgery, this was not necessarily taken to refute the existence of Hermes Trismegistus, much less the entire tradition of pagan wisdom embodied in the *prisca theologia*. More than a century after Casaubon, William Warburton lamented the persistence of belief in Hermetic wisdom: "Now though that imposture [the Hermetic Corpus] hath been sufficiently exposed, yet on pretence, that the writers of those books took the substance of them from the ancient Egyptian physiology, they preserve, I do not know how, a certain authority amongst the learned, by no means due unto them."[36] The relegation of occult philosophy to the margins of European scholarship was underway, but

---

34. Gale, *Court of the Gentiles* (1676), 7, 35. On Gale see Malusa, "Cambridge Platonists" (1993), 292–330.
35. Blackwell, "Thales Philosophus" (1997).
36. Warburton, *Divine Legation of Moses* (1846), 391.

far from complete, when Kircher published *Egyptian Oedipus* in 1655. It was not until the eighteenth century that the underlying premise of perennial philosophy, in both its Adamic and Hermetic forms, decisively gave way to the sustained attacks of authors espousing an evolutionary model of the progress of language and knowledge from primitive beginnings.

THE IDOL OF AMBIGUITY

As we saw in chapter 1, the idea of hieroglyphs as symbolic wisdom was widely shared in the sixteenth century and remained current in the seventeenth. Kircher subscribed to what I have called the strong theory of symbolism, rooted in Neoplatonic metaphysics. The fundamental Platonic distinction between the many and the one—between a lower realm of transience, multiplicity, and finitude and an ultimate realm of unity, eternity, and infinity—sustained the idea of the mystical symbol as the best possible representation of absolute truth in the world of change. Likewise, the idea that reality was structured in hierarchical levels underwrote allegorism by implying that the full significance of any single representation, be it a linguistic or natural object, was distributed across numerous layers of meaning.[37] During the seventeenth century this kind of semantics, which had always been at odds with the Aristotelian-rhetorical tradition, was increasingly disputed by philosophical and scientific reformers. The debate played out in discussions about the possibility of a universal language, in which Chinese ideograms, understood by Europeans as an example of a "real character," displaced Egyptian hieroglyphs as the model of linguistic perfection.[38]

Like hieroglyphs, "real characters" were defined as referring to ideas rather than words, and thus had the potential to be understood by speakers of any language. But whereas a hieroglyph was a symbol (or "emblem," as Francis Bacon, who popularized the distinction, put it) whose meaning had a natural relationship to its form, a real character was an arbitrary sign established by convention. Thus, to understand real characters it was not enough to be wise; one needed a key or dictionary. In addition, real characters were defined by a strict one-to-one relationship between sign and meaning. From a Neoplatonic perspective, the highly constrained semantic range of the real character was a limitation. If ultimate reality was located in the realm of the infinite, the

---

37. For an excellent discussion, see Lamberton, *Homer the Theologian* (1986), especially about Plotinus, 88. See also Eco, *Search for the Perfect Language* (1995), 162.

38. On the European idea of Chinese "ideograms," see Saussy, *Great Walls of Discourse* (2001), ch. 3.

job of a perfect or "philosophical" language was to mediate between that ultimate reality and the finite world of experience with minimal loss of meaning; hence, the excellence of the inexhaustible hieroglyph. By contrast, in the view of Bacon's successors, like John Wilkins, author of *An Essay towards a Real Character* (London, 1668), a philosophical language should limit rather than proliferate meaning. It was precisely the overfecundity of Neoplatonic semantics that annoyed the new philosophers and their Enlightenment heirs, who identified equivocal language as the cause of many errors.[39]

The decline of allegorism and the strong theory of symbols—what William Ashworth has dubbed the "emblematic worldview"—was a complex process, and a full account of its multiple causes lies beyond the scope of this discussion.[40] Eventually, the rejection of Neoplatonic hermeneutics combined with the historical critique of the *prisca theologia* to definitively undermine the scholarly status of occult philosophy. But change was gradual and piecemeal. Francis Bacon, for example, continued to believe that ancient pagans had communicated profound truths through their myths, even as he opposed the idea of the mystical hieroglyphic symbol. In the eighteenth century, attacks on the allegorical interpretation of ancient myth by Pierre Bayle and Bernard Fontenelle found traction that previous criticism, like that by Rabelais, had not.[41] But as late as 1740, Warburton observed that the notion that Egyptian hieroglyphs were sacred symbols concealing sublime mysteries, which he considered the root of Kircher's error, was the common opinion of most modern writers. Although he was one of Kircher's harshest critics, from the vantage point of the mid-eighteenth century Warburton viewed him as an earnest victim of the prevailing ideas of the age.[42]

SACRED HISTORY, PAGAN WISDOM, AND THE ORIGIN OF LANGUAGE

In 1655, the year Kircher published *Egyptian Oedipus*, another book dealing with remote antiquity and the relationship between biblical history and

---

39. On real characters vs. hieroglyphs, see Singer, "Hieroglyphs" (1989); Eco, *Search for the Perfect Language* (1995). On the importance of univocality in the Scientific Revolution, see Funkenstein, *Theology and the Scientific Imagination* (1986).

40. Ashworth, "Natural History" (1990). See also Harrison, *The Bible* (1998); Killeen and Forshaw, eds., *The Word and the World* (2007); and ch.1, n. 55.

41. On Bayle and Fontenelle, see Manuel, *Eighteenth Century Confronts the Gods* (1959), 15–53. Rabelais mocked the allegorical interpretation of Homer in the prologue to the first book of *Gargantua and Pantagruel*.

42. Warburton, *Divine Legation of Moses* (1846), vol. 2, 365.

pagan civilization appeared. Kircher heard about it in a letter from his friend Barthold Nihus in Erfurt. Writing in February 1656, Nihus told Kircher that he had just gotten his hands on a "little book" published anonymously in Holland and entitled *Praeadamitae*. Skimming its contents, he discovered outrageous claims about the existence of human beings before Adam. Nihus asked Kircher if the work was known in Rome, where it was sure to be condemned.[43] He was right. *Men before Adam*, whose author was later identified as Isaac La Peyrère, met condemnation among Catholics and Protestants alike, becoming one of the era's most notorious works.[44] La Peyrère argued that the first chapters of Genesis described two separate creations of humanity—the first generating all gentile nations while the second, of Adam and Eve, spawned only the Jews—and he demoted Noah's flood to a local Near Eastern event. Although primarily based on unorthodox scriptural interpretation, La Peyrère also appealed to evidence from Egyptian and Chinese annals that reckoned thousands of years of continuous history before the Flood.[45]

In La Peyrère's wake, questions about pagan civilizations' antiquity and relationship to the biblical tradition took on new importance and sensitivity, a situation that was amplified by the appearance of Spinoza's *Theological-Political Treatise* in 1670, as well as the diffusion of the ideas of Thomas Hobbes. Kircher composed his hieroglyphic studies before this development, but the reception of *Egyptian Oedipus*—and the fate of occult philosophy more generally—must be understood in the context of the new controversies over sacred history that emerged in the second half of the seventeenth century. In order to defend the authority of the Bible against arguments based on the chronological priority of pagan traditions, some Christian apologists inveighed against the claim that nations like Egypt possessed profound wisdom in remote antiquity. These polemics spawned new visions of the origins of civilization and the evolution of language which would undermine the entire historical and hermeneutic edifice of occult philosophy.[46]

About the supposed wisdom of the Egyptians, in 1666 the Anglican bishop Samuel Parker wrote: "What childish fooleries their Hieroglyphicks were Learned Men now prove from the lost labour and fruitless industry of *Kirchers Oedipus Aegyptiacus*. Certainly, if they had design'd to abuse

43. Nihus to Kircher, Erfurt, 28 February 1656, APUG 557, fol. 167^r.
44. Popkin, *Isaac La Peyrère* (1987); Grafton, *Defenders of the Text* (1994), 204–13.
45. La Peyrère, *Praeadamitae* (1655); La Peyrère, *Systema theologicum* (1655).
46. Rossi, *Dark Abyss of Time* (1984).

and debauch this humour, they could scarce have contrived more fond and extravagant Emblems; and indeed their courseness and unlikeness to the things they should resemble, sufficiently discover them to have been but the rude Essays of a barbarous and undisciplined Fancy."[47] A fellow of the Royal Society with linguistic views akin to Wilkins, Parker gave the criticism of hieroglyphic symbolism a historical rationale.[48] He dismissed the very possibility of ancient Egyptians possessing profound knowledge because their use of symbolic writing was typical of an immature stage of cultural development. This notion, which called into question the most basic premise of Kircher's interpretation of the hieroglyphs, was further elaborated by eighteenth-century thinkers who developed naturalistic accounts of the origin of language. Its most influential exponent was another Anglican bishop, William Warburton, whose sprawling, apologetic treatise *The Divine Legation of Moses Demonstrated* (1738–41) contained an original theory about the origin and development of writing, including a devastating attack on Kircher's theory of the hieroglyphs.

The source of all misunderstanding, according to Warburton, was the nearly universal misconception that hieroglyphs had been invented by Egyptian priests to hide arcane wisdom from the masses. Instead, he explained, they represented an early and defective stage in the development of language. Warburton laid out a progressive scheme in which all societies naturally developed a succession of increasingly perfect forms of spoken and written communication. The earliest and crudest form of writing was "simple painting," as used by the Mexicans. The next stage was the development of "pictured characters," which communicated more efficiently thanks to various means of compression, such as using a part to represent the whole, but whose forms maintained a nonarbitrary relationship to their referents. The Egyptians were the first and most famous to use such characters. In the third stage, represented by Chinese ideograms, resemblance of sign and referent was abandoned for a system of arbitrary signs or "real characters" in which "every distinct idea has its characteristic mark." Warburton used the term "hieroglyphic" in a broad sense to designate all three of these stages, which were characteristic of "all the barbarous nations upon earth, before the invention or introduction of letters." Finally, ideograms were transformed into alphabetic writing, "invented to express *sounds* instead of *things*," which had the inestimable advantage of avoiding ambi-

---

47. Parker, *Free and Impartial Censure* (1666), 97.
48. Parkin, *Science, Religion, and Politics* (1999), 37.

guity.⁴⁹ (Kircher believed in a similar sequence, but saw it as a process of corruption, not progress.)

Believing hieroglyphs to be the expression of people at a primitive stage of development with limited ability to think abstractly, Warburton utterly rejected the notion that their original function was to record profound wisdom. The hieroglyphic inscriptions on Egyptian monuments such as obelisks, he insisted, had been used to record, "openly and plainly," banal affairs of state. Inevitably, Kircher stood square in the path of Warburton's demolition effort. "It is pleasant," Warburton wrote of the Jesuit's hieroglyphic studies, "to see him labouring through half a dozen folios with the writings of late Greek Platonists, and the forged books of Hermes, which contain a philosophy, not Egyptian, to explain and illustrate old monuments, not philosophical." Warburton accepted the traditional claim for Egypt's extreme antiquity, as well as the idea that its priests had used a symbolic mode of writing, referring to things rather than sounds, and considered Horapollo an accurate guide. Where he differed from Kircher was in his judgment of the implications of extreme antiquity. It did not signify that, being closer to Creation, the Egyptians had been closer to the pristine source of perennial philosophy and therefore had used symbols and allegory to communicate sacred wisdom. It meant that they were a semiliterate "infant nation" who wrote in crude images for want of better understanding.⁵⁰

A similar verdict was pronounced by the Neapolitan philosopher Giambattista Vico. The rejection of the *prisca theologia* was one of the key tenets of Vico's *New Science*. His famous axiom, "the conceit of scholars," which stated that learned men falsely "assert that what they know is as old as the world," was supposed to refute "all the scholars who have praised the incomparable wisdom of the ancients. It proves," wrote Vico, "that the oracles of Zoroaster the Chaldaean, the lost oracles of Anacharsis the Scythian, the *Pimander* of Hermes Trismegistus, the *Orphica* (or *Hymns* of Orpheus), and the *Golden Verses* of Pythagoras were impostures, as all astute critics agree. It also exposes the absurdity of all the mystical senses which scholars have read into Greek myths." Philological criticism corroborated, but was by no means necessary, to invalidate the *Hermetica* and other testimonies of ancient occult wisdom which followed from Vico's first principles. The claim "that philosophers invented hieroglyphics to conceal the mysteries of their esoteric wisdom" was a fundamental error that Vico would eradi-

---

49. Warburton's theory of the hieroglyphs occupies book 4, section 4: Warburton, *Divine Legation of Moses* (1846), vol. 2, 172–246, quotations at 178 and 183.

50. Ibid., vol. 2, 193–96.

cate. Like Warburton, he found no better target than Kircher. Regarding the Jesuit's treatment of the hieroglyph of the winged globe and serpent, Vico declared: "The conceit of scholars led to such madness that in [*Pamphilian Obelisk*] Athanasius Kircher says that it signifies the Holy Trinity."[51]

Thus Vico consigned the Neoplatonic interpretation of the hieroglyphs to the dustbin of his new science of history. But he did not doubt the hieroglyphs' great antiquity or symbolic nature. He was no precursor of Champollion, predicting that the Egyptian writing system simply recorded a normal language. For Vico, like Warburton, hieroglyphs were the language of the first age of the world, the universal mode of expression of primitive men, incapable of more subtle cogitation, who "shared a natural need to speak by means of hieroglyphic symbols."[52] The hieroglyphs were symbolic but crude, expressing the same simple ideas as the bodily gestures that functioned as the spoken form of "hieroglyphic" communication.

After the metaphysical bombast of *Egyptian Oedipus*, it can be refreshing to read these critics of hieroglyphic wisdom, especially Warburton, who so clearly located the source of Kircher's error in the syncretic, Platonizing culture of the early Christian era. But one must be wary of overestimating the modernity of occult philosophy's early modern enemies. No less than more traditional critics, who attacked the wisdom of pagan *prisci theologi* while upholding the antediluvian science of Adam, the thinkers who first challenged the idea of perennial philosophy at its root were motivated by apologetic goals. An idiosyncratic Catholic, Vico designed his new science, with its absolute separation of Jewish and gentile history, to safeguard sacred history and its compressed chronology from the threat that he detected in authors who extolled the antiquity and accomplishments of ancient pagan nations like Egypt.[53] Bishop Warburton developed his theory about the evolution of language in order to defend the revelatory authority of the Bible against Spinoza and his English disciples, such as John Toland. He belonged to a liberal Protestant tradition reaching back to Samuel Parker and including contemporaries like the German historian of philosophy Jacob Brucker, which aimed to purge Christianity of the corrupting influence of Neoplatonism.[54] A Lutheran pastor, Brucker believed that disentangling the history of philosophy from the history of revealed wisdom would strengthen

---

51. Vico, *New Science* (1999), 77, 175, 269.
52. Ibid., 175.
53. Rossi, *Dark Abyss of Time* (1984); Momigliano, "La nuova storia romana" (1980).
54. See Pocock, "Edward Gibbon in History" (1988), esp. 338–55. On the destruction of Christian Platonism, see Mulsow, *Moderne aus dem Untergrund* (2002), 261–307; Lehmann-Brauns, *Weisheit in der Weltgeschichte* (2004).

both rational inquiry and Christian faith.⁵⁵ Brucker, Warburton, and likeminded scholars built on Casaubon's philological criticism, but they went much further, systematically deconstructing Neoplatonic philosophy, its infiltration of Christianity, and its latter-day resurgence in the form of Renaissance occult philosophy, which Brucker diagnosed as "Pythagorean-Platonic-Kabbalistic philosophy."⁵⁶ But their motivations, like Casaubon's a century earlier, remained conservative.

Critiques of ancient wisdom were also made outside of Christian apologetics, most notably by Fontenelle and Bayle.⁵⁷ But it would be some time before the attack on occult philosophy was associated primarily with thinkers who challenged sacred history. In the late seventeenth and early eighteenth centuries, radical critics of the Christian tradition embraced their own versions of the *prisca theologia*, arguing that the wise men of ancient nations like Egypt, Persia, and Greece had followed an esoteric, natural religion based on reason. Thus, John Toland confirmed Vico's conceit of the scholars by declaring Spinozism the oldest knowledge in the world.⁵⁸ In this context, as Martin Mulsow has shown, it was quite conceivable for a German freethinker to constructively merge material from Kircher's *Egyptian Oedipus* and Bayle's treatise on the comet.⁵⁹ Isaac Newton was neither a freethinker nor a Neoplatonist occult philosopher, but he was an unconditional believer in a form of the *prisca theologia*, attributing the theory of universal gravitation to ancient pagan and biblical wise men, a genealogy that had considerable influence on the first generations of Newtonians, both Christian and Deist.⁶⁰

It was a century after the publication of *Egyptian Oedipus* before Enlightenment theorists of progress absorbed and secularized the evolutionary model of the origin of language and civilization. Diderot and d'Alembert's *Encyclopedia* (1751–72) promulgated Warburton's theory, which had been translated into French in 1744, in a series of articles on hieroglyphs, writing, Egyptian writing, and Chinese writing, all compiled by Jaucourt. In another entry, "Antediluvian," Diderot took aim at belief in Adamic philosophy, cribbing most of his argument from Brucker's *Critical History of Philos-*

---

55. Schmidt-Biggemann and Stammen, eds., *Jacob Brucker* (1998); Longo, "Storia 'critica' della filosofia" (1979).
56. Brucker, *Historia critica philosophiae* (1742–44), vol. 4, 353–448.
57. See above, n. 41.
58. Harrison, *"Religion" and the Religions* (1990); Champion, *Pillars of Priestcraft* (1992), esp. 131.
59. Mulsow, *Moderne aus dem Untergrund* (2002), 214.
60. Haycock, "Long-Lost Truth" (2004).

ophy while jettisoning its Christian framework.[61] Around the same time, authors like David Hume and Charles de Brosses put forth psychological theories about the origin of religion according to which irrational traditions such as Egyptian animal worship were not to be explained away by means such as allegory, but taken literally as expressions of primitive man's abject ignorance.[62] With Adamic science, Hermetic wisdom, and sacred history all exposed as fables of an unenlightened age, it became possible to read Kircher's hieroglyphic studies in a modern way.

This is not to say that belief in ancient symbolic wisdom disappeared from European intellectual history. But even among those late-eighteenth-century scholars who revived the grand tradition of allegorical mythography, the differences are as telling as the similarities. Take, for example, Antoine Court de Gébelin's eight-volume study, *The Primitive World Analyzed and Compared to the Modern World* (1773–82), which repurposed the explanatory frameworks of seventeenth-century interpreters of ancient paganism, including Kircher. An active player in French Masonic lodges and a renowned antiquary, Court believed in a version of the *prisca theologia*, arguing that the first authors of ancient religion had been monotheists who used allegorical communication to convey their teachings. Even while valorizing the prudence and morality of the earliest men, however, he followed Warburton and Vico in viewing the primordial language of symbols as characteristic of a primitive mentality incapable of complexity and abstraction. For Court, the wisdom that the "genius of the primitive world" had encoded in allegory was Enlightenment utilitarianism, based on a cult of nature and devoid of any transcendental or mystical dimension.[63] Much the same can be said for the allegorical naturalism of Charles Dupuis, whose *Origin of All Religious Worship* (1795) offered a sexological key to all mythologies, as well as lesser-known authors such as Sir William Drummond, whose scandalous treatise *Oedipus Judaicus* (1811) explained enigmatic biblical passages as esoteric allegories of an ancient solar cult.[64]

Kircher's uncritical defense of the Hermetic Corpus was obsolete on arrival. But the larger historical argument of *Egyptian Oedipus* still breathed life in 1655. Casaubon's philological criticism had not spelled the end of

---

61. Diderot, "Antédiluvienne" (1751–65).
62. Manuel, *Eighteenth Century Confronts the Gods* (1959), 168–209.
63. Court de Gébelin, *Monde primitif* (1773–81); Mercier-Faivre, "Court de Gébelin" (2005); Manuel, *Eighteenth Century Confronts the Gods* (1959), 250–58; Edelstein, "Egyptian French Revolution" (2010).
64. On Dupuis, see Manuel, *Eighteenth Century Confronts the Gods* (1959), 259–70. On Drummond, see Godwin, *Theosophical Enlightenment* (1994), 39–48.

occult philosophy's role in European scholarship, although it was an important factor in its eventual demise. The real watershed came later, with the triumph of theories of the gradual development of language, civilization, and philosophy from primitive origins. As we have seen, these theories themselves evolved gradually over the course of a century. In the early stages, their significance was ambiguous. It is hard to say, for example, whether Parker was more "modern" than Cudworth; both were moderate Anglicans devoted to defending the faith by articulating a form of rational Christianity. This common goal led Cudworth to embrace occult philosophy, and Parker to explode it.

More unusual than Kircher's beliefs about Egyptian wisdom was the secondary role that Christian apologetics played in his study of the hieroglyphs. Because the scholarly discourses in which he partook were steeped in religious controversy, this dimension was by no means absent from his work. But in contrast to many works treating similar topics—for example, those of Wits, Gale, Parker, Cudworth, More, Bouvet, Huet, Lafitau, La Peyrère, Warburton, Vico, and Brucker, to refer only to authors mentioned in this chapter—*Egyptian Oedipus* was not primarily an argument about the relationship between Christianity and pagan traditions, undertaken to defend true religion. Rather, it presumed a certain position about that relationship as its starting point, and on that foundation explored the details of ancient pagan beliefs and practices—which is another way of saying that *Egyptian Oedipus* didn't advocate occult philosophy so much as put it to use to explain other things. Kircher didn't defend the textual sources of the *prisca theologia* in order to prove the affinity of pagan theology and Christianity. He believed in that affinity, to be sure, but the goal that drove him was to solve the great hieroglyphic enigma and reap the glory.

IMPERISHABLE ENCYCLOPEDISM

The unexpectedly long shelf life of Kircher's hieroglyphic studies was partly due, then, to the endurance of its historical framework and the ongoing salience of debates over the relationship between ancient paganism and biblical tradition. But it was also possible to read Kircher without regard to his larger arguments. It was the nature of works written in baroque erudition's digressive, encyclopedic style that bits of information were easily separated from an author's interpretative framework. Even the eighteenth-century *philosophes*, who heaped scorn on the erudite Latin compendia of the preceding century, lived off their legacy, pillaging their contents for useful information to repackage in a more polite form. Kircher was frequently cited, if not substantively engaged, in Diderot and d'Alembert's *Encyclopedia*.[65] Consider the case of Abenephius, a.k.a. Barachias Nephi, the reputed but apparently nonexistent author of a mysterious Arabic manuscript on Egypt and the hieroglyphs, known only from Kircher's testimony, as discussed in chapter 2. Following Kircher's inclusion of quotations from his work in *Egyptian Oedipus*, Abenephius entered European scholarship's intertextual bloodstream and never left—turning up, for example, in the *Encyclopedia*'s article on chemistry.[66]

Oriental philologists and biblical scholars were especially intrigued by passages in which Abenephius discussed the Egyptian origin of various Jewish rites described in the Old Testament. In *Smegma Orientale* (1658), Johann Heinrich Hottinger, a Calvinist theologian at Heidelberg, borrowed a quotation by Abenephius from *Egyptian Oedipus* in support of the argument that the Hebrew *theraphim* derived from the Egyptian *Serapis*. A decade later, the English scholar John Spencer quoted the same Abenephius passage, although he disagreed with the proposed derivation.[67] In another lengthy quotation, Abenephius described how the Mosaic law provided the Jews with pious imitations of idolatrous Egyptian ceremonies to which they had grown accustomed. As Kircher noted, the quotation strongly resembled a famous passage in Maimonides' *Guide for the Perplexed*.[68] After John Spencer's revival of this theory, a decades-long controversy raged over the Egyptian origin of Jewish rituals, in which Abenephius, via *Egyptian Oedipus*, played a small supporting role. Hermann Wits, for example, seems to have recognized Kircher's treatise as a precursor to Spencer's thesis, which he was at pains to refute in his *Aegyptiaca*, and thus discussed Abenephius at some length in order to cast doubt on his credibility.[69]

By this time, Abenephius had infiltrated the bibliographies. In 1668, Gottlieb Spitzel's sample of a proposed universal bibliography included an entry on "Abenephius, Arabic Historian." Spitzel noted that "Athanasius Kircher mentioned him repeatedly in *Egyptian Oedipus*, where . . . he showed how many treasures unknown to the Latins lie hidden in the monuments of the Arabs and the Jews." A year later, Christoph Hendreich, librarian to the Elector of Brandenburg, borrowed from Spitzel to compose his own entry on Abenephius, printed in the first volume of his unfinished uni-

---

65. See Edelstein, "Humanism" (2009), appendix 2.
66. Venel, "Chymie" (1751–65), 425.
67. Hottinger, *Smegma orientale* (1658), 90; Spencer, *De Urim* (1669), 52, 92 and passim.
68. See above, ch. 2, n. 40.
69. Witsius, *Aegyptiaca* (1739), 33, 53–57, 225.

versal bibliography. By the early eighteenth century, Abenephius's dubious identity had become a lively topic among scholars of Oriental languages. In his *Antiquarian and Exegetical Dictionary* (1724–25), Peter Zorn, yet another German bibliographer, devoted twenty-one pages to an entry on Abenephius, which attacked the Jewish-Arabic author's credibility as a source of ancient knowledge. In 1727, J. C. Wolf, whom Zorn had admonished for neglecting Abenephius in his bibliography of Jewish literature, made amends by adding an entry for "אבן נעפי ABEN NEPHI, vel ABENEPHI Arabs" in a new appendix.[70]

Another Abenephius quotation, proffering a Trinitarian explanation for the hieroglyph of the winged globe with serpent, also garnered notoriety as it became caught up in the debate over pagan anticipations of Christian mysteries.[71] Along with other passages from Abenephius, it continued to be discussed in the works of biblical scholars and Orientalists into the nineteenth century.[72] As that century progressed, however, Abenephius appeared more often in works by spiritualists and theosophists, like Madame Blavatsky, than in those of more sober scholars of antiquity. This represented a significant shift in Kircher's readership, but it is noteworthy how late it occurred.[73] As recently as 1975, *The Journal of Theological Studies* published an article refuting a theory about Saint John's vision of the heavenly Jerusalem, put forward by two twentieth-century biblical scholars who relied on the authority of Abenephius.[74]

Kircher meant *Egyptian Oedipus* to be a definitive solution to the mystery of the hieroglyphs. In this endeavor he certainly failed. But he also explicitly envisioned the work as a contribution to erudite historical research, a showcase of newly discovered Oriental texts about antiquity. The trail of Abenephius shows how Kircher lived on as an "authority" whose works could be mined for isolated factoids and opinions, just as he had done to other authors in composing his own text. In this manner, European scholars of the seventeenth and eighteenth centuries frequently cited Kircher in the

---

70. Spitzel, *Sacra bibliothecarum* (1668), k5^v; Hendreich, *Pandectae Brandenburgicae* (1699), 12; Zorn, *Bibliotheca antiquaria* (1724–25), 133–54; Wolf, *Bibliotheca Hebraea* (1715–33), vol. 3, 10–13.

71. See above, ch. 7, n. 16.

72. For example: Shaw, *Travels* (1808; originally published 1738), 171; Walsh, *Sketches* (1793), 226; Maurice, *Indian Antiquities* (1800), 325; TW, "Remains of Sanchoniatho" (1829); Russell, *A Connection of Sacred and Profane History* (1865), vol. 1, 24; Gutschmid, "War Ibn Wahshijjah ein Nabatäischer Herodot?" (1890), 726.

73. Blavatsky, *Secret Doctrine* (1895), 388. See also Finney, *Skeletons of a Course* (1840), Lecture 17; Ragon, *Orthodoxie Maçonnique* (1853), 566, 574.

74. Glasson, "Order of Jewels" (1975).

company of the great names of early modern erudition, such as Joseph Scaliger, Isaac Casaubon, G. J. Vossius, and John Selden.[75]

The seventeenth- and eighteenth-century responses to Kircher's hieroglyphic studies, negative and positive and in various contexts, had one thing in common: they treated his work as historical scholarship. Brucker, for example, did not include Kircher in his analysis of "Pythagorean-Platonic-Kabbalistic philosophy," alongside authors like Agrippa and Patrizi. On occasion, Brucker cited *Egyptian Oedipus*, but always as a secondary source, treating Kircher as a historian of Egyptian and Chaldean traditions. (When Brucker did identify Kircher with a school of philosophy, he did so in the context of experimental science, listing him as a "magnetic philosopher" alongside William Gilbert and Niccolò Cabeo.[76]) Although Kircher would have been dismayed—but perhaps not so very surprised—that his interpretation of the hieroglyphs did not carry the day, his work was received as intended: an erudite investigation of Egyptian antiquities and a showcase of the application of new Oriental sources to historical research.

---

75. For example, in the *Encyclopedia* Kircher was cited more often than Selden, but less often than Scaliger; Edelstein, "Humanism" (2009), appendix 2. Haugen, *Richard Bentley* (2011), 49, names "Joseph Scaliger, Claude Saumaise, Isaac Casaubon, Richard Montagu, William Spencer, G. J. Vossius, Isaac Vossius, Gilles Menage, Athanasius Kircher, and Hermann Conring" as authors frequently cited by scholars in Restoration-era Cambridge.

76. Brucker, *Historia critica philosophiae* (1742–44), vol. 5, 615–16.

EPILOGUE

The Twilight of Tradition and the Clear Light of History

It remained, then, virtually true, as it had been for two thousand years, that for all that we could learn of the history of the Old Orient in preclassical days, we must go solely to the pages of the Bible and to a few classical authors, notably Herodotus and Diodorus. A comparatively few pages summed up, in language often vague and mystical, all that the modern world had been permitted to remember of the history of the greatest nations of antiquity. To these nations the classical writers had ascribed a traditional importance, the glamour of which still lighted their names, albeit revealing them in the vague twilight of tradition rather than in the clear light of history. It would have been a bold, not to say a reckless, dreamer who dared predict that any future researches could restore to us the lost knowledge that had been forgotten more than two milleniums.
—"Chronology," *Encyclopaedia Britannica* (1911)

When the eleventh edition of the *Encyclopaedia Britannica* invoked the opposition between "the clear light of history" and "the vague twilight of tradition" to describe the historiographic revolution brought about by the decipherment of hieroglyphics and cuneiform, the metaphor was already a literary cliché. Nineteenth-century writers commonly described reliable knowledge of the past based on credible evidence as the "clear light of history," which they contrasted with various phrases such as "the mist of tradition," "mythological fiction and poetry," the "faint gleam of tradition and fable," "the dimness of legendary fable and myth," the "impenetrable veil of tradition, mystery, and silence," or simply "tradition."[1]

---

1. More examples can easily be found by searching Google Books. The quotations are from Antranig Azhderian, *The Turk and the Land of Haig* (New York, 1898), 111; *The Calcutta*

This was the language of disenchantment, of science displacing myth, and historicism inexorably dissolving tradition. Indeed, could there be a more vivid instantiation of the idea of "the disenchantment of the world" than the deciphering of hieroglyphics? In Champollion's wake, the occult philosophy of Hermes Trismegistus would seem to have melted into air, and in its place appeared more solid records of a Bronze Age civilization. But the narrative of disenchantment, with its implication that occult philosophy ("legendary fable and myth") simply gave way beneath the weight of archeological evidence ("history's clear light") fails to capture the complexity of modern scholarship's evolution.

In this book I have argued that the fusion of occult philosophy and empirical historical research found in Kircher's hieroglyphic studies was not anomalous. The opposition between "history" and "tradition," which came so naturally to the nineteenth century, was foreign to the seventeenth. Kircher trusted that the progress of historical scholarship, based on the investigation of material evidence and the discovery of new Oriental texts, would illuminate, not vaporize, the symbolic wisdom of ancient Egypt, as surely as Isaac Casaubon believed that the critical philology he unleashed on the Hermetic Corpus in 1614 would fortify biblical authority, undermining only religion's enemies. Because of Casaubon, critical philology is firmly established in the narrative of occult philosophy's demise. But, since Frances Yates put the "Hermetic tradition" on the agenda in the 1960s, research on early modern Neoplatonism and occult sciences such as magic, astrology, and alchemy has mostly concentrated on the realms of natural science and philosophy.[2] The case of Kircher's hieroglyphic studies points to another crucial context: historical scholarship on ancient religion, philosophy, and magic. Even after Casaubon's critique, European erudition was by no means a uniformly hostile environment for occult philosophy. On the contrary, historical research was an important factor in occult philosophy's ongoing vitality.

Kircher was uncommon in his monomaniacal determination to wring a solution from the hieroglyphic monuments, and I know no other example of an antiquarian project that marshaled the tradition of Renaissance occult philosophy on such a massive, encyclopedic scale. But the elements of

---

Monthly Journal, July 1836, 325–26; A. H. L. Heeren, *Reflections on the Politics, Intercourse, and Trade of the Ancient Nations of Africa* (Oxford, 1832), 294; *Proceedings of the Suffolk Institute of Archeological Statistics and Natural History* 2 (1859), 171; *Bibliotheca Sacra and Theological Review* 3 (1846), 436.

2. Yates, *Giordano Bruno* (1964). See Hanegraaff, "Beyond the Yates Paradigm" (2001), 13–15.

his method were not unusual. To many seventeenth-century scholars, the genealogy of the *prisca theologia*, ideas preserved in esoteric literature, and Neoplatonic hermeneutic techniques seemed like plausible and efficacious resources for accomplishing one of their era's greatest aspirations: the reconciliation of Christian and pagan traditions.[3] In the seventeenth century occult philosophy was not detritus waiting to be jettisoned, but a creative intellectual force.

By focusing on occult philosophy's nexus with erudition, the preceding chapter has also offered new insights into its eventual marginalization within European scholarship. Historians of science have linked the demise of Neoplatonic magic, astrology, and other occult sciences to the overthrow of the traditional physics and cosmology that provided their theoretical underpinnings.[4] But debates about ancient history played a key role as well. The Renaissance Neoplatonic tradition, which I have been calling occult philosophy, was one of several currents that many scholars now refer to under the rubric of "Western esotericism."[5] Occult philosophy was arguably the most important such current in early modern Europe, because its legitimating history about the transmission of primordial knowledge served as a master narrative for other occult sciences. The unraveling of that master narrative spelled the end of occult philosophy within the realm of mainstream European historical scholarship. But the process that produced this outcome was not due to an essential opposition between occult philosophy and rational inquiry (or between "occult" and "scientific mentalities," as Brian Vickers would have it).[6] To a significant extent, occult philosophy was collateral damage in a larger battle that ultimately brought about the collapse of sacred history based on the literal interpretation of the Bible. The history of these developments was complex and anything but linear, since, beginning in the second half of the seventeenth century, the enemies of Neoplatonism who launched the frontal attack on occult philosophy were Christian apologists engaged in a rearguard attempt to shore up some version of traditional biblical authority. Indeed, Wouter Hanegraaff argues that it was anti-Platonist Protestant authors like Daniel Colberg who first articulated a coherent vision of something like Western esotericism as an au-

---

3. Ralf Häfner's monumental, panoramic study, although framed in somewhat different terms, provides extensive evidence of this phenomenon. Häfner, *Götter im Exil* (2003).

4. Copenhaver, "Occultist Tradition" (1998); Copenhaver, "Tale of Two Fishes" (1991); Copenhaver, "Astrology and Magic" (1988).

5. See introduction, n. 65.

6. Vickers, "Analogy Versus Identity" (1984).

tonomous, heterodox tradition under banners such as "Platonic-Hermetic Christianity."[7]

As a case study of the conjuncture of esotericism and empiricism in the seventeenth century, this investigation of Kircher's hieroglyphic studies provides confirmation of Hanegraaff's argument that, before the Enlightenment, the currents of Western esotericism did not constitute an autonomous tradition, but were integral elements of mainstream European intellectual and religious culture. Only after the eighteenth century did something like an esotericist "counterculture" start to emerge, defined first by critics, then by adherents, in opposition to Enlightenment epistemology.[8] But, even then, the relationship between esotericism and empiricism was hardly clear-cut. Despite Enlightenment efforts to demarcate a realm of legitimate knowledge, grounded in reason and experience, from the domain of unfounded superstition and magic, a growing body of research makes clear how difficult it proved for eighteenth-century thinkers to define and maintain such boundaries.[9] I have argued that the particular tradition of Renaissance occult philosophy came to be marginalized within the realm of historical scholarship in a gradual process spanning the seventeenth and eighteenth centuries. But one should not extrapolate this argument into a general claim about the influence of other esoteric currents, or even of occult philosophy itself, which continued to play significant roles in other domains of modern European culture.

Describing the state of historical knowledge about the ancient Near East before the discovery of the Rosetta Stone and the excavation of Nineveh, the 1911 *Encyclopaedia Britannica* pronounced that only a "bold, not to say reckless, dreamer" could have predicted that new research would one day supplement the scanty evidence preserved in the Bible and a handful of Greek texts. Kircher was, indeed, both bold and reckless, as his translations of hieroglyphic inscriptions prove. But his dream that new discoveries of Oriental records would unlock lost ancient worlds was shared by other seventeenth-century scholars, including many of a less "visionary" stripe. The pessimism referred to in the *Britannica* article belonged to eighteenth-century scholars who discarded that dream, abandoned the project of deci-

---

7. Hanegraaff, "Forbidden Knowledge" (2005), 242 (building on Antoine Faivre and Monica Neugebauer-Wölk).

8. Hanegraaff, *New Age Religion* (1996); Hanegraaff, "Beyond the Yates Paradigm" (2001); Hanegraaff, "Forbidden Knowledge" (2005). See also Stuckrad, *Locations of Knowledge* (2010).

9. Neugebauer-Wölk, ed., *Aufklärung und Esoterik* (1999); Edelstein, ed., *Super-Enlightenment* (2010); Reill, *Vitalizing Nature* (2005).

pherment, and declared classical literature the unique, if inadequate, source of hieroglyphic knowledge. Thus, Thomas Young, surveying the previous two centuries of research in 1823, wrote: "Among the works of more modern authors, who had employed themselves in the study of the hieroglyphics, it is difficult to say whether those were the more discouraging, which, like the productions of Father Kircher . . . professed to contain explanations of every thing, or which, like the ponderous volume of Zoëga on the Obeliscs, confessed, after collecting all that was really on record, that the sum and substance of the whole amounted absolutely to nothing."[10] One imagines that an eighteenth-century Egyptologist like Zoega, who had already disenchanted the Egyptian past on theoretical grounds, would not have been terribly surprised by the content of many of nineteenth-century Egyptology's discoveries. But the fact that such momentous discoveries about the ancient world could still be made—the *Britannica* author described the decipherment of cuneiform as "miraculous"—would have astonished him. By contrast, Kircher and other seventeenth-century Orientalists and antiquaries had predicted and tried to bring about just such a breakthrough.

The rejection of occult philosophy by eighteenth-century scholars, coincided with declining interest in Near Eastern philology.[11] The two developments were related, although only indirectly: the fate of occult philosophy, like that of early modern Oriental studies, was tied to the fate of sacred history. Kircher's principal contributions to Orientalist research proved insubstantial: his interpretation of the hieroglyphs was unpersuasive in the short term and decisively refuted in the long term; his exploration of Near Eastern literature, most notably Arabic treatises on Egyptian history, generated more excitement but did not provide genuine access to the Orient's deep past as he imagined it did. The greatest achievements of seventeenth-century Oriental studies took place elsewhere: in the realm of biblical scholarship, where sacred philologists transformed the interpretation of scripture through the study of Near Eastern literature. Here, too, Kircher was active, but the most significant accomplishments belonged to other scholars. When the apologetic agendas that powered sacred philology lost steam in the eighteenth century, Orientalist research fell off, until the rise of Indology and comparative philology at the turn of the nineteenth century reinvigorated the study of what came to be known as Semitic languages.

10. Young, *Account of some recent discoveries* (1823), 2.

11. Irwin, *Dangerous Knowledge* (2006), 109–29; Sutcliffe, *Judaism and Enlightenment* (2003), 23–41; Feingold, "Decline and Fall: Arabic Science in Seventeenth-Century England" (1996).

Perhaps more significant than the positive results of seventeenth-century Orientalist scholars were their aspirations and the new global conditions that made them thinkable. The world of early modern erudition was a Mediterranean world, dependent on the circulation of texts and people between East and West. Kircher and other scholars believed that by mastering Near Eastern languages and recovering ancient manuscripts hidden away in the famous libraries of Cairo and Damascus, they would bring about a revolution in historical knowledge. In crucial respects it would turn out to be a dream deferred. But for all their allegiance to traditional frameworks like sacred history and the *prisca theologia*, Kircher and other seventeenth-century Orientalists were attempting nothing less than to reinvent the study of antiquity by moving beyond the Western canon and seeking knowledge in the heritage of non-European cultures.

ACKNOWLEDGMENTS

My first debt is to Paula Findlen, who supervised my research during the book's early stages and followed its evolution with crucial advice and support. At Stanford I also had the pleasure of working with Brad Gregory and Haun Saussy, who shared their knowledge and enthusiasm. Tony Grafton has supported this project since its beginning with extraordinary generosity for which I am most appreciative. A postdoctoral fellowship at the Max Planck Institute for the History of Science in Berlin provided valuable time to let Kircher ripen; I am grateful to Raine Daston, whose research group offered a nearly utopian environment for thinking about early modern knowledge, and whose expansive vision of *Wissenschaftgeschichte* gave me confidence that a project that turned out to be about the study of the past more than of nature remains a contribution to the history of science. Another postdoctoral position at the Michigan Society of Fellows allowed me to pursue new projects while Kircher fermented. Since 2007 the history department at the University of California, Davis, has provided a stimulating and collegial intellectual home, as well as a powerful motivation to bring my work on Kircher to completion.

    I was fortunate to conduct research in Italy while Antonella Romano organized a series of roundtables on early modern Roman scientific culture at the École française de Rome; I thank her for inviting me to participate and for taking an ongoing interest in my project. Joan Cadden, Dan Edelstein, Jeremy Mumford, Neil Safier, Jonathan Sheehan, and Jake Soll read parts of the penultimate manuscript, offering valuable feedback. I also thank the readers for the University of Chicago Press. The book is much improved thanks to Will Stenhouse's meticulous reading and thoughtful criticisms. I'm grateful as well to Moti Feingold, who commented on the penultimate draft. At an earlier stage, Okasha El Daly and Kevin van Bladel generously read Arabic passages in Kircher's work and shared their evaluations. I also wish to thank David Biale, Brian Copenhaver, Allison Coudert, Brian Curran, Michael John Gorman, Katie Harris, Bernard Heyberger, Angela Mayer-Deutsch, Peter Miller, Susan Miller, Giovanni Pizzorusso, Ingrid Rowland, Mike Saler, and Nick Wilding. Among the many libraries and archives where I have worked, I owe a particular debt to the staff at Stanford's Department of Special Collections, especially John Mustain and Becky Fischbach. At the Press, it has been a pleasure to work with Karen Darling, who guided the book to publication, as well as Renaldo Migaldi and Abby Collier. Financial support has come from the Fulbright Foundation, the National Science Foundation (SES-0080717), the Mellon Foundation, the UC Davis Hellman Fellowship, the UC Davis Publication Assistance Fund, and the Dean's Office of the Division of Social Sciences.

I would be remiss not also to acknowledge the essential contributions of an impersonal historical force and a central Italian urban agglomeration. First, I owe an incalculable debt to the zeitgeist which lifted Athanasius Kircher from oblivion shortly after I undertook this project. Pondering a dissertation topic in the last days of the twentieth century, I worried that the eccentric Jesuit polymath might be too obscure a subject. Within two years he had become a postmodern cultural icon, his quatrocentenary reported by the *New York Times*, National Public Radio, the *New Yorker* magazine, and other media while his enormous engraved likeness emblazoned the sides of Roman buses and a permanent exhibit celebrated his genius at Los Angeles's Museum of Jurassic Technology. More unexpectedly, a decade on, he has not returned to obscurity. If this book sells, I thank the strange times we live in. The city of Rome has left a profound mark on the book and its author in ways both tangible and intangible. I thank Athanasius Kircher (and the Fulbright and Mellon Foundations) for the opportunity to spend two splendid years amid the city's unhurried chaos, tangled layers of history, and indelible beauty. I thank Rome for bringing a scholarly research project to life. Among my Roman friends I am especially grateful to Wendy Artin and Bruno Boschin, who welcomed me into their family and introduced me to the living city.

Friends and family have sustained me over the years of this project. In particular I thank Judith Levine, Nomi Stolzenberg, and especially Gabriel Stolzenberg, who read the manuscript at key stages, lending his keen ear for language. Above all, Mara and, more recently, Milo have kept me going on a daily basis. Strange to say, neither of them has ever known me not working on Athanasius Kircher. We all have something to look forward to. As a work of scholarship, this book advances arguments, based on original research, intended to contribute to early modern intellectual history. I have also tried to tell a good story. The story is for Mara.

BIBLIOGRAPHY

ARCHIVES

Archivio della Pontificia Università Gregoriana, Rome
Archivum Romanum Societatis Iesu, Rome
    Fondo Gesuitico (FG)
        Censurae Librorum
        Lugd. (Catalogi)
        Rhen. Inf. (Catalogi)
        Rhen. Sup. (Catalogi, Indiam petentes)
        Rom. (Catalogi)
        Sic. (Catalogi)
Archivio Segreto Vaticano, Vatican City
    Fondo Della Valle–Del Bufalo
Biblioteca Apostolica Vaticana, Vatican City
    Barberini Latini
    Neofiti
    Vaticani Latini
Biblioteca nazionale centrale di Roma
    Fondo Gesuitico
Bibliothèque interuniversitaire (Montpellier), Section Médecine (formerly: Montpellier, Bibliothèque de l'École de Médecine)
Bibliothèque nationale de France, Paris
    Fonds Dupuy
    Fonds français (FF)
    Nouvelles acquistions françaises (NAF)
    Nouvelles acquistions latines (NAL)

PRINTED WORKS COMPOSED BEFORE 1850

Addison, Joseph. "Dialogues upon the Usefulness of Medals." In *The Works of the Right Honourable Joseph Addison*, vol. 1, 447–539. London: Printed for Jacob Tonson, 1721.

Agrippa von Nettesheim, Heinrich Cornelius. *De occulta philosophia libri tres*. Edited by V. Perrone Compagni. Leiden: Brill, 1992. Original publication, 1533.

Alciati, Andrea. *Emblemata cum commentariis Claudii Minois I.C. Francisci Sanctii Brocensis, & notis Laurentii Pignorii Patavini*. Patavii: Typis Pauli Frambotti bibliopolae, 1661. Original publication, 1612.

Aleandro, Giorlamo. *Antiquae tabulae marmoreae solis effigie symbolisque exculptae accurata explicatio*. Romae: Ex Typographia Bartholomaei Zannetti, 1616.

Allacci, Leone. *Apes Urbanae, sive de viris illustribus, qui ab anno MDCXXX per totum MDCXXXII Romae adfuerunt, ac typis aliquid evulgarunt*. Romae: Ludovicus Grignanus, 1633.

Alphabetum Cophtum sive Aegyptiacum. [Rome]: [Congregatio de Propaganda Fide], [1630].

Antiquitates ecclesiae orientalis, clarissimorum virorum . . . Dissertationibus epistolicis enucleatae. Londini: Prostant apud Geo. Wells, 1682.

Arcangelo da Borgonovo. *Cabalistarum selectiora, obscurioraque dogmata, a Ioanne Pico ex eorum commentationibus pridem excerpta*. Venetiis: Apud Franciscum Franciscium Senensem, 1569.

Assemani, Giuseppe Simonio. *Bibliotheca orientalis Clementino-vaticana, I, De scriptoribus Syris Orthodoxis*. Romae: Typis Sacrae Congregationis de Propaganda Fide, 1719.

Assemani, Simone. *Catologo de' codici manoscritti della Biblioteca Naniana*. 2 vols. Padua: Nella Stamperia del Seminario, 1787–92.

Assemani, Stefano Evodio. *Bibliothecae Mediceae Laurentianae et Palatinae codicum mms. orientalium catalogus*. Florentiae: Ex Typographio Albiziniano, 1742.

Bandini, Angelo Maria. *Dell'obelisco di Cesare Augusto scavato dalle rovine del Campo Marzo*. Roma: Appresso Niccolò e Marco Pagliarini, 1750.

Bartolocci, Giulio. *Bibliotheca magna rabbinica de scriptoribus, & scriptis Hebraicis, ordine alphabetico Hebraicè, & Latinè digestis*. 4 vols. Romae: Ex Typographia Sacrae Congregationis de Propaganda Fide, 1675–93.

Boas, George. *The Hieroglyphics of Horapollo*. Princeton, NJ: Princeton University Press, 1993. Original publication, 1950.

Bonfrère, Jacques. *Pentateuchus Moysis commentario illustratus; praemissis, quae ad totius Scripturae intelligentiam manuducant, praeloquiis perutilibus*. Antwerpiae: Ex Officina Plantiniana, Apud Balthasarem Moretum, et Viduam Ioannis Moreti, et Io. Meurisium, 1625.

Bouchard, Jean-Jacques, ed. *Monumentum Romanum Nicolao Claudio Fabricio Perescio senatori Aquensi doctrinae virtutisque causa factum*. Romae: Typis Vaticanis, 1638.

Brucker, Jacob. *Historia critica philosophiae a mundi incunabulis ad nostram usque aetatem deducta*. 6 vols. Lipsiae: Apud Bernh. Christoph Breitkopf, 1742–44.

Bujunda, J. M. de, ed. *Index de Rome 1590, 1593, 1596, avec étude des index de Parme 1580 et Munich 1582*. 9 vols. Sherbrooke: Centre d'études de la Renaissance, 1993.

Buxtorf, Johannes. *Bibliotheca rabbinica nova.* Basilae: Typis Conradi Waldirchi, 1613.
Cancellieri, Francesco. *Il mercato, il lago dell'Acqua Vergine ed il Palazzo Panfiliano nel Circo Agonale detto voglarmente Piazza Navona.* Rome: Francesco Bourlie, 1811.
Carmignac, Jean, ed. *Evangiles de Matthieu et de Marc traduits en hébreu en 1668 par Giovanni Battista Iona retouchés en 1805 par Thomas Yeates.* Turnhout: Brépols, 1982.
Cartari, Vincenzo. *Imagini delli Dei de gl'Antichi.* Milan: Luni, 2004. Original publication, 1647.
Casali, Giovanni Battista. *De veteribus Aegyptiorum ritibus.* Romae: Ex Typographia Andreae Phaei, 1644.
Casaubon, Isaac. *De rebus sacris et ecclesiasticis exercitationes XVI.* Londini: Ex officina Nortoniana apud I. Billium, 1614.
Cassian, John. *John Cassian: The Conferences.* Translated by Boniface Ramsey. New York: Paulist Press, 1997.
Caussin, Nicolas. *De symbolica Aegyptiorum sapientia, in qua symbola, historiae selectae, quae ad omnem emblematum, aenigmatum, hieroglyphicorum cognitionem viam praestat.* Coloniae Agrippinae: Apud Ioannem Kinckium, 1631. Original publication, 1618.
Champollion, Jean-François. *Dictionnaire égyptien.* Arles: Solin-Actes Sud, 2000.
——— *Lettre à M. Dacier.* Paris: Chez Firmin Didot pére et fils, 1822.
———. *Grammaire égyptienne; ou, Principes généraux de l'écriture sacrée égyptienne.* Paris: Typographie de Firmin Didot frères, 1836.
Ciantes, Joseph. *De sanctissima Trinitate ex antiquorum Hebraeorum testimoniis evidenter comprobata.* Romae: Typis Varesii, 1667.
Court de Gébelin, Antoine. *Monde primitif, analysé et comparé avec le monde moderne, consideré dans son génie allégorique et dans les allégories auxquelles conduisit ce génie.* 9 vols. Paris: Durand, 1773–81.
Cudworth, Ralph. *The True Intellectual System of the Universe.* 2nd edition. London: Printed for J. Walthoe, 1743. Original publication, 1678.
D'Alembert, Jean Le Rond. "Erudition." In *Encyclopédie, ou dictionnaire raisonné des sciences, des arts et des métiers,* edited by Denis Diderot and Jean le Rond d'Alembert, vol. 5, 914–18. Paris: Briasson, 1751–65.
———. *Preliminary Discourse to the Encyclopedia of Diderot.* Translated by Richard Schwab. Chicago: University of Chicago Press, 1995. Original publication, 1751.
Della Valle, Pietro. *Viaggi di Pietro Della Valle il Pellegrino con minuto raguaglio di tutte le cose notabili osservate in essi discritti da lui medesimo in 54. lettere familiari.* Roma: Apresso Vitale Mascardi, 1650–1663.
Descartes, René. *Discourse on Method and Meditations on First Philosophy.* Translated by Donald A. Cress. 2nd ed. Indianapolis: Hackett, 1980. Original publication, 1637, 1641.
Diderot, Denis. "Antédiluvienne." In *Encyclopédie, ou dictionnaire raisonné des sciences, des arts et des métiers,* edited by Denis Diderot and Jean Le Rond d'Alembert, vol. 1, 493–95. Paris: Briasson, 1751–65.
Dindorf, Ludwig August, ed. *Chronicon Paschale.* Preussen: E. Weber, 1832.
Du Choul, Guillaume. *Discorso della religione antica de romani.* In Lione: Appresso Guglielmo Rovillio, 1569.

Ecchellensis, Abraham. *Semita sapientiae sive ad scientias comparandas methodus.* Parisiis: Apud Adrianum Taupinart, 1646.

Ficino, Marsilio. *Opera Omnia.* 2 vols. Turin: Bottega d'Erasmo, 1962. Original publication, 1576.

———. *Three Books on Life.* Translated by Carol V. Kaske and John R. Clark. Binghamton: Medieval & Renaissance Texts & Studies, 1989. Original publication, 1489.

Finney, Charles G. *Skeletons of a Course in Theological Lectures.* Oberlin: James Steele, 1840.

Fulvio, Andrea. *L'antichità di Roma.* Edited by Giorlamo Ferrucci. In Venetia: Per Girolamo Francini Libraro in Roma, 1588.

Gale, Theophilus. *The Court of the Gentiles; or, A Discourse Touching the Traduction of Philosophie from the Scriptures and Jewish Church. Part II, Of Barbaric and Grecanic Philosophie.* London: Printed by J. Macock for Thomas Gilbert, 1676.

Gassendi, Pierre. *The Mirrour of True Nobility & Gentility. Being the Life of the Renowned Nicolaus Claudius Fabricius Lord of Peiresk, Senator of the Parliament at Aix.* Translated by W. Rand. London: J. Streater for Humphrey Moseley, 1657.

Gigli, Giacinto. *Diario Romano (1608–1670).* Edited by Giuseppe Ricciotti. Rome: Tumminelli, 1958.

Gorp, Jean van. *Hieroglyphica. Opera.* Antwerpiae: Ex officina Christophori Plantini, 1580.

Hendreich, Christoph. *Pandectae Brandenburgicae.* Berolini: Typis Viduae Salfeldianis, 1699.

Herwart von Hohenburg, Johann Friedrich. *Admiranda ethnicae theologiae mysteria propalata.* Monachii: Formis Nicolai Henrici, 1626. Original publication, 1623.

Herwart von Hohenburg, Johann Georg. *Thesaurus Hieroglyphicorum.* [Munich?]: [1610].

Holstenius, Lucas. *Epistolae ad diversos, quas ex editis et ineditis codicibus collegit atque illustravit Jo. Franc. Boissonade.* Edited by J. F. Boissonade. Parisiis: J. Gratiot, 1817.

———. *Vetus pictura nymphaeum referens commentariolo explicata.* Romae: Typis Barberinis, 1676.

Hottinger, Johann Heinrich. *Smegma orientale: Sordibus barbarismi, contemtui praesertim linguarum orientalium oppositum.* Heidelbergae: Typis & impensis Adriani Wyngaerden, Academ. Bibliopolae & typographi, 1658.

Huet, Pierre Daniel. *Demonstratio Evangleica, ad Serenissimum Delphinum. Sexta editio ab auctore recognita, castigata et amplificata.* Francofurti: Sumptibus Thomae Fritschii, 1722. Original publication, 1679.

Iamblichus. *On the Mysteries.* Translated by Emma C. Clarke, John M. Dillon, and Jackson Hershbell. Atlanta: Society of Biblical Literature, 2003.

Jacobs, Fr., and F. A. Ukert. *Beiträge zur ältern Litteratur oder Merkwürdigkeiten der Herzogl. öffentlichen Bibliothek zu Gotha.* Leipzig: Dyk'sche Buchhandlung, 1835.

Kircher, Athanasius. *Ad Alexandrum VII Pont. Max. obelisci Aegyptiaci, nuper inter Isaei Romani rudera effossi interpretatio hieroglyphica.* Romae: Ex typographia Varesii, 1666.

———. *Ars magna lucis et umbrae in decem libros digesta.* Romae: Sumptibus Hermanni Scheus ex typographia Ludovici Grignani, 1646.

———. *Ars magna sciendi in XII libros digesta*. Amstelodami: Apud Joannem Janssonium à Waesberge, & viduam Elizei Weyerstraet, 1669.

———. *Ars magnesia, hoc est, disquisitio bipartita-empirica seu experimentalis, physico-mathematica de natura, viribus, et prodigiosis effectibus magnetis*. Herbipoli: Typis Eliae Michaelis Zinck, 1631.

———. *China monumentis, quà sacris quà profanis, nec non variis naturae & artis spectaculis, aliarumque rerum memorabilium argumentis illustrata*. Amstelodami: Apud Joannem Janssonium à Waesberge & Elizeum Weyerstraet, 1667.

———. *Itinerarium exstaticum quo mundi opificium, id est coelestis expansi, siderumque tam errantium, quàm fixorum natura, vires, proprietates, singulorumque compositio & structura, ab infimo telluris globo, usque ad ultima mundi confinia, per ficti raptus integumentum explorata, nova hypothesi exponitur ad veritatem interlocutoribus Cosmiele et Theodidacto*. Romae: Typis Vitalis Mascardi, 1656.

———. *Lingua Aegyptiaca restituta, opus tripartitum, quo linguae Coptae sive idiomatis illius primaevi Aegyptiorum pharaonici, vetustate temporum paene collapsi, ex abstrusis Arabum monumentis, plena instauratio continetur, cui adnectitur supplementum earum rerum, quae in Prodromo Copto, & opere hoc tripartito, vel omissa, vel obscurius tradita sunt, nova, & peregrina eruditione contextum, ad instauratae linguae usum, speciminis loco declarandum*. Romae: Sumptibus Hermanni Scheus apud Ludovicum Grignanum, 1643.

———. *Magnes, sive, de arte magnetica opus tripartitum*. Romae: Sumptibus Hermanni Scheus. Ex typographia Ludovici Grignani, 1641.

———. *Mundus Subterraneus, in XII libros digestos, quo divinum subterrestris mundi opificium, mira ergasteriorum naturae in eo distributio, verbo παντἀμορφον Protei regnum, universae denique naturae majestas & divitiae summa rerum varietate exponuntur*. 2 vols. Amstelodami: Apud Joannem Janssonium & Elizeum Weyerstraten, 1665.

———. *Musurgia universalis, sive, ars magna consoni et dissoni in X libros digesta*. Romae: Ex typographia haeredum Francisci Corbelletti, 1650.

———. *Obeliscus Pamphilius, hoc est, interpretatio nova & hucusque intentata obelisci hieroglyphici quem non ita pridem ex veteri hippodromo Antonini Caracallae Caesaris, in Agonale Forum transtulit, integritati restituit & in urbis aeternae ornamentum erexit Innocentius X Pont. Max.: In quo post varia Aegyptiacae, Chaldaicae, Hebraicae, Graecanicae antiquitatis, doctrinaeque quà sacrae, quà profanae monumenta, veterum tandem theologia, hieroglyphicis involuta symbolis, detecta è tenebris in lucem asseritur*. Romae: Typis Ludovici Grignani, 1650.

———. *Oedipus Aegyptiacus: Hoc est, Universalis hieroglyphicae veterum doctrinae temporum iniuria abolitae instauratio: Opus ex omni Orientalium doctrina & sapientia conditum, nec non viginti diversarum linguarum, authoritate stabilitum, felicibus auspiciis Ferdinandi III, Austriaci sapientissimi & invictissimi Romanorum imperatoris semper augusti è tenebris erutum, atque bono reipublicae literariae consecratum*. 3 in 4 vols. Romae: Ex typographia Vitalis Mascardi, 1652–54.

———. *Primitiae gnomonicae catoptricae, hoc est, horologiographiae novae specularis*. Avenione: Ex typographia I. Piot S. Officii typographi, 1635.

———. *Prodromus Coptus sive Aegyptiacus . . . in quo cùm linguae Coptae, sive Aegypti-

acae, quondam Pharaonicae, origo, aetas, vicissitudo, inclinatio, tùm hieroglyphicae literaturae instauratio: Uti per varia variarum eruditionum, interpretationumque difficillimarum specimina, ita nova quoque & insolita methodo exhibentur. Romae: Typis S. Cong. de Propag. Fide, 1636.

———. *Specula Melitensis encyclica, Hoc est, Syntagma novum instrumentorum Physico-Mathematicorum.* Neapoli: typis Secundini Roncagliolo, 1638.

———. *Sphinx mystagoga, sive Diatribe hieroglyphica qua mumiae, ex Memphiticis pyramidum adytis erutae & non ita pridem in Galliam transmissae, juxta veterum hieromystarum mentem, intentionemque, plena fide & exacta exhibetur interpretatio.* Amstelodami: Ex Officina Janssonio-Waesbergiana, 1676.

———. *Turris Babel, sive archontologia.* Amstelodami: Ex officina Janssonio-Waesbergiana, 1679.

———. *Vita Admodum Reverendi P. Athanasii Kircheri, Societ. Jesu Viri toto orbe celebratissimi.* Fasciculus Epistolarum, edited by Hieronymus Langenmantel. Augusta Vinedlicorum: Typis Utzschneiderianis, 1684.

"Kircher, Athanase." In *Encyclopédie; ou, Dictionnaire universel raisonné des connoissances humaines*, edited by Fortuné Barthélemy de Félice, vol. 25, 365. Yverdon, 1770-75.

Kollár, Adám Ferencz. *Ad Petri Lambecii Commentariorum de Augusta Bibliotheca Caes. Vindobonensi Libros VIII Supplementorum liber primus posthumus.* Vindobonae: Typis et Sumpt. Joan. Thomae Nob. de Trattern, 1790.

La Peyrère, Isaac de. *Praeadamitae, sive, exercitatio super versibus duodecimo, decimotertio, & decimoquarto, capitis quinti Epistolae D. Pauli ad Romanos.* [Amsterdam]: 1655.

———. *Systema theologicum, ex praeadamitarum hypothesi.* [Amsterdam]: 1655.

Lafitau, Joseph-François. *Moeurs des sauvages ameriquains, comparées aux moeurs des premiers temps.* 2 vols. A Paris: Chez Saugrain . . . [et] Hochereau, 1724.

Lambeck, Peter. *Petri Lambecii Hamburgensis Commentariorum de Augustissima Bibliotheca Caesarea Vindobonensi liber primus [-octavus].* Edited by Adám Ferencz Kollár. Editio altera. 8 vols. Vindobonae: Typis et Sumptibus Joan. Thomae Nob. de Trattnern, 1766-82.

Liceti, Fortunio. *De lucernis antiquorum reconditis libb. sex.* Utini: Ex typographia Nicolai Schiratii; Prostant Patavii: apud Franciscum Bolzettam, 1652. Original publication, 1621.

———. *Hieroglyphica sive antiqua schemata gemmarum anularum.* Patavii: Typis Sebastiani Sardi, 1653.

Macrobius. *The Saturnalia.* Translated by Percival Vaughan Davies. New York: Columbia University Press, 1969.

Maimonides, Moses. *Doctor perplexorum.* Translated by Johannes Buxtorf. Westmead: Gregg, 1969. Original publication, 1629.

———. *Dux seu director dubita[n]tium aut perplexorum.* Frankfurt: Minerva, 1964. Original publication, 1520.

Marracci, Ludovico. *Prodromus ad refutationem Alcorani.* 4 vols. Romae: Typis Sac. Cong. de Prop. Fide, 1691.

Maurice, Thomas. *Indian Antiquities.* London: 1800.

Mencken, Johann Burckhard. *The Charlatanry of the Learned.* Translated by F. Litz. With notes and an introduction by H. L. Mencken. New York: Knopf, 1937.

Mercati, Michele. *De gli obelischi di Roma.* In Roma: Domenico Basa, 1589.

Montfaucon, Dom Bernard de. *L'Antiquité expliquée et représentée en figures.* 5 in 10 vols. Paris: F. Delaulne, 1719.

"Neofito." In *Dizionario di erudizione storico-ecclesiastica,* edited by Gaetano Romano, vol. 47, 267–76. Venezia: Tipografia Emiliana, 1840–61.

Parker, Samuel. *A free and impartial censure of the Platonick philosophie.* Oxford: Printed by W. Hall for Richard Davis, 1666.

Patrizi, Francesco. *Nova de universis philosophia.* Ferrariae: apud Benedictum Mammarellum, 1591.

Peiresc, Nicolas-Claude Fabri de. *Lettres à Cassiano dal Pozzo, 1626–1637.* Edited by Jean-François Lhote. Clermont-Ferrand: Adosa, 1989.

———. *Lettres à Claude Saumaise et à son entourage.* Edited by Agnes Bresson. Florence: Olschki, 1992.

Philo of Byblos. *The Phoenician History.* Introduction, critical text, translation, and notes by Harold W. Attridge and Robert A. Oden, Jr. Washington: Catholic Biblical Association of America, 1981.

Pignoria, Lorenzo. *Mensa Isiaca, qua sacrorum apud Aegyptios ratio & simulachra subiectis tabulis aeneis simul exhibentur & explicantur. Accessit eiusdem authoris de magna deum matre discursus, & sigillorum, gemmarum, amuletorum aliquot figurae & earundem ex Kirchero Chifletoque interpretatio.* Amstelodami: Sumptibus Andreae Frisii, 1669.

———. *Symbolarum epistolicarum Liber.* Patavii: Ex. Typograph. Ioannis Baptistae de Martinis, 1628.

———. *Vetustissimae tabulae aeneae sacris Aegyptiorum simulachris coelate accurata explicatio.* Venetiis: Apud Io: Anto: Rampazaettum . . . Sumptibus Iacobi Franco, 1605.

Pistorius, Johannes. *Artis cabalisticae, hoc est, reconditae theologiae et philosophiae, scriptorum, tomus I.* Basilae: Per Sebastianum Henricpetri, [1587].

Plantavit de la Pause, Jean. *Planta Vitis seu thesaurus synonymicus Hebraico-Chaldaico-Rabbinicus.* Lodovae: Typis Arnaldi Colomerii, 1644.

Plotinus. *Enneads.* With an English translation by A. H. Armstrong. 7 vols. Cambridge: Harvard University Press, 1966–88.

Possevino, Antonio. *Apparatus sacer ad scriptores veteris, & novi Testamenti.* 3 vols. Venetiis: Apud Societatem Venetam, 1606. Original publication, 1603.

Quatremère, Étienne. "Mémoire sur les Nabatéens, Troisième Section." *Nouveau Journal Asiatique* 15 (1835): 209–71.

———. *Recherches critiques et historiques sur la langue et la littérature de l'Égypte.* Paris: De l'imprimerie impériale, 1808.

Ricius, Paulus. *De coelesti agricultura.* Excusus Augustae Vindelicorum: per Henricum Stayner, 1541.

Rittangel, Joannes Stephan. *[Sefer Yetzirah] id est Liber Iezirah qui Abrahamo Patriarchae adscribitur, unà cum commentario Rabi Abraham F. D. super 32 semitis sapientiae, à quibus liber Iezirah incipit.* Amstelodami: Apud Joannem & Jodocum Janssonios, 1642.

Sacy, Silvestre de. "Review of Schnurrer's *Bibliotheca Arabica.*" *Magasin Encyclopédique, ou Journal des sciences, des lettres, et des arts* 1 (1814): 192–211.

Saumaise, Claude. *Epistolarum liber primus.* Lugduni Batavorum: Ex typographia Adriani Wyngaerden, 1656.

Schnurrer, Christian Frederic. *Bibliotheca Arabica.* Halae ad Salam: Typis et sumtu I. C. Hendelii, 1811.

Schott, Kaspar. *Technica curiosa, sive mirabilia artis.* Norimbergae: Sumptibus Johannis Andreae Endteri, & Wolfgangi Junioris Haeredum, excudebat Jobus Hertz, typographus Herbipol., 1664.

Selden, John. *De dis Syris syntagmata II, adversaria nempe de numinibus commentitiis in Vetere Instrumento memoratis: Accedunt fere quae sunt reliqua Syrorum, prisca porro Arabum, Aegyptiorum, Persarum, Afrorum, Europaeorum item theologia, subinde illustratur.* Editio altera, emendatior & tertia parte auctior. Lugduni Batavorum: Ex Officinâ Bonaventurae & Abrahami Elsevir, 1629. Original publication, 1617.

Serarius, Nicolaus. *Iosue, ab utero ad ipsum usque tumulum.* 2 vols. Moguntiae: Ex officina typographica Ioannis Albini, 1609–10.

———. *Rabbini et Herodes.* Moguntiae: E typographeo Balthasaris Lippii, 1607.

Serarius, Nicolaus, and Christian Joannes Georgius. *Moguntiacarum rerum libri quinque.* Moguntiae: typis Balthasaris Lippii, 1722. Original publication, 1604.

Shaw, Thomas. *Travels; or, Observations, Relating to Several Parts of Barbary and the Levant.* Edinburgh: 1808. Original publication, 1738.

Spencer, John. *De Urim et Thummim.* Cantabrigiae: Impensis Timoth. Garthwait, 1669.

Spitzel, Gottlieb. *Sacra bibliothecarum illustriorum arcana retecta.* Augustae Vindelicorum: Apud Gottlieb Goebelium, 1668.

Stukeley, William. *Abury, a temple of the British druids, and some others described.* London: Printed for the author, and sold by W. Innys R. Manby B. Dod, 1743.

Suárez, Francisco. *Disputationes metaphysicae.* 2 vols. Hildesheim: Georg Olms, 1965. Original publication, 1866.

Tamizey de Larroque, Philippe, ed. *Les Correspondants de Peiresc, fasc. III. Jean-Jacques Bouchard. Lettres inédites, Écrites à Rome, à Peiresc (1633–1637).* Paris: Alphonse Picard, 1881.

———. *Les Correspondants de Peiresc, fasc. V. Claude de Saumaise. Lettre inédites ecrites de Dijon, de Paris et de Leyde à Peiresc.* Extrait des Mémoires de l'Academie de Dijon. Dijon: Imprimerie Darantiere, 1882.

———. *Les Corréspondants de Peiresc, fasc. XIV. Samuel Petit. Lettres inédites écrites, de Nimes et de Paris, a Peiresc.* Nimes: Chastanier, 1887.

———. *Les Corréspondants de Peiresc, fasc. XV. Thomas d'Arcos. Lettres inédites écrites de Tunis . . . (1633–1636).* Algiers: A. Jourdan, 1889.

———. *Lettres de Peiresc.* 7 vols, Collections de documents inédits sur l'histoire de France. Paris: Imprimerie Nationale, 1888–98.

Tenison, Thomas. *Of Idolatry.* London: Printed for Francis Tyton, 1678.

Tomasini, Giacomo Filippo. *Manus aenae Cecropii Votum referentis dilucidatio.* Patavii: Typis Sebastiani Sardi, 1649.

TW. "Remains of Sanchoniatho." *The Classical Journal* 40 (1829): 157–59.

Valence, Apollinaire de, ed. *Correspondance de Peiresc avec plusieurs missionnaires et religieux de l'ordre des Capucins, 1631–1637*. Paris: Alphonse Picard, 1892.
Valeriano, Pierio. *Hieroglyphica, sive de sacris Aegyptiorum literis commentarii*. Basileae: [M. Isengrin], 1556.
Venel, Gabriel-François. "Chymie ou Chimie." In *Encyclopédie, ou dictionnaire raisonné des sciences, des arts et des métiers*, edited by Denis Diderot and Jean Le Rond d'Alembert, vol. 3, 408–37. Paris: Briasson, 1751–1765.
Vico, Giambattista. *New Science*. Translated by David Marsh. New York: Penguin, 1999. Original publication, 1730.
Walsh, John. *Sketches of Hebrew and Egyptian Antiquity*. Dublin: 1793.
Warburton, William. *The Divine Legation of Moses Demonstrated*. 10th ed., 3 vols. London: Thomas Tegg, 1846. Original publication 1737–41.
Witsius, Herman. *Aegyptiaca, et ΔΕΚΑΥΛΟΝ, sive de Aegyptiacorum sacrorum cum Hebraicis collatione libri tres*. Basilae: Apud Joh. Rudolphum Im-Hoff, 1739. Original publication, 1683.
Wolf, Johann Christoph. *Bibliotheca Hebraea*. 4 vols. Hamburgi: Impensis C. Liebezeit, 1715–33.
Wyttenbach, Daniel. *Animadversiones in Plutarchi libri*. Oxoni: E typographeo Clarendonio, 1821.
Young, Thomas. *An Account of Some Recent Discoveries in Hieroglyphical Literature, and Egyptian Antiquities*. London: John Murray, 1823.
Zenker, Julius Theodor. *Bibliotheca orientalis*. 2 vols. Leipzig: G. Engelmann, 1846–61.
Zoega, Georg. *De origine et usu obeliscorum ad Pium Sextum, Pontificem Maximum*. Romae: Typis Lazzarini Typographi Cameralis, 1797.
Zorn, Peter. *Bibliotheca antiquaria et exegetica*. Francofurti & Lipsiae: Apud Christoph. Gottl. Nicolai, 1724–25.

PRINTED WORKS COMPOSED AFTER 1850

Allen, Don Cameron. *Mysteriously Meant: The Rediscovery of Pagan Symbolism and Allegorical Interpretation in the Renaissance*. Baltimore: Johns Hopkins University Press, 1970.
———. "The Predecessors of Champollion." *Proceedings of the American Philosophical Society* 104 (1960), no. 5: 527–47.
Allen, Michael J. B. "At Variance: Marsilio Ficino, Platonism, and Heresy." In *Platonism at the Origins of Modernity: Studies on Platonism and Early Modern Philosophy*, edited by Douglas Hedley and Sarah Hutton, 31–44. Dordecht: Springer, 2008.
———. *Synoptic Art: Marsilio Ficino on the History of Platonic Interpretation*. Florence: Olschki, 1998.
Ashworth, William. "Natural History and the Emblematic World View." In *Reappraisals of the Scientific Revolution*, edited by David Lindberg and Robert Westman, 303–33. Cambridge: Cambridge University Press, 1990.
Assmann, Jan. *Moses the Egyptian: The Memory of Egypt in Western Monotheism*. Cambridge, MA: Harvard University Press, 1997.
Aufrère, Sydney. *La momie et la tempête: Nicolas-Claude Fabri de Peiresc et la curiosité*

égyptienne en Provence au début du XVIIe siècle. Avignon: Editions A. Barthélemy, 1990.

Baker, Keith Michael. *Inventing the French Revolution: Essays on French Political Culture in the Eighteenth Century.* Cambridge: Cambridge University Press, 1990.

Baldini, Ugo. "Christoph Clavius and the Scientific Scene in Rome." In *Gregorian Reform of the Calendar: Proceedings of the Vatican Conference to Commemorate its 400th Anniversary,* edited by George V. Coyne, Michael A. Hoskin, and Olaf Petersen. Vatican City: Specola Vaticana, 1983.

———. *Legem impone subactis. Studi su filosofia e scienza dei Gesuiti in Italia 1540–1632.* Rome: Bulzoni, 1992.

Baldwin, Martha. "Pious Ambition: Natural Philosophy and the Jesuit Quest for Patronage of Printed Books in the Seventeenth Century." In *Jesuit Science and the Republic of Letters,* edited by Mordechai Feingold, 285–329. Cambridge, MA: MIT Press, 2003.

———. "Reverie in a Time of Plague: Athanasius Kircher and the Plague Epidemic of 1656." In *Athanasius Kircher: The Last Man Who Knew Everything,* edited by Paula Findlen, 63–77. London: Routledge, 2004.

Barkan, Leonard. *Unearthing the Past: Archaeology and Aesthetics in the Making of Renaissance Culture.* New Haven: Yale University Press, 1999.

Bartòla, Alberto. "Alessandro VII e Athanasius Kircher S.J.: Ricerche ed appunti sulla loro corrispondenza erudita e sulla storia di alcuni codici chigiani." In *Miscellanea Bibliothecae Apostolicae Vaticanae, III,* 7–105. Vatican City: Biblioteca Apostolica Vaticana, 1989.

Beinlich, Horst. "Athanasius Kircher und die Kenntnis vom Alten Ägypten." In *Magie des Wissens. Athanasius Kircher 1602–1680: Universalgelehrter, Sammler, Visionär,* edited by Horst Beinlich, Christoph Daxelmüller, Hans-Joachim Vollrath and Klaus Wittstadt, 85–98. Dettelbach: J. H. Röll, 2002.

Beinlich, Horst, Christoph Daxelmüller, Hans-Joachim Vollrath, and Klaus Wittstadt, eds. *Magie des Wissens. Athanasius Kircher 1602–1680: Universalgelehrter, Sammler, Visionär.* Dettelbach: J. H. Röll, 2002.

Beinlich, Horst, Hans-Joachim Vollrath, and Klaus Wittstadt, eds. *Spurensuche. Wege zu Athanasius Kircher.* Dettelbach: J. H. Röll, 2002.

Ben-Schlomo, J. "Cordovero, Moses." In *Encyclopaedia Judaica,* vol. 5, 967–70. New York: Keter, 1971–72.

Benin, Stephen D. "The 'Cunning of God' and Divine Accommodation." *Journal of the History of Ideas* 45, no. 2 (1984): 179–91.

Bepler, Jill. "*Vicissitudo Temporum*: Some Sidelights on Book Collecting in the Thirty Years' War." *Sixteenth Century Journal* 32: 4 (2001): 953–68.

Biagioli, Mario. *Galileo, Courtier: The Practice of Science in the Culture of Absolutism.* Chicago: University of Chicago Press, 1993.

Bidez, Joseph, and Franz Cumont. *Les Mages Hellenisés: Zoroastre, Ostanès et Hystaspe.* 2 vols. Paris: Les Belles Lettres, 1938.

Bignami Odier, Jeanne. *La Bibliothèque Vaticane de Sixte IV à Pie XI. Recherches sur l'histoire des collections et manuscrits,* Studi e Testi 272. Vatican City: Biblioteca Apostolica Vaticana, 1973.

Billings, Timothy. "Jesuit Fish in Chinese Nets: Athanasius Kircher and the Translation of the Nestorian Tablet." *Representations*, no. 87 (2004): 1–42.

Bizzocchi, Roberto. *Genalogie incredibili: Scritti di storia nell'Europa moderna*. Bologna: Il Mulino, 1995.

Blackwell, Constance. "*Thales Philosophus*: The Beginning of Philosophy as a Discipline." In *History and the Disciplines: The Reclassification of Knowledge in Early Modern Europe*, edited by Donald R. Kelley, 61–82. Rochester: University of Rochester Press, 1997.

Blair, Ann. "Humanist Methods in Natural Philosophy: The Commonplace Book." *Journal of the History of Ideas* 53, no. 4 (1992): 541–51.

———. *The Theater of Nature: Jean Bodin and Renaissance Science*. Princeton, NJ: Princeton University Press, 1997.

Blavatsky, H. P. *The Secret Doctrine: The Synthesis of Science, Religion, and Philosophy*. 3rd revised edition. London: Theosophical Publishing House, 1895.

Bloch, Marc. *The Historian's Craft*. New York: Vintage Books, 1953.

Borg, Vincent. *Fabio Chigi, Apostolic Delegate in Malta (1634–1639): An Edition of His Correspondence*, Studi e Testi 249. Vatican City: Biblioteca Apostolica Vaticana, 1967.

Bottin, Francesco, Luciano Malusa, Giuseppe Micheli, Giovanni Santinello, and Ilario Tolomīo. *Models of the History of Philosophy I: From Its Origins in the Renaissance to the "Historia Philosophica."* Edited by Giovanni Santinello and Constance Blackwell. Dordecht: Kluwer, 1993.

Bouwsma, William. *Concordia Mundi: The Career and Thought of Guillaume Postel (1510–1581)*. Cambridge, MA: Harvard University Press, 1957.

Bredekamp, Horst. *The Lure of Antiquity and the Cult of the Machine*. Translated by Allison Brown. Princeton,NJ: Markus Wiener, 1995.

Brisson, Luc. *How Philosophers Saved Myths: Allegorical Interpretations and Classical Mythology*. Translated by Catherine Tihanyi. Chicago: University of Chicago Press, 2004. Original publication, 1996.

Brunet, Jacques-Charles. *Manuel du libraire et de l'amateur de livres*. 5th edition. 6 vols. Paris: Didot Frères, 1860–65.

Burke, Peter. "Images as Evidence in Seventeenth-Century Europe." *Journal of the History of Ideas* 64, no. 2 (2003): 273–96.

———. *The Renaissance Sense of the Past*. New York: St. Martin's Press, 1969.

———. "Rome as Center of Information and Communication for the Catholic World, 1550–1650." In *From Rome to Eternity: Catholicism and the Arts in Italy, ca. 1550–1650*, edited by Pamela M. Jones and Thomas Worcester. Leiden: Brill, 2002.

Burnett, Charles. "The Establishment of Medieval Hermeticism." In *The Medieval World*, edited by Peter Linehan and Janet L. Nelson, 111–30. London: Routledge, 2001.

———. "The Second Revelation of Arabic Philosophy and Science: 1492–1562." In *Islam and the Italian Renaissance*, edited by Charles Burnett and Anita Contadini, 185–98. London: Warburg Institute, 1999.

Burnett, Stephen G. *From Christian Hebraism to Jewish Studies: Johannes Buxtorf (1564–1629) and Hebrew Learning in the Seventeenth Century*. Leiden: Brill, 1996.

Burstein, Stanley M. "Images of Egypt in Greek Historiography." In *Ancient Egyptian Literature. History and Forms*, edited by Antonio Loprieno, 590–604. Leiden: Brill, 1996.
Busi, Giulio. *Qabbala visiva*. Turin: Einaudi, 2005.
Cahen-Salvador, Georges. *Un grand humaniste: Peiresc, 1580–1637*. Paris: Michel, 1951.
Cappelletti, Francesca, and Caterina Volpi. "New Documents Concerning the Discovery and Early History of the *Nozze Aldobrandini*." *Journal of the Warburg and Courtauld Institutes* 56 (1993): 274–81.
Casciato, Maristella, Maria Grazia Ianniello, and Maria Vitale, eds. *Enciclopedismo in Roma barocca: Athanasius Kircher e il museo del Collegio Romano tra Wunderkammer e museo scientifico*. Venice: Marsilio, 1986.
Cavarra, Angela A., ed. *Hebraica: il mondo ebraico nell'interpretazione cristiana nei secoli XV–XVIII*. Rome: Aisthesis, 2000.
Celenza, Christopher S. "The Search for Ancient Wisdom in Early Modern Europe: Reuchlin and the Late Ancient Esoteric Paradigm." *The Journal of Religious History* 25, no. 2 (2001): 115–33.
Cerruti, Simona, and Gianna Pomata, eds. *Fatti: Storie dell'evidenza empirica. Theme Issue. Quaderni Storici*. Vol. 108, no. 3 (2001).
Chaine, M. "Une composition oubliée du Père Kircher en honneur de Peiresc." *Revue de l'Orient Chrétien* 3e série, IX (1933–34): 207–8.
Champion, J. A. I. *The Pillars of Priestcraft Shaken: The Church of England and Its Enemies, 1660–1730*. Cambridge: Cambridge University Press, 1992.
Ciampi, Ignazio. *Della vita e delle opere di Pietro Della Valle Il Pellegrino. Monografia illustrata con nuovi documenti*. Rome: Tipografia Barbèra, 1880.
Cipriani, Giovanni. *Gli obelischi egizi: Politica e cultura nella Roma barocca*. Florence: Olschki, 1993.
Claridge, Amanda. "Archaeologies, Antiquaries, and the *Memorie* of Sixteenth- and Seventeenth-Century Rome." In *Archives and Excavations: Essays on the History of Archaeological Excavations in Rome and Southern Italy from the Renaissance to the Nineteenth Century*, edited by Ilaria Bignamini, 33–53. London: British School of Rome, 2004.
Clark, Stuart. *Thinking with Demons: The Idea of Witchcraft in Early Modern Europe*. Oxford: Oxford University Press, 1997.
Codici ebraici della Pia Casa dei Neofiti a Roma. Rome: Tipographia della Accademia dei Lincei, 1893.
Cohen, Thomas V. *Words and Deeds in Renaissance Rome*. Toronto: Toronto University Press, 1993.
Collani, Claudia von. *P. Joachim Bouvet S.J. Sein Leben und sein Werk*, Monumenta Serica Monograph Series XVII. Nettetal: Steyler, 1985.
Collins, Jeffrey. "Obelisks as Artifacts in Early Modern Rome: Collecting the Ultimate Antiquities." *Ricerche di storia dell'arte*, no. 71 (2000): 49–68.
"Congregationes Cardinalices." *Analecta Juris Pontificii. Droit canonique, liturgie, théologie et histoire*, no. 227 (1886): 887–89.
Contini, Riccardo. "Gli inizi della linguistica siriaca nell'Europa rinascimentale." In *Italia ed Europa nella linguistica del Rinascimento*, edited by Mirko Tavoni, vol. 2, 483–502. Modena: Franco Cosimo Panini, 1996.

Cook, Michael. "Pharaonic History in Medieval Egypt." *Studia Islamica* 57 (1983): 67–103.
Copenhaver, Brian. "Astrology and Magic." In *The Cambridge History of Renaissance Philosophy*, edited by Charles Schmitt and Quentin Skinner, 264–300. Cambridge: Cambridge University Press, 1988.
———. "Did Science Have a Renaissance?" *Isis* 83 (1992): 387–407.
———. "Hermes Trismegistus, Proclus, and the Question of a Philosophy of Magic in the Renaissance." In *Hermeticism and the Renaissance: Intellectual History and the Occult in Early Modern Europe*, edited by Ingrid Merkel and Allen G. Debus, 79–110. Washington: Folger Shakespeare Library, 1988.
———. "Iamblichus, Synesius and the 'Chaldean Oracles' in Marsilio Ficino's 'De Vita Libri Tres': Hermetic Magic or Neoplatonic Magic?" In *Supplementum Festivum: Studies in Honor of Paul Oskar Kristeller*, edited by James Hankins, John Monfasani, and Frederick Purnell, 441–55. Binghamton, NY: Medieval & Renaissance Texts & Studies, 1987.
———. "Natural Magic, Hermetism, and Occultism in Early Modern Science." In *Reappraisals of the Scientific Revolution*, edited by David Lindberg and Robert Westman, 261–302. Cambridge: Cambridge University Press, 1990.
———. "The Occultist Tradition and Its Critics." In *The Cambridge History of Seventeenth-Century Philosophy*, edited by Daniel Garber and Michael Ayers, 454–512. Cambridge: Cambridge University Press, 1998.
———. "A Tale of Two Fishes: Magical Objects in Natural History from Antiquity through the Scientific Revolution." *Journal of the History of Ideas* 52 (1991): 373–98.
Copenhaver, Brian, ed. *Hermetica: The Greek Corpus Hermeticum and the Latin Asclepius in a New English Translation, with Notes and Introduction*. Cambridge: Cambridge University Press, 1992.
Copenhaver, Brian, and Charles Schmitt. *Renaissance Philosophy*. Oxford: Oxford University Press, 1992.
Coudert, Allison P. *The Impact of the Kabbalah in the Seventeenth Century: Francis Mercury van Helmont (1614–1698)*. Leiden: Brill, 1999.
Coudert, Allison P., and Jeffrey S. Shoulson, eds. *Hebraica Veritas? Christian Hebraists and the Study of Judaism in Early Modern Europe*. Philadelphia: University of Pennsylvania Press, 2004.
Cropper, Elizabeth, Giovanna Perini, and Francesco Solinas, eds. *Documentary Culture: Florence and Rome from Grand-Duke Ferdinand I to Pope Alexander VII*. Bologna: Nuova Alfa Editoriale, 1992.
Cunnally, John. *Images of the Illustrious: The Numismatic Presence in the Renaissance*. Princeton, NJ: Princeton University Press, 1999.
Curran, Brian A. "'De sacrorum litterarum Aegyptiarum interpretatione.' Reticence and Hubris in Hieroglyphic Studies of the Renaissance: Pierio Valeriano and Annius of Viterbo." *Memoirs of the American Academy of Rome* 43/44 (1998/1999): 139–82.
———. *The Egyptian Renaissance: The Afterlife of Ancient Egypt in Early Modern Italy*. Chicago: University of Chicago Press, 2007.
Curran, Brian A., Anthony Grafton, Pamela O. Long, and Benjamin Weiss. *Obelisk: A History*. Cambridge, MA: Burndy Library, 2009.

Daly, Okasha El. *Egyptology: The Missing Millennium. Ancient Egypt in Medieval Arabic Writings.* London: UCL Press, 2005.
Dan, Joseph, ed. *The Christian Kabbalah: Jewish Mystical Books and their Christian Interpreters.* Cambridge, MA: Harvard College Library, 1997.
Dannenfeldt, Karl. "Egypt and Egyptian Antiquities in the Renaissance." *Studies in the Renaissance* 6 (1959): 7–27.
———. "The Pseudo-Zoroastrian Oracles in the Renaissance." *Studies in the Renaissance* 4 (1957): 8–30.
———. "The Renaissance Humanists and the Knowledge of Arabic." *Studies in the Renaissance* 2 (1955): 96–117.
Daston, Lorraine. "The Language of Strange Facts in Early Modern Science." In *Inscribing Science: Scientific Texts and the Materiality of Communication*, edited by Timothy Lenoir, 20–38. Stanford, CA: Stanford University Press, 1998.
"Marvelous Facts and Miraculous Evidence in Early Modern Europe." *Critical Inquiry* 18 (1991): 93–124.
———. "The Moral Economy of Science." *Osiris* 10 (1995): 3–24.
Daston, Lorraine, and Katharine Park. *Wonders and the Order of Nature, 1150–1750.* Cambridge, MA: Zone Books, 1998.
Dauphinais, Michael, Barry David, and Matthew Levering, eds. *Aquinas the Augustinian.* Washington: Catholic University of America Press, 2007.
David, Madeleine V. *Le débat sur les écritures et l'hiéroglyphe aux XVIe et XVIIe siècles et l'application de la notion de déchiffrement aux écritures mortes.* Paris: S.E.V.P.E.N., 1965.
Davies, W. V. *Egyptian Hieroglyphs.* Berkeley: University of California Press, 1987.
Daxemüller, Christoph. "Die Welt als Einheit: Eine Annäherung an das Wissenschaftskonzept des Athanasius Kircher." In *Magie des Wissens. Athanasius Kircher 1602–1680: Universalgelehrter, Sammler, Visionär*, edited by Horst Beinlich, Christoph Daxelmüller, Hans-Joachim Vollrath and Klaus Wittstadt, 27–48. Dettelbach: J. H. Röll, 2002.
Dear, Peter. *Discipline and Experience: The Mathematical Way in the Scientific Revolution.* Chicago: University of Chicago Press, 1995.
Dempsey, Charles. "The Classical Perception of Nature in Poussin." *Journal of the Warburg and Courtauld Institutes* 29 (1966): 219–49.
Dew, Nicholas. *Orientalism in Louis XIV's France.* Oxford: Oxford University Press, 2009.
Dietz, Luc. "Space, Light, and Soul in Patrizi's *Nova de universis philosophia* (1591)." In *Natural Particulars: Nature and the Disciplines in Renaissance Europe*, edited by Anthony Grafton and Nancy Siraisi, 139–69. Cambridge, MA: MIT Press, 1999.
Ditchfield, Simon. *Liturgy, Sanctity and History in Tridentine Italy: Pietro Maria Campi and the Preservation of the Particular.* Cambridge: Cambridge University Press, 1995.
Donadoni, Sergio. "I geroglifici di Athanasius Kircher." In *Athanasius Kircher: Il museo del mondo*, edited by Eugenio Lo Sardo, 101–10. Rome: Edizioni De Luca, 2001.
Donato, Maria Pia, and Jill Kraye, eds. *Conflicting Duties: Science, Medicine and Religion*

in Rome 1550–1750, Warburg Institute Colloquia 15. London: Warburg Institute, 2009.

D'Onofrio, Cesare. *Gli obelischi di Roma: Storia e urbanistica di una città dall'età antica al XX secolo.* 3rd revised edition. Rome: Romana Società Editrice, 1992.

Dooley, Brendan. *Morandi's Last Prophecy and the End of Renaissance Politics.* Princeton, NJ: Princeton University Press, 2002.

Duits, Rembrandt, and François Quiviger, eds. *Images of the Pagan Gods: Papers of a Conference in Memory of Jean Seznec*, Warburg Institute Colloquia. London: Warburg Institute, 2009.

Eamon, William. *Science and the Secrets of Nature: Books of Secrets in Medieval and Early Modern Culture.* Princeton, NJ: Princeton University Press, 1994.

Ebeling, Florian. *The Secret History of Hermes Trismegistus: Hermeticism from Ancient to Modern Times.* Translated by David Lorton. Ithaca, NY: Cornell University Press, 2007.

Eco, Umberto. *The Search for the Perfect Language.* Translated by J. Fentress. Oxford: Blackwell, 1995.

Edelstein, Dan. "The Egyptian French Revolution: Antiquarianism, Freemasonry and the Mythology of Nature." In *The Super-Enlightenment: Daring to Know Too Much*, edited by Dan Edelstein, 215–41. Oxford: Voltaire Foundation, 2010.

———. *The Enlightenment: A Genealogy.* Chicago: University of Chicago Press, 2010.

———. "Humanism, l'Esprit Philosophique, and the Encyclopédie." *Republics of Letters: A Journal for the Study of Knowledge, Politics, and the Arts* 1, no. 1 (2009).

Edelstein, Dan, ed. *The Super-Enlightenment: Daring to Know Too Much.* Oxford: Voltaire Foundation, 2010.

Elukin, Jonathan. "Maimonides and the Rise and Fall of the Sabians: Explaining Mosaic Law and the Limits of Scholarship." *Journal of the History of Ideas* 63 (2002): 619–37.

Englmann, Felicia. *Sphärenharmonie und Mikrokosmos: Das politische Denken des Athanasius Kircher (1602–1680).* Köln: Böhlau, 2006.

Ernst, Germana. "Astrology, Religion and Politics in Counter-Reformation Rome." In *Science, Culture and Popular Belief in Renaissance Europe*, edited by Stephen Pumfrey, Paolo Rossi and Maurice Slawinski, 249–73. Manchester, UK: Manchester University Press, 1991.

Evans, R. J. W. *The Making of the Habsburg Monarchy 1550–1700: An Interpretation.* Oxford: Clarendon Press, 1979.

Fabre, Pierre-Antoine, and Antonella Romano, eds. *Les jésuites dans le monde moderne. Nouvelles approches historiographiques.* Theme Issue. *Revue de Synthèse.* Vol. 120 (2–3), 1999.

Faivre, Antoine. *Access to Western Esotericism.* Buffalo: State University of New York Press, 1994.

Faivre, Antoine, and Frédérick Tristan, eds. *Kabbalistes Chrétiens*, Cahiers de l'Hermétisme. Paris: Albin Michel, 1979.

Fehrenbach, Frank. *Compendia Mundi: Gianlorenzo Berninis Fontana dei Quattro Fiumi (1648–1651) und Nicola Salvis Fontana dei Trevi (1732–1762).* Munich: Deutscher Kunstverlag, 2008.

Feingold, Mordechai. "Decline and Fall: Arabic Science in Seventeenth-Century England." In *Tradition, Transmission, Transformation: Proceeedings of Two Conferences on Pre-Modern Science Held at the University of Oklahoma*, edited by F. Jamil Ragep and Sally P. Ragep, 441–70. Leiden: Brill, 1996.

———. "The Grounds of Conflict: Grienberger, Grassi, Galileo, and Posterity." In *The New Science and Jesuit Science: Seventeenth Century Perspectives*, edited by Mordechai Feingold, 121–57. Dordecht: Kluwer, 2003.

Feingold, Mordechai, ed. *Jesuit Science and the Republic of Letters*. Cambridge, MA: MIT Press, 2003.

———. *The New Science and Jesuit Science: Seventeenth Century Perspectives*, Archimedes, New Studies in the History and Philosophy of Science and Technology, Volume 6. Dordecht: Kluwer, 2003.

Feldhay, Rivka. *Galileo and the Church: Political Inquisition or Critical Dialogue?* Cambridge: Cambridge University Press, 1995.

———. "The Use and Abuse of Mathematical Entities: Galileo and the Jesuits Revisited." In *The Cambridge Companion to Galileo*, edited by Peter Machamer, 80–146. Cambridge: Cambridge University Press, 1998.

Festugière, A. J. *La révélation d'Hermès Trismégiste*. 4 vols. Paris: Les Belles lettres, 1981. Original publication, 1949–54.

Findlen, Paula. "The Janus Faces of Science in the Seventeenth Century: Athanasius Kircher and Isaac Newton." In *Rethinking the Scientific Revolution*, edited by Margaret Osler, 221–46. Cambridge: Cambridge University Press, 2000.

———. "The Last Man Who Knew Everything . . . Or Did He? Athanasius Kircher, S. J. (1602–80) and His World." In *Athanasius Kircher: The Last Man Who Knew Everything*, edited by Paula Findlen, 1–48. London: Routledge, 2004.

———. "Living in the Shadow of Galileo: Antonio Baldigiani (1647–1711), a Jesuit Scientist in Late Seventeenth-Century Rome." In *Conflicting Duties: Science, Medicine and Religion in Rome, 1550–1750*, edited by Maria Pia Donato and Jill Kraye. London: Warburg Institute, 2009.

———. *Possessing Nature: Museums, Collecting and Scientific Culture in Early Modern Italy*. Berkeley: University of California Press, 1994.

———. "Science, History, and Erudition: Athanasius Kircher's Museum at the Collegio Romano." In *The Great Art of Knowing: The Baroque Encyclopedia of Athanasius Kircher*, edited by Daniel Stolzenberg, 17–26. Stanford, CA: Stanford University Libraries, 2001.

———. "Scientific Spectacle in Baroque Rome: Athanasius Kircher and the Roman College Museum." *Roma Moderna e Contemporanea* 3 (1995): 625–65.

Findlen, Paula, ed. *Athanasius Kircher: The Last Man Who Knew Everything*. London: Routledge, 2004.

Fiorani, Luigi. "Astrologi, superstiziosi e devoti nella società Romana del seicento." *Ricerche per la storia religiosa di Roma: Studi, documenti, inventari* 2 (1978): 97–162.

Fletcher, John. *A Study of the Life and Works of Athanasius Kircher, "Germanus Incredibilis," with a Selection of His Unpublished Correspondence and an Annotated Translation of His Autobiography*. Edited by Elizabeth Fletcher. Leiden: Brill, 2011.

———. "Claude Fabri de Peiresc and the Other French Correspondents of Athanasius Kircher." *Australian Journal of French Studies* 9 (1972): 250–73.
———. "Johann Marcus von Marci Writes to Athanasius Kircher." *Janus* 59 (1972): 95–118.
Fletcher, John, ed. *Athanasius Kircher und seine Beziehungen zum gelehrten Europa seiner Zeit*. Wiesbaden: Otto Harrassowitz, 1988.
Flint, Valerie. *The Rise of Magic in Early Medieval Europe*. Princeton, NJ: Princeton University Press, 1991.
Fodor, A. "The Origins of the Arabic Legends of the Pyramids." *Acta Orientalia Academiae Scientiarium Hungaricae* 23 (1970): 335–63.
Fosi, Irene. *All'ombra dei Barberini: fedeltà e servizio nella Roma barocca*. Rome: Bulzoni, 1997.
Foucault, Michel. *The Order of Things: An Archaeology of the Human Sciences*. Translated by Alan Sheridan. New York: Random House, 1980.
Fowden, Garth. *The Egyptian Hermes: A Historical Approach to the Late Pagan Mind*. Princeton, NJ: Princeton University Press, 1986.
Franklin, Julian. *Jean Bodin and the Sixteenth-Century Revolution in the Methodology of Law and History*. New York: Columbia University Press, 1963.
Freedberg, David. *The Eye of the Lynx: Galileo, His Friends, and the Beginnings of Modern Natural History*. Chicago: University of Chicago Press, 2002.
Frigo, Gian Franco. "Il ruolo della sapienza egizia nella rappresentazione del sapere di Athanasius Kircher." In *Athanasius Kircher: L'idea di scienza universale*, edited by Federico Vercellone and Alessandro Bertinetto, 91–105. Milan: Mimesis, 2007.
Funkenstein, Amos. *Perceptions of Jewish History*. Berkeley: University of California Press, 1993.
———. *Theology and the Scientific Imagination from the Middle Ages to the Seventeenth Century*. Princeton, NJ: Princeton University Press, 1986.
García-Arenal, Mercedes, and Fernando Rodríguez Mediano. *Un Oriente español: Los moriscos y el Sacromonte en tiempos de Contrareforma*. Madrid: Marcial Pons Historia, 2010.
Gentile, Sebastiano, and Carlos Gilly, eds. *Marsilio Ficino e il ritorno di Ermete Trismegisto*. Florence: Centro Di, 1999.
Geoffroy, E. "al-Suyuti, Abu 'l-Fadl 'Abd al-Rahman b. Abi Bakr b. Muhammad Djalal al-Din al-Khudayri." In *Encyclopaedia of Islam*, vol. 9, 913–16. Leiden: Brill, 1960–2005.
Giard, Luce, ed. *Les jésuites à la Renaissance: Système éducatif et production de savoir*. Paris: Presses Universitaires de France, 1995.
Giehlow, Karl. "Die Hieroglyphenkunde des Humanismus in der Allegorie der Renaissance, besonders der Ehrenpforte Kaisers Maximilian I." *Jahrbuch der Kunsthistorischen Sammlungen des allerhöchsten Kaiserhauses* 32, no. 1 (1915): 1–232.
Gigli, Giacinto. *Diario Romano (1608–1670)*. Edited by Giuseppe Ricciotti. Rome: Tumminelli, 1958.
Ginzburg, Carlo. *History, Rhetoric, and Proof*. Hanover, NH: University Press of New England, 1999.
Glasson, T. Francis. "The Order of Jewels in Revelation XXI.19–20: A Theory Eliminated." *The Journal of Theological Studies* 26, no. 1 (1975): 95–100.

Godman, Peter. *The Saint as Censor: Robert Bellarmine between Inquisition and Index.* Leiden: Brill, 2000.

Godwin, Joscelyn. *Athanasius Kircher's Theatre of the World: The Life and Work of the Last Man to Search for Universal Knowledge.* Rochester, VT: Inner Traditions, 2009.

———. *The Theosophical Enlightenment.* Albany: State University of New York Press, 1994.

Gombrich, E. H. "*Icones Symbolicae*: Philosophies of Symbolism and their Bearing on Art." In *Gombrich on the Renaissance. Volume 2: Symbolic Images*, 123–95. London: Phaidon, 1985. Original publication, 1972.

Gorman, Michael John. "The Angel and the Compass: Athanasius Kircher's Geographical Project." In *Athanasius Kircher: The Last Man Who Knew Everything*, edited by Paula Findlen. London: Routledge, 2004.

———. "Mathematics and Modesty in the Society of Jesus: The Problems of Christoph Grienberger." In *The New Science and Jesuit Science: Seventeenth Century Perspectives*, edited by Mordechai Feingold, 1–120. Dordecht: Kluwer, 2003.

———. "A Matter of Faith? Christoph Scheiner, Jesuit Censorship, and the Trial of Galileo." *Perspectives on Science* 4 (1996): 283–320.

———. "The Scientific Counter-Revolution: Mathematics, Natural Philosophy, and Experimentalism in Jesuit Culture, 1580–c. 1670." PhD dissertation, European University Institute, 1999.

Grafton, Anthony. *Defenders of the Text: The Traditions of Scholarship in an Age of Science, 1450–1800.* Cambridge, MA: Harvard University Press, 1994.

———. *The Footnote: A Curious History.* Cambridge, MA: Harvard University Press, 1999.

———. *Forgers and Critics: Creativity and Duplicity in Western Scholarship.* Princeton, NJ: Princeton Univesity Press, 1990.

———. *Joseph Scaliger: A Study in the History of Classical Scholarship.* 2 vols. Oxford: Oxford University Press, 1983–93.

———. "Kircher's Chronology." In *Athanasius Kircher: The Last Man Who Knew Everything*, edited by Paula Findlen, 171–87. London: Routledge, 2004.

———. "Momigliano's Method and the Warburg Institute: Studies in his Middle Period." In *Momigliano and Antiquarianism: Foundations of the Modern Cultural Sciences*, edited by Peter N. Miller, 97–126. Toronto: University of Toronto Press, 2007.

———. *What Was History? The Art of History in Early Modern Europe.* Cambridge: Cambridge University Press, 2007.

———. "Where Was Salomon's House? Ecclesiastical History and the Intellectual Origin of Bacon's *New Atlantis*." In *Worlds Made By Words: Scholarship and Community in the Modern West*, 98–113. Cambridge, MA: Harvard University Press, 2009.

———. "The World of the Polyhistors." *Central European History* 18, no. 1 (1985): 31–47.

Grafton, Anthony, ed. *Rome Reborn: The Vatican Libary and Renaissance Culture.* Washington: Library of Congress, 1993.

Grafton, Anthony, and Joanna Weinberg. *"I Have Always Loved the Holy Tongue": Isaac Casaubon, the Jews, and a Forgotten Chapter in Renaissance Scholarship.* Cambridge, MA: Belknap Press of Harvard University, 2011.

Gravit, Francis West. *The Peiresc Papers*. Ann Arbor: University of Michigan Press, 1950.

Grégoire, Réginald. "Costituzioni, Visite apostoliche e atti ufficiali nella storia del Collegio Maronita di Roma." *Ricerche per la storia religiosa di Roma: Studi, documenti, inventari* 1 (1977): 175–229.

Griggs, Tamara. "Antiquaries and the Nile Mosaic: The Changing Face of Erudition." *Ricerche di storia dell'arte*, no. 71 (2000): 37–48.

Gutschmid, Alfred von. "War Ibn Wahshijjah ein Nabatäischer Herodot?" In *Kleine Schriften*, vol. 2, 717–53. Leipzig: B. G. Teubner, 1890. Original publication, 1862.

Haarmann, Ulrich. "Medieval Muslim Perceptions of Pharaonic Egypt." In *Ancient Egyptian Literature. History and Forms*, edited by Antonio Loprieno, 605–27. Leiden: Brill, 1996.

———. "Misalla." In *Encyclopaedia of Islam*, vol. 7, 140–41. Leiden: Brill, 1960–2005.

Häfner, Ralf. *Götter im Exil: Frühneuzeitliches Dichtungsverständnis im Spannungsfeld christlicher Apologetik und philologischer Kritik (ca. 1590–1736)*. Tübingen: Max Niemeyer Verlag, 2003.

Hämeen-Antilla, Jaako. *The Last Pagans of Iraq: Ibn Wahshiyya and his Nabatean Agriculture*. Leiden: Brill, 2006.

Hamilton, Alastair. *The Copts and the West, 1439–1822: The European Discovery of the Egyptian Church*. Oxford: Oxford University Press, 2006.

———. "Eastern Churches and Western Scholarship." In *Rome Reborn: The Vatican Libary and Renaissance Culture*, edited by Anthony Grafton, 225–49. Washington: Library of Congress, 1993.

———. *William Bedwell the Arabist 1563–1632*. Leiden: Brill, 1985.

Hamilton, Alastair, Maurits Van Den Boogert, and Bart Westerweel, eds. *The Republic of Letters and the Levant*. Leiden: Brill, 2005.

Hammond, Frederick. *Music and Spectacle in Baroque Rome: Barberini Patronage under Urban VIII*. New Haven: Yale University Press, 1994.

Hanegraaff, Wouter J. "Beyond the Yates Paradigm: The Study of Western Esotericism between Counterculture and New Complexity." *Aries* 1, no. 1 (2001): 5–37.

———. "Esotericism." In *Dictionary of Gnosis and Western Esotericism*, edited by Wouter J. Hanegraaff, Antoine Faivre, Roelof van den Broek, and Jean-Pierre Brach, vol. 1, 336–40. Leiden: Brill, 2005.

———. "Forbidden Knowledge: Anti-Esoteric Polemics and Academic Research." *Aries* 5, no. 2 (2005): 225–54.

———. *New Age Religion and Western Culture: Esotericism in the Mirror of Secular Thought*. Leiden: Brill, 1996.

———. "Occult / Occultism." In *Dictionary of Gnosis and Western Esotericism*, edited by Wouter J. Hanegraaff, Antoine Faivre, Roelof van den Broek, and Jean-Pierre Brach, vol. 2, 884–89. Leiden: Brill, 2005.

———. "Tradition." In *Dictionary of Gnosis and Western Esotericism*, edited by Wouter J. Hanegraaff, Antoine Faivre, Roelof van den Broek, and Jean-Pierre Brach, vol. 2, 1125–35. Leiden: Brill, 2005.

Hankins, James. *Plato in the Italian Renaissance*. 2 vols. Leiden: Brill, 1990.

Harkness, Deborah E. *John Dee's Conversations with Angels: Cabala, Alchemy, and the End of Nature*. Cambridge: Cambridge University Press, 1999.

Harris, A. Katie. "Forging History: The *Plomos* of the Sacromonte de Granada in Francisco Bermúdez de Pedraza's *Historia Ecclesiastica*." *Sixteenth Century Journal* 30 (1999): 945–66.

Harris, Steven J. "Confession-Building, Long-Distance Networks, and the Organization of Jesuit Science." *Early Science and Medicine* 1 (1996): 287–318.

Harrison, Peter. *The Bible, Protestantism, and the Rise of Modern Science*. Cambridge: Cambridge University Press, 1998.

———. *The Fall of Man and the Foundations of Science*. Oxford: Oxford University Press, 2007.

———. *"Religion" and the Religions in the English Enlightenment*. Cambridge: Cambridge University Press, 1990.

Haskell, Francis. *History and Its Images: Art and the Interpretation of the Past*. New Haven: Yale University Press, 1993.

Haugen, Kristine. *Richard Bentley: Poetry and Enlightenment*. Cambridge, MA: Harvard University Press, 2011.

Haycock, David Boyd. "'The Long-Lost Truth': Sir Isaac Newton and the Newtonian Pursuit of Ancient Knowledge." *Studies in the History and Philosophy of Science* 35 (2004): 605–23.

———. *William Stukeley: Science, Religion and Archaeology in Eighteenth-Century England*. Woodbridge, UK: Boydell Press, 2002.

Hellyer, Marcus. "'Because the Authority of My Superiors Commands': Censorship, Physics and the German Jesuits." *Early Science and Medicine* 1 (1996): 319–54.

———. *Catholic Physics: Jesuit Natural Philosophy in Early Modern Germany*. Notre Dame, IN: University of Notre Dame Press, 2005.

———. "The Construction of the *Ordinatio Pro Studiis Superioribus* of 1651." *Archivum Historicum Societatis Iesu* 72 (2003): 3–44.

Henkel, Willy. "The Polyglot Printing-office of the Congregation." In *Sacrae Congregationis de Propaganda Fide Memoria Rerum*, edited by Josef Metzler, vol. 1, 335–50. Rome: Herder, 1971.

Herklotz, Ingo. "Arnaldo Momigliano's 'Ancient History and the Antiquarian': A Critical Review." In *Momigliano and Antiquarianism: Foundations of the Modern Cultural Sciences*, edited by Peter N. Miller, 127–53. Toronto: University of Toronto Press, 2007.

———. *Cassiano dal Pozzo und die Archäologie des 17. Jahrhunderts*, Römische Forschungen der Bibliotheca Hertziana 28. Munich: Hirmer, 1999.

———. "Excavations, Collectors and Scholars in Seventeenth-Century Rome." In *Archives and Excavations: Essays on the History of Archaeological Excavations in Rome and Southern Italy from the Renaissance to the Nineteenth Century*, edited by Ilaria Bignamini, 55–88. London: The British School of Rome, 2004.

Heyberger, Bernard. "Abraham Ecchellensis dans la République des Lettres." In *Orientalisme, Science et Controverse: Abraham Ecchellensis (1605–1664)*, edited by Bernard Heyberger, 9–51. Turnhout, Belgium: Brepols, 2010.

———. *Les Chrétiens du Proche-Orient au temps de la Réforme Catholique (Syrie, Liban, Palestine, XVIIe–XVIIIe siècles)*. Rome: École Française de Rome, 1994.
Heyberger, Bernard, ed. *Orientalisme, Science et Controverse: Abraham Ecchellensis (1605–1664)*. Turnhout, Belgium: Brepols, 2010.
Idel, Moshe. *Kabbalah: New Perspectives*. New Haven: Yale University Press, 1988.
———. "Major Currents in Italian Kabbalah between 1560 and 1660." In *Essential Papers on Jewish Culture in Renaissance and Baroque Italy*, edited by David B. Ruderman, 345–68. New York: New York Univesity Press, 1992.
Infelise, Mario. *I libri proibiti da Gutenberg all'Encyclopédie*. Rome and Bari: Editori Laterza, 1999.
Institutum Societatis Iesu. 3 vols. Florence: Ex typographia a SS. Conceptione, 1892–93.
Irwin, Robert. *Dangerous Knowledge: Orientalism and its Discontents*. Woodstock, NY: Overlook Press, 2006.
Iversen, Erik. *The Myth of Egypt and its Hieroglyphs in European Tradition*. Princeton, NJ: Princeton University Press, 1993. Original publication, 1961.
———. *Obelisks in Exile, I: The Obelisks of Rome*. Copenhagen: G. E. C. Gad, 1968.
Jacks, Philip. *The Antiquarian and the Myth of Antiquity: The Origins of Rome in Renaissance Thought*. Cambridge: Cambridge University Press, 1993.
Jäger, Berthold. "Athanasius Kircher, Geisa, und Fulda." In *Spurensuche: Wege zu Athanasius Kircher*, edited by Horst Beinlich, Hans-Joachim Vollrath, and Klaus Wittstadt. Wiesbaden: Harrassowitz, 2002.
Jones, Robert. "The Medici Oriental Press (Rome 1584–1614) and the Impact of Its Arabic Publications on Northern Europe." In *The "Arabick" Interest of the Natural Philosophers in Seventeenth-Century England*, edited by G. A. Russell, 88–108. Leiden: Brill, 1994.
Katchen, Aaron L. *Christian Hebraists and Dutch Rabbis: Seventeenth Century Apologetics and the Study of Maimonides' Mishneh Torah*. Cambridge, MA: Harvard University Press, 1984.
Kelley, Donald. *Foundations of Modern Historical Scholarship: Language, Law, and History in the French Renaissance*. New York: Columbia University Press, 1970.
Kelley, Donald R. "Writing Cultural History in Early Modern Europe: Christophe Milieu and his Project." *Renaissance Quarterly* 52 (1999): 342–65.
Killeen, Kevin, and Peter J. Forshaw, eds. *The Word and the World: Biblical Exegesis and Early Modern Science*. New York: Palgrave Macmillan, 2007.
Kleinhans, Arduino. *Historia Studii Linguae Arabicae et Collegii Missionum Ordinis Fratrum Minorum in Conventu ad S. Petrum in Monte Aureo Romae Erecti*, Biblioteca Bio-Bibliografica della Terra Santa e dell'Oriente Francescano, Nuova Serie—Documenti 13. Florence: Quaracchi, 1930.
Klutstein, Ilana. *Marsilio Ficino et la theologie ancienne: Oracles Chaldaïques—Hymnes Orphiques—Hymnes de Proclus*. Florence: Olschki, 1987.
Kraye, Jill. "The Pseudo-Aristotelian *Theology* in Sixteenth- and Seventeenth-Century Europe." In *Pseudo-Aristotle in the Middle Ages: The Theology and Other Texts*, edited by Jill Kraye, W. F. Ryan, and C. B. Schmitt, 265–86. London: Warburg Institute, 1986.

Kraye, Jill, W. F. Ryan, and C. B. Schmitt, eds. *Pseudo-Aristotle in the Middle Ages: The Theology and Other Texts*. London: Warburg Institute, 1986.
Kristeller, Paul Oskar. *Eight Philosophers of the Italian Renaissance*. Stanford, CA: Stanford University Press, 1964.
Lamberton, Robert. *Homer the Theologian: Neoplatonist Allegorical Reading and the Growth of the Epic Tradition*. Berkeley: University of California Press, 1986.
Lantschoot, Arnold van. *Un précurseur d'Athanase Kircher: Thomas Obicini et la Scala Vat. Copte 71*, Bibliothèque du Muséon 22. Louvain: Bureaux du Muséon, 1948.
Lehmann-Brauns, Sicco. *Weisheit in der Weltgeschichte: Philosophiegeschichte zwischen Barock und Aufklärung*. Tübingen: Max Niemayer Verlag, 2004.
Leijenhorst, Cees. "Francesco Patrizi's Hermetic Philosophy." In *Gnosis and Hermeticism from Antiquity to Modern Times*, edited by Roelof van den Broek and Wouter J. Hanegraaff, 125–47. Albany: State University of New York Press, 1998.
Leinkauf, Thomas. "Interpretation und Analogie. Rationale Strukturen im Hermetismus der Frühen Neuzeit." In *Antike Weisheit und kulturelle Praxis*, edited by Anne-Charlott Trepp and Hartmut Lehmann, 41–61. Göttingen: Vandenhoeck & Ruprecht, 2001.
———. *Mundus Combinatus: Studien zur Struktur der barocken Universalwissenschaft am Beispiel Athanasius Kircher SJ (1602–1680)*. Berlin: Akademie Verlag, 1993.
Leospo, Enrica. *La Mensa Isiaca di Torino*. Leiden: Brill, 1978.
Leospo, Enrichetta. "La collezione egizia del Museo Kircheriano." In *Athanasius Kircher: Il museo del mondo*, edited by Eugenio Lo Sardo, 125–30. Rome: Edizioni De Luca, 2001.
Levi della Vida, Giorgio. "Abramo Ecchellense." In *Dizionario Biografico degli Italiani*, vol. 1. Rome: Istituto della Enciclopedia Italiana, 1960.
———. *Ricerche sulla formazione del più antico fondo dei manoscritti orientali della Biblioteca Vaticana*, Studi e Testi 92. Vatican City: Biblioteca Apostolica Vaticana, 1939.
Longo, Mario. "Storia 'critica' della filosofia e primo illuminismo: Jakob Brucker." In *Storia delle storie generali della filosofia, Vol. 2, Dall' età cartesiana a Brucker*, edited by Francesco Bottin, Mario Longo, and Gregorio Piaia, 527–635. Brescia: Editrice La Scuola, 1979.
Lo Sardo, Eugenio, ed. *Athanasius Kircher: Il museo del mondo*. Rome: Edizioni De Luca, 2001.
Mahé, Jean-Pierre. *Hermès en Haute-Egypte: Les textes hermétiques de Nag Hammadi et leurs paralleles grecs et latins*. Québec: Presses de l'Université Laval, 1978.
Majercik, Ruth. *The Chaldean Oracles: Text, Translation, and Commentary*. Leiden: Brill, 1989.
Malusa, Luciano. "The Cambridge Platonists and the History of Philosophy." In Bottin et al., *Models of the History of Philosophy from Its Origins in the Renaissance to the "Historia Philosophica,"* 279–370. Dordecht: Kluwer, 1993.
———. "Renaissance Antecedents to the Historiography of Philosophy." In Bottin et al., *Models of the History of Philosophy from Its Origins in the Renaissance to the "Historia Philosophica,"* 3–65. Dordecht: Kluwer, 1993.
Manning, John. *The Emblem*. London: Reaktion, 2002.

Manuel, Frank E. *The Broken Staff: Judaism through Christian Eyes*. Cambridge, MA: Harvard University Press, 1993.

———. *The Eighteenth Century Confronts the Gods*. Cambridge, MA: Harvard University Press, 1959.

———. *Issac Newton, Historian*. Cambridge, MA: Belknap Press of Harvard University, 1963.

Marchand, Suzanne L. *German Orientalism in the Age of Empire*. Cambridge: Cambridge University Press, 2009.

Marquet, Jean-François. "La quête isiaque d'Athanase Kircher." *Études Philosophiques*, nos. 2–3 (1987): 227–41.

Marrone, Caterina. *I geroglifici fantastici di Athanasius Kircher*. Viterbo: Nuovi Equilibri, 2002.

Marucchi, Orazio. *Gli obelischi egizi di Roma*. Rome: Loescher & Co., 1898.

Mastroianni, Aldo. "Kircher e l'Oriente nel Museo del Collegio Romano." In *Athanasius Kircher: Il museo del mondo*, edited by Eugenio Lo Sardo, 65–75. Rome: Edizioni De Luca, 2001.

Mayer-Deutsch, Angela. "Iconographia Kircheriana." In *Athanasius Kircher: Il museo del mondo*, edited by Eugenio Lo Sardo, 353–63. Rome: Edizioni De Luca, 2001.

———. *Das Musaeum Kircherianum. Kontemplative Momente, historische Rekonstruktion, Bildrhetorik*. Zurich: Diaphanes, 2010.

McCuaig, William. *Carlo Sigonio: The Changing World of the Late Renaissance*. Princeton, NJ: Princeton University Press, 1989.

McGuire, J. E., and P. M. Rattansi. "Newton and the 'Pipes of Pan.'" *Notes and Records of the Royal Society of London* 21, no. 2 (1966): 108–43.

Mercier, Raymond. "English Orientalists and Mathematical Astronomy." In *The "Arabick" Interest of the Natural Philosophers in Seventeenth-Century England*, edited by G. A. Russell, 158–214. Leiden: Brill, 1994.

Mercier-Faivre, Anne-Marie. "Court de Gébelin, Antoine." In *Dictionary of Gnosis and Western Esotericism*, edited by Wouter J. Hanegraaff, Antoine Faivre, Roelof van den Broek and Jean-Pierre Brach, vol. 1, 279–81. Leiden: Brill, 2005.

Meserve, Margaret. *Empires of Islam in Renaissance Historical Thought*. Cambridge, MA: Harvard University Press, 2008.

Miller, Peter N. "The 'Antiquarianization' of Biblical Scholarship and the London Polyglot Bible (1653–57)." *Journal of the History of Ideas* 62 (2001): 463–82.

———. "An Antiquary between Philology and History: Peiresc and the Samaritans." In *History and the Disciplines*, edited by Donald Kelley, 163–84. Rochester, NY: University of Rochester Press, 1997.

———. "The Antiquary's Art of Comparison: Peiresc and *Abraxas*." In *Philologie und Erkenntnis: Beiträge zu Begriff und Problem frühneuzeitlicher "Philologie,"* edited by Ralph Häfner, 57–94. Tübingen: Max Niemeyer, 2001.

———. "Copts and Scholars: Kircher in Peiresc's Republic of Letters." In *Athanasius Kircher: The Last Man Who Knew Everything*, edited by Paula Findlen, 103–48. London: Routledge, 2004.

———. "Introduction: Momigliano, Antiquarianism, and the Cultural Sciences." In

Momigliano and Antiquarianism: Foundations of the Modern Cultural Sciences, edited by Peter N. Miller, 3–65. Toronto: University of Toronto Press, 2007.

———. "Making the Paris Polygot Bible: Humanism and Orientalism in the Early Seventeenth Century." In *Die europäische Gelehrtenrepublik im Zeitalter des Konfessionalismus: The European Republic of Letters in the Age of Confessionalism*, edited by Herbert Jaumann, 59–85. Wiesbaden: Harrassowitz, 2001.

———. "The Mechanics of Christian-Jewish Intellectual Collaboration in Seventeenth-Century Provence: N.-C. Fabri de Peiresc and Salomon Azubi." In *Hebraica Veritas? Christian Hebraists and the Study of Judaism in Early Modern Europe*, edited by Allison P. Coudert and Jeffrey S. Shoulson, 71–101. Philadelphia: University of Pennsylvania Press, 2004.

———. "Peiresc, the Levant and the Mediterranean." In *The Republic of Letters and the Levant*, edited by Alastair Hamilton, 103–22. Leiden: Brill, 2005.

———. *Peiresc's Europe: Learning and Virtue in the Seventeenth Century*. New Haven: Yale University Press, 2000.

———. *Peiresc's Orient: Antiquarianism as Cultural History*. Aldershot: Ashgate Variorum, 2012.

———. "A Philologist, a Traveller and an Antiquary Rediscover the Samaritans in Seventeenth-Century Paris, Rome and Aix: Jean Morin, Pietro della Valle, and N.-C. de Peiresc." In *Die Praktiken der Gelehrsamkeit in der Frühen Neuzeit*, edited by Helmut Zedelmaier and Martin Mulsow, 123–46. Tübingen: Max Niemeyer Verlag, 2001.

———. "Taking Paganism Seriously: Anthropology and Antiquarianism in Early Seventeenth-Century Histories of Religion." *Archiv für Religionsgeschichte* 3 (2001): 182–209.

Miller, Peter N., ed. *Momigliano and Antiquarianism: Foundations of the Modern Cultural Sciences*. Toronto: University of Toronto Press, 2007.

Momigliano, Arnaldo. "Ancient History and the Antiquarian." *Journal of the Warburg and Courtauld Institutes* 13, no. 3/4 (1950): 285–315.

———. *The Classical Foundations of Modern Historiography*. Berkeley: University of California Press, 1990.

———. "La nuova storia romana di G. B. Vico." In *Sesto contributo alla storia degli studi classici e del mondo antico*, vol. 1, 192–210. Rome: Edizioni di storia e letteratura, 1980. Original publication, 1965.

———. "The Rise of Antiquarian Research." In *The Classical Foundations of Modern Historiography*, 54–79. Berkeley: University of California Press, 1990.

Moss, Ann. "The *Politica* of Justus Lipsius and the Commonplace-Book." *Journal of the History of Ideas* 59 (1998): 421–36.

———. *Printed Commonplace-Books and the Structuring of Renaissance Thought*. Oxford: Clarendon Press, 1996.

Mulsow, Martin. "Ambiguities of the *Prisca Sapientia* in Late Renaissance Humanism." *Journal of the History of Ideas* 65, no. 1 (2004): 1–13.

———. "Antiquarianism and Idolatry: The *Historia* of Religions in the Seventeenth Century." In *Historia: Empiricism and Erudition in Early Modern Europe*, edited by Gianna Pomata and Nancy G. Siraisi, 181–209. Cambridge, MA: MIT Press, 2005.

———. "John Seldens *De Diis Syris:* Idolatriekritik und vegleichende Religionsgeschichte im 17. Jahrhundert." *Archiv für Religionsgeschichte* 3 (2001): 1–24.

———. *Moderne aus dem Untergrund: Radikale Frühaufklärung in Deutschland, 1680–1720.* Hamburg: Felix Meiner, 2002.

———. "Das schnelle und langsame Ende des Hermetismus." In *Das Ende des Hermetismus: Historische Kritik und neue Naturphilosophie in der Spätrenaissance,* edited by Martin Mulsow, 305–10. Tübingen: Mohr Siebeck, 2002.

Mulsow, Martin, ed. *Das Ende des Hermetismus: Historische Kritik und neue Naturphilosophie in der Spätrenaissance.* Tübingen: Mohr Siebeck, 2002.

Mungello, David E. *Curious Land: Jesuit Accommodation and the Origins of Sinology.* Wiesbaden: F. Steiner, 1985.

Nemoy, Leon. "The Treatise on the Egyptian Pyramids by Jalal al-Din al-Suyuti: Edited with Introduction, Translation and Notes." *Isis* 30, no. 1 (1939): 2–37.

Neugebauer-Wölk, Monika, ed. *Aufklärung und Esoterik.* Hamburg: Meiner, 1999.

Newman, William, and Lawrence Principe. *Alchemy Tried in the Fire: Starkey, Boyle and the Fate of Helmontian Chymistry.* Chicago: University of Chicago Press, 2002.

Newman, William R., and Anthony Grafton. "Introduction: The Problematic Status of Astrology and Alchemy in Premodern Europe." In *Secrets of Nature: Astrology and Alchemy in Early Modern Europe,* edited by William R. Newman and Anthony Grafton, 1–37. Cambridge, MA: MIT Press, 2001.

O'Malley, John W., Gauvin Alexander Bailey, Steven J. Harris, and T. Frank Kennedy. *The Jesuits: Cultures, Sciences, and the Arts, 1540–1773.* Toronto: University of Toronto Press, 1999.

———. *The Jesuits II: Cultures, Sciences, and the Arts, 1540–1773.* Toronto: University of Toronto Press, 2006.

O'Neill, Charles, and Joaquin Domínguez, eds. *Diccionario Histórico de la Compañia de Jesús Biográfico-Temático.* 3 vols. Rome: Institutum Historicum, S. I., 2001.

Ogilvie, Brian. *The Science of Describing: Natural History in Renaissance Europe.* Chicago: University of Chicago Press, 2006.

Onori, Lorenza Mochi, Sebastian Schütze, and Francesco Solinas, eds. *I Barberini e la cultura europea del seicento.* Rome: De Luca Editori d'Arte, 2007.

Osorio Romero, Ignacio, ed. *La luz imaginaria: Epistolario de Atanasio Kircher con los novohispanos.* Mexico City: Universidad Nacional Autónoma de México, 1993.

Parkin, Jon. *Science, Religion, and Politics in Restoration England: Richard Cumberland's De Legibus Naturae.* Woodbridge, UK: Boydell Press, 1999.

Pastine, Dino. *La nascita dell'idolatria: L'Oriente religioso di Athanasius Kircher.* Florence: La Nuova Italia, 1978.

Perkins, Franklin. *Leibniz and China: A Commerce of Light.* Cambridge: Cambridge University Press, 2004.

Pfeiffer, Heinrich. "Il concetto di simbolo nell 'Obeliscus Alexandrinus' di Kircher." In *Enciclopedismo in Roma barocca: Athanasius Kircher e il museo del Collegio Romano tra Wunderkammer e museo scientifico,* edited by Maristella Casciato, Maria Grazia Ianniello, and Maria Vitale, 39–45. Venice: Marsilio, 1986.

Piemontese, Angelo Michele. "Grammatica e lessicografia araba in Italia dal XVI al XVII secolo." In *Italia ed Europa nella linguistica del Rinascimento*, edited by Mirko Tavoni, vol. 2, 519–32. Modena: Franco Cosimo Panini, 1996.

Pizzorusso, Giovanni. "Agli antipodi di Babele: Propaganda Fide tra imagine cosmopolita e orrizonti romani (xvii–xix secolo)." In *Roma, la città del papa: Vita civile e religiosa dal giubileo di Bonficacio VIII al giubileo di papa Wojtyla*, edited by Luigi Fiorani and Adriano Prosperi, 477–518. Turin: Einaudi, 2000.

———. "Les écoles de langue arabe et le milieu orientaliste autour de la congrégation De Propaganda Fide au temps d'Abraham Ecchellensis." In *Orientalisme, Science et Controverse: Abraham Ecchellensis (1605–1664)*, edited by Bernard Heyberger, 59–80. Turnhout, Belgium: Brepols, 2010.

———. "Filippo Guadagnoli, i Caracciolini e lo studio delle lingue orientali e della controversia con l'Islam a Roma nel XVII secolo." In *L'Ordine dei Chierici Regolari minori (Caracciolini): Religione e cultura in età postridentina: Atti del Convegno (Chieti, 11–12 aprile 2008). Theme Issue. Studi medievali e moderni*, edited by Irene Fosi and Giovanni Pizzorusso, vol. 14. no. 1 (2010): 245–78.

———. "L'indagine geo-etnografica nelle istruzioni ai missionari della Congregazione 'De Propaganda Fide' (secoli XVII–XIX)." In *Viaggi e scienza: Le istruzioni scientifiche per i viaggiatori nei secoli XVII–XIX*, edited by Maurizio Bossi and Claudio Greppi, 287–308. Florence: Olschki, 2005.

———. "I satelliti di Propaganda Fide: Il Collegio Urbano e la Tipografia Poliglotta." *Mélanges de l'école française de Rome: Italie et Méditerranée* 116, no. 2 (2004): 471–98.

Plessner, M. "Hermes Trismegistus and Arabic Science." *Studia Islamica* 2 (1954): 45–59.

———. "Hirmis." In *Encyclopaedia of Islam*, vol. 3, 463–64. Leiden: Brill, 1960–2005.

Plutarch. *Plutarch's De Iside et Osiride*. Edited with an introduction, translation, and commentary by J. Gwyn Griffiths. Cardiff: University of Wales Press, 1970.

Pocock, J. G. A. *The Ancient Constitution and the Feudal Law: A Study of English Historical Thought in the Seventeenth Century*. Cambridge: Cambridge University Press, 1957.

———. *Barbarism and Religion, Volume One: The Enlightenments of Edward Gibbon*. Cambridge: Cambridge University Press, 1999.

———. "Edward Gibbon in History: Aspects of the Text in 'The History of the Decline and Fall of the Roman Empire.'" In *The Tanner Lectures in Human Values*, edited by Grethe B. Petersen, vol. 11, 289–364. Salt Lake City: University of Utah Press, 1988.

Pomata, Gianna. "*Praxis Historialis:* The Uses of *Historia* in Early Modern Medicine." In *Historia: Empiricism and Erudition in Early Modern Europe*, edited by Gianna Pomata and Nancy G. Siraisi. Cambridge, MA: MIT Press, 2005.

Pomata, Gianna, and Nancy G. Siraisi, eds. *Historia: Empiricism and Erudition in Early Modern Europe*. Cambridge, MA: MIT Press, 2005.

Popkin, Richard. *The History of Scepticism from Erasmus to Spinoza*. Berkeley: University of California Press, 1979.

———. *Isaac La Peyrère (1596–1676): His Life, Work, and Influence*. Leiden: Brill, 1987.

Popper, Nicholas. "'Abraham, Planter of Mathematics': Histories of Mathematics and Astrology in Early Modern Europe." *Journal of the History of Ideas* 67, no. 1 (2006): 87–106.

Porter, David. *Ideographia: The Chinese Cipher in Early Modern Europe*. Stanford, CA: Stanford University Press, 2001.

Praz, Mario. *Studies in Seventeenth-Century Imagery*. 2nd edition, Sussidi Eruditi 16. Rome: Edizioni di Storia e Letteratura, 1964.

Prosperi, Adriano. *L'inquisizione romana: Letture e ricerche*. Rome: Edizioni di Storia e Letteratura, 2003.

Proverbio, Delio V. "Alle origini delle collezioni librarie orientali." In *Le origini della Biblioteca Vaticana tra umanesimo e rinascimento*, edited by Antonio Manfredi, 467–85. Vatican City: Biblioteca Apostolica Vaticana, 2010.

Purnell, Frederick. "Francesco Patrizi and the Critics of Hermes Trismegistus." *Journal of Medieval and Renaissance Studies* 6 (1976): 155–74.

Ragon, Jean Marie. *Orthodoxie Maçonnique*. Paris: 1853.

Rankin, Oliver Shaw. *Jewish Religious Polemic of Early and Later Centuries: A Study of Documents Here Rendered in English*. New York: Ktav, 1970. Original publication, 1953.

Rappe, Sarah. *Reading Neoplatonism: Non-Discursive Thinking in the Texts of Plotinus, Proclus, and Damascius*. Cambridge: Cambridge University Press, 2000.

Redondi, Pietro. *Galileo Heretic*. Translated by Raymond Rosenthal. Princeton, NJ: Princeton University Press, 1987.

Reill, Peter Hanns. *Vitalizing Nature in the Enlightenment*. Berkeley: University of California Press, 2005.

Reilly, Conor. *Athanasius Kircher: A Master of a Hundred Arts, 1602–1680*. Wiesbaden: Edizioni del Mondo, 1974.

Reitzenstein, R. *Poimandres: Studien zur griechisch-ägyptischen und frühchristlichen Literatur*. Leipzig: B. G. Teubner, 1904.

Richter, Gerhard. "Athanasius Kircher und seine Vaterstadt Geisa." *Fuldaer Geschichtsblätter* 20, no. 4 (1927): 49–59.

Rietbergen, Peter. "A Maronite Mediator between Seventeenth-Century Mediterranean Cultures: Ibrahim al Hakilani, or Abraham Ecchellense (1605–1664)." *Lias* 16 (1989): 13–42.

———. *Power and Religion in Baroque Rome: Barberini Cultural Policies*. Leiden: Brill, 2006.

Rivosecchi, Valerio. *Esotismo in Roma barocca: Studi sul Padre Kircher*. Rome: Bulzoni, 1982.

Rizza, Cecilia. *Peiresc e l'Italia*. Turin: Giappichelli, 1965.

Rolet, Stéphane. "D'étranges objets hiéroglyphiques: Les monnaies antiques dans les *Hieroglyphica* de Pierio Valeriano (1556)." In *Polyvalenz und Multifunktionalität der Emblematik. Multivalence and Multifunctionality of the Emblem*, edited by Wolfgang Harms and Dietmar Peil, vol. 2, 813–44. Frankurt am Main: Peter Lang, 2002.

———. "Invention et exégèse symbolique à la Renaissance: Le 'tombeau d'Aureolus' dans les *Antiquitates Mediolanenses* d'Alciat et les *Hieroglyphica* de Valeriano." *Albertiana* 5 (2002): 109–40.

Romano, Antonella. *La contre-réforme mathématique: Constitution et diffusion d'une culture mathématique jésuite à la Renaissance (1540–1640)*. Rome: École Française de Rome, 1999.

———. "Il mondo della scienza." In *Roma Moderna*, edited by Giorgio Ciucci, 275–305. Rome and Bari, 2002.

———. "Rome, un chantier pour les savoirs de la Catholicité post-tridentine." *Revue d'histoire moderne et contemporaine* 55, no. 2 (2008): 101–20.

———. "Sciences, activités scientifiques, et acteurs de la science, dans la Rome de la Renaissance." *Mélanges de l'école française de Rome* 114, no. 2 (2002): 467–75.

Romano, Antonella, ed. *Rome et la science moderne: Entre Renaissance et Lumières*. Rome: École française de Rome, 2008.

Rooden, Peter T. van. *Theology, Biblical Scholarship, and Rabbinical Studies in the Seventeenth Century: Constantijn L'Empereur (1591–1648), Professor of Hebrew and Theology at Leiden*. Leiden: Brill, 1989.

Rossi, Paolo. *The Dark Abyss of Time: The History of the Earth and the History of Nations From Hooke to Vico*. Translated by Lydia Cochrane. Chicago: University of Chicago Press, 1984.

———. *Logic and the Art of Memory: The Quest for a Universal Language*. Translated by Stephen Clucas. Chicago: University of Chicago Press, 2000. Original publication, 1983.

Roullet, Anne. *The Egyptian and Egypticizing Monuments of Rome*. Leiden: Brill, 1972.

Rowbotham, Arnold H. "The Jesuit Figurists and Eighteenth-Century Religious Thought." *Journal of the History of Ideas* 17, no. 4 (1956): 471–85.

Rowland, Ingrid. "Athanasius Kircher, Giordano Bruno, and the *Panspermia* of the Infinite Universe." In *Athanasius Kircher: The Last Man Who Knew Everything*, edited by Paula Findlen, 191–205. New York: London, 2004.

———. *The Ecstatic Journey: Athanasius Kircher in Baroque Rome*. Chicago: University of Chicago Press, 2000.

———. *Giordano Bruno: Philosopher/Heretic*. New York: Farrar, Straus and Giroux, 2008.

———. "Kircher Trismegisto." In *Athanasius Kircher: Il museo del mondo*, edited by Eugenio Lo Sardo, 113–21. Rome: Edizioni De Luca, 2001.

———. "'Th' United Sense of the Universe': Athanasius Kircher in Piazza Navona." *Memoirs of the American Academy in Rome* 46 (2001): 154–81.

Ruska, Julius. *Tabula Smagdarina: Ein Beitrag zur Geschichte der hermetischen Literatur*. Heidelberg: Carl Winter's Universitätsbuchhandlung, 1926.

Russell, Daniel. "Illustration, Hieroglyph, Icon: The Status of the Emblem Picture." In *Polyvalenz und Multifunktionalität der Emblematik: Multivalence and Multifunctionality of the Emblem*, edited by Wolfgang Harms and Dietmar Peil, vol. 1, 73–90. Frankurt am Main: Peter Lang, 2002.

Russell, G. A., ed. *The "Arabick" Interest of the Natural Philosophers in Seventeenth-Century England*. Leiden: Brill, 1994.

Russell, Michael. *A Connection of Sacred and Profane History*. Edited by J. Talboys Wheeler. London: William Tegg, 1865.

Russell, Susan. "Pirro Ligorio, Cassiano Dal Pozzo and the Republic of Letters." *Papers of the British School at Rome* 75 (2007): 239–74.

Saussy, Haun. *Great Walls of Discourse: Adventures in Cultural China*. Cambridge, MA: Harvard University Asia Center, 2001.

Scharlau, Ulf. *Athanasius Kircher (1601–1680) als Musikschriftsteller.* Marburg: Görich and Weiershaüser, 1969.

Schmidt-Biggemann, Wilhelm. "Cattolicesimo e Cabala: L'esempio di Athanasius Kircher (1602–1680)." In *Athanasius Kircher: L'idea di scienza universale,* edited by Federico Vercellone and Alessandro Bertinetto, 59–89. Milan: Mimesis, 2007.

———. "Hermes Trismegistos, Isis und Osiris in Athanasius Kirchers *Oedipus Aegyptiacus.*" *Archiv für Religionsgeschichte* 3 (2001): 67–88.

———. *Philosophia Perennis: Historical Outlines of Western Spirituality in Ancient, Medieval and Early Modern Thought.* Dordecht: Springer, 2004.

———. *Topica Universalis: Eine Modellgeschichte humanistischer und barocker Wissenschaft.* Hamburg: Felix Meiner, 1983.

Schmidt-Biggemann, Wilhelm, and Theo Stammen, eds. *Jacob Brucker (1696–1770): Philosoph und Historiker der europäischen Aufklärung.* Berlin: Akademie Verlag, 1998.

Schmitt, Charles B. "Perennial Philosophy: From Agostino Steuco to Leibniz." *Journal of the History of Ideas* 27 (1966): 505–32.

Schnapp, Alain. *The Discovery of the Past.* Translated by Ian Kinnes and Gillian Varndell. New York: Harry N. Abrams, 1997.

Scholem, Gershom. *Bibliographia Kabbalistica: die jüdische Mystik (Gnosis, Kabbala, Sabbatianismus, Frankismus, Chassidismus) behandelnden Bücher und Aufsätze von Reuchlin bis zur Gegenwart. Mit einem Anhang: Bibliographie des Zohar und seiner Kommentare.* Berlin: Schocken, 1933.

Scott, Walter, and A. S. Ferguson, eds. *Hermetica: The Ancient Greek and Latin Writings Which Contain Religious or Philosophical Teachings Ascribed to Hermes Trismegistus.* 4 vols. Oxford: Clarendon Press, 1926–1936.

Secret, François. "Guillaume Postel et les études arabes à la renaissance." *Arabica* 9 (1962): 21–36.

———. *Hermétisme et Kabbale.* Naples: Bibliopolis, 1992.

———. "Les Jésuites et le kabbalisme chrétien à la renaissance." *Bibliothèque d'humanisme et renaissance* 20 (1958): 543–55.

———. *Les Kabbalistes Chrétiens.* Nouvelle édition mise à jour et augmentée. Milan: Archè, 1985.

———. "Notes sur quelques kabbalistes chrétiens." *Bibliothèque d'humanisme et renaissance* 36 (1974): 67–82.

Seznec, Jean. *The Survival of the Pagan Gods: The Mythological Tradition and its Place in Renaissance Humanism and Art.* Princeton, NJ: Princeton University Press, 1995. Original publication, 1953.

Shalev, Zur. "Measurer of All Things: John Greaves (1602–1652), the Great Pyramid, and Early Modern Metrology." *Journal of the History of Ideas* 63, no. 4 (2002): 555–75.

Shapiro, Barbara. *Probability and Certainty in Seventeenth-Century England: A Study of the Relationships between Natural Science, Religion, History, Law, and Literature.* Princeton, NJ: Princeton University Press, 1983.

Sheehan, Jonathan. *The Enlightenment Bible: Translation, Scholarship, Culture.* Princeton, NJ: Princeton University Press, 2005.

———. "Sacred and Profane: Idolatry, Antiquarianism, and the Polemics of Distinction in the Seventeenth Century." *Past and Present,* no. 192 (2006): 35–66.

Shelford, April G. "Thinking Geometrically in Pierre-Daniel Huet's *Demonstratio evangelica* (1679)." *Journal of the History of Ideas* 63, no. 4 (2002): 599–617.

———. *Transforming the Republic of Letters: Pierre-Daniel Huet and European Intellectual Life, 1650–1720.* Rochester, NY: University of Rochester Press, 2007.

Sider, Sandra. "Horapollo." In *Catalogus Translationum et Commentariorum: Mediaeval and Renaissance Latin Translations and Commentaries*, edited by Paul Kristeller, vol. 6, 15–29 and vol. 7, 325. Washington: Catholic University Press, 1960–1992.

Siebert, Harald. *Die grosse kosmologische Kontroverse: Rekonstruktionsversuche anhand des Itinerarium exstaticum von Athanasius Kircher SJ (1602–1680).* Stuttgart: Steiner, 2006.

Singer, Thomas C. "Hieroglyphs, Real Characters, and the Idea of Natural Language in English Seventeenth-Century Thought." *Journal of the History of Ideas* 50, no. 1 (1989): 49–70.

Siraisi, Nancy G. *History, Medicine, and the Traditions of Renaissance Learning.* Ann Arbor: University of Michigan Press, 2007.

Solinas, Francesco, ed. *Cassiano dal Pozzo: Atti del seminario internazionale di studi.* Rome: De Luca, 1989.

Soll, Jacob. "The Antiquary and the Information State: Colbert's Archives, Secret Histories and the Affair of the Régale." *French Historical Studies* 31, no. 1 (2008): 3–28.

———. *Publishing the Prince: History, Reading, and the Birth of Political Criticism.* Ann Arbor: University of Michigan Press, 2005.

Spitz, Lewis W. "The Significance of Leibniz for Historiography." *Journal of the History of Ideas* 13, no. 3 (1952): 333–48.

Stausberg, Michael. *Faszination Zarathushtra: Zoroaster und die Europäische Religionsgeschichte der Frühen Neuzeit.* Berlin: Walter de Gruyter, 1998.

Stenhouse, William. "Classical Inscriptions and Antiquarian Scholarship in Italy, 1600–1650." In *The Afterlife of Inscriptions: Reusing, Rediscovering and Reinventing Ancient Inscriptions*, edited by Alison E. Cooley, 77–89. London: Institute of Classical Studies, 2000.

———. *Reading Inscriptions and Writing Ancient History: Historical Scholarship in the Late Renaissance.* London: Institute of Classical Studies, 2005.

———. "Visitors, Display, and Reception in the Antiquity Collections of Late-Renaissance Rome." *Renaissance Quarterly* 58 (2005): 397–484.

Stephens, Walter. "Berosus Chaldaeus: Counterfeit and Fictive Editors of the Early Sixteenth Cetnury." PhD dissertation, Cornell University, 1979.

———. "*Livres de haulte gresse:* Bibliographic Myth from Rabelais to Du Bartas." *MLN* 120, Supplement (2005): S60–S83.

Stolzenberg, Daniel. "The Connoisseur of Magic." In *The Great Art of Knowing: The Baroque Encyclopedia of Athanasius Kircher*, edited by Daniel Stolzenberg, 49–57. Stanford, CA: Stanford University Libraries, 2001.

———. "Four Trees, Some Amulets, and the Seventy-Two Names of God: Kircher Reveals the Kabbalah." In *Athanasius Kircher: The Last Man Who Knew Everything*, edited by Paula Findlen, 149–69. London: Routledge, 2004.

———. "John Spencer and the Perils of Sacred Philology." *Past and Present* 214 (2012): 129–63.

———. "Oedipus Censored: *Censurae* of Athanasius Kircher's Works in the Archivum Romanum Societatis Iesu." *Archivum Historicum Societatis Iesu* 73 (2004): 3–52.

———. "The Universal History of the Characters of Letters and Languages: An Unknown Manuscript by Athanasius Kircher." *Memoirs of the American Academy in Rome* 56/57 (2011/2012): 305–21.

———. "Utility, Edification, and Superstition: Jesuit Censorship and Athanasius Kircher's *Oedipus Aegyptiacus*." In *The Jesuits, II: Cultures, Sciences, and the Arts, 1540–1773*, edited by John O'Malley, Gauvin Bailey, Steven Harris, and T. Frank Kennedy, 336–54. Toronto: University of Toronto Press, 2006.

Stolzenberg, Daniel, ed. *The Great Art of Knowing: The Baroque Encyclopedia of Athanasius Kircher*. Stanford, CA: Stanford University Libraries, 2001.

Strasser, Gerhard. "La contribution d'Athanase Kircher à la tradition humaniste hiéroglyphique." *XVIIe Siècle* 40 (1988): 72–92.

———. "Das Sprachdenken Athanasius Kirchers." In *The Language of Adam*, edited by Allison Coudert. Wiesbaden: Harrassowitz, 1999.

Stroumsa, Guy. *A New Science: The Discovery of Religion in the Age of Reason*. Cambridge, MA: Harvard University Press, 2010.

Stroumsa, Sarah. "Sabéens de Harran et Sabéens de Maïmonide." In *Maïmonide. Philosophe et Savant (1138–1204)*, edited by Tony Lévy and Roshdi Rashed, 335–52. Leuven: Peeters, 2004.

Struck, Peter. *Birth of the Symbol: Ancient Readers at the Limits of Their Texts*. Princeton, NJ: Princeton University Press, 2004.

Stuckrad, Kocku von. *Locations of Knowledge in Early Modern Europe: Esoteric Discourse and Western Identities*. Leiden: Brill, 2010.

Sutcliffe, Adam. *Judaism and Enlightenment*. Cambridge: Cambridge University Press, 2003.

Tamani, Giuliano. "Gli studi di aramaico giudaico nel sec. XVI." In *Italia ed Europa nella linguistica del Rinascimento*, edited by Mirko Tavoni, vol. 2, 503–16. Modena: Franco Cosimo Panini, 1996.

Thorndike, Lynn. *A History of Magic and Experimental Science*. 8 vols. New York: Columbia University Press, 1923–58.

Toomer, G. J. *Eastern Wisedome and Learning: The Study of Arabic in Seventeenth-Century Learning*. Oxford: Oxford University Press, 1996.

———. *John Selden: A Life in Scholarship*. 2 vols. Oxford: Oxford University Press, 2009.

Totaro, Giunia. *L'autobiographie d'Athanasius Kircher: L'écriture d'un jésuite entre vérité et invention au seuil de l'oeuvre. Introduction et traduction française et italienne*. Bern: Peter Lang, 2009.

Ullmann, Manfred. *Die Natur- und Geheimwissenschaften in Islam*, Handbuch der Orientalistik. Leiden: Brill, 1972.

Vasoli, Cesare. "Francesco Patrizi e la Tradizione Ermetica." *Nuova Rivista Storica* 64 (1980): 25–40.

Vercellone, Federico, and Alessandro Bertinetto, eds. *Athansius Kircher e l'idea di scienza universale*. Milan: Mimesis, 2007.

Vervliet, H. D. L. "Robert Granjon à Rome (1578-1589)." *Bulletin de l'institut historique de Belge de Rome* 38 (1967): 177-231.
Vickers, Brian. "Analogy Versus Identity: The Rejection of Occult Symbolism, 1580-1680." In *Occult and Scientific Mentalities in the Renaissance*, edited by Brian Vickers, 95-163. Cambridge: Cambridge University Press, 1984.
Völkel, Markus. *"Pyrrhonismus historicus" und "fides historica." Die Entwicklung der deutschen historischen Methodologie unter dem Gesichtspunkt der historischen Skepsis*. Frankfurt am Main: Peter Lang, 1987.
———. *Römische Kardinalshaushalte des 17. Jahrhunderts: Borghese, Barberini, Chigi*, Bibliothek des Deutschen Historischen Instituts in Rom 74. Tübingen: Niemeyer, 1993.
Walker, D. P. *The Ancient Theology: Studies in Christian Platonism from the Fifteenth to the Eighteenth Century*. Ithaca, NY: Cornell University Press, 1972.
Wear, Andrew. "English Medical Writers and their Interest in Classical Arabic Medicine in the Seventeenth Century." In *The 'Arabick Interest' of the Natural Philosophers in Seventeenth-Century England*, edited by G. A. Russell, 266-77. Leiden: Brill, 1994.
Weiss, Roberto. *The Renaissance Discovery of Classical Antiquity*. Oxford: Basil Blackwell, 1969.
Whitehouse, Helen. "Towards a Kind of Egyptology: The Graphic Documentation of Ancient Egypt, 1587-1666." In *Documentary Culture: Florence and Rome from Grand-Duke Ferdinand I to Pope Alexander VII*, edited by Elizabeth Cropper, Giovanna Perini, and Francesco Solinas, 63-79. Bologna: Nuova Alfa, 1992.
Wilding, Nick. "'If You Have a Secret, Either Keep It, or Reveal It': Cryptography and Universal Language." In *The Great Art of Knowing: The Baroque Encyclopedia of Athanasius Kircher*, edited by Daniel Stolzenberg, 93-103. Stanford, CA: Stanford University Libraries, 2001.
Wirszubski, Chaim. "Francesco Giorgi's Commentary on Pico's Kabbalistic Theses." *Journal of the Warburg and Courtauld Institutes* 37 (1974): 145-56.
Wüstenfeld, Ferdinand. "Die älteste Aegyptische Geschichte nach den Zauber- und Wunderzählungen der Araber." *Orient und Occident insbesondere in ihren gegenseitigen Beziehungen. Forschungen und Mittheilungen* 1 (1861): 326-40.
———. "Histoire de l'Égypte d'après les légendes Arabes." *Revue Germanique* 16 (1861): 275-85.
Yates, Frances. *Giordano Bruno and the Hermetic Tradition*. Chicago: University of Chicago Press, 1964.
Zammit Ciantar, Joe. "Athanasius Kircher in Malta." *Studi Magrebini* 23 (1991): 23-65.
Ziller Camenietzki, Carlos. "L'extase interplanetaire d'Athanasius Kircher: Philosophie, cosmologie et discipline dans la Compagnie de Jèsus au XVIIe siècle." *Nuncius* 10 (1995): 3-32.
Zinguer, Ilana. *L'hébreu au temps de la Renaissance*. Leiden: Brill, 1992.

INDEX

Page numbers in italics refer to figures.

Abela, D., 126
Abenephius (Barachias Nephi), 71, 74–84, *86*, 127, 140, 146, 156, 162, 177, 205; content of the treatise, 84–88, 159; identity and authenticity of the author and his text, 83–88, 206; references by scholars after Kircher, 251–52
Abenvaschia. *See* Ibn Wahshiyya
Abraham, 136, 154, 155, 217
Abū Ma'shar, 155, 156
Abyssinians. *See* Ethiopian language and literature
Ad Alexandrum obelsici interpretatio (Kircher), 33n77
Adam, 30, 37, 44, 141, 161, 241, 244. *See also* Adamic wisdom
Adamic wisdom, 30, 37, 63, 67, 135, 136, 137, 154, 160, 236, 240–42, 247–49. *See also* Adam; antediluvian knowledge; perennial philosophy
Addison, Joseph, 36, 57
Agrippa, Cornelius, 25, 55, 67, 132, 134, 135, 151, 178, 180, 185, 253
Akiba (rabbi), 164
Alberti, Leon Battista, 43
Albertus Magnus, 223
alchemy, 26n64, 37, 132, 142n36, 149, 222–23, 255
Alciati, Andrea, 60
al-Dimishqi, 156
Aleandro, Girolamo, 60–63, *61*, 67
Alexander VII (pope), 3

Alexandria, 23
Alexandrine Chronicle, 160
Allacci, Leone, 105, 112, 126
allegory and allegorism, 6, 24, 25, 26, 37, 41, 57–66, 69, 112, 131, 133, 139, 199, 213, 242, 243, 246, 249. *See also* mythography; occult philosophy; symbols and symbolism
Allen, Don Cameron, 69
al-Suyuti, Jalal al-Din (Gelaldinus), 85, 109, 126, 127, 140, 156–57, 159, 161, 170
America, 65, 68, 95, 238; its discovery as model of intellectual novelty, 24, 223
Ammianus Marcellinus, 43, 77, 80, 140n32
Amphion, 139
amulets, 47, 51–54, *52*, 67, 85, 88, 158. *See also* Kabbalah: kabbalistic amulets
anachronism (in textual criticism), 217–18, 221, 238
ancient theology. *See prisca theologia*
Angeloni, Francesco, 126
Annius of Viterbo, 45, 153, 162n36
antediluvian knowledge, 37–38, 136–37, 149, 154, 155, 159, 161, 236, 248–49. *See also* Adamic wisdom; perennial philosophy
Antenor, 65
Antinori, Giovanni, 230f
antiquarianism, 6, 20–21, 29, 55, 56, 67–70; defined, 20; relation to erudition and philology, 21, 53; relation to history, 20n42; study of ancient imagery, 57–63, *63*, 66–69. *See also* antiquities, collections and

297

antiquarianism (continued)
    museums of; Egypt: Egyptian antiquities;
    erudition
antiquities, collections and museums of, 24,
    53, 72, 100, 106–7, 111, 112, 119, 124,
    126. See also Egypt: Egyptian antiquities;
    Musaeum Kircherianum
Apollo, 61
apologetics, Christian,: 31, 49, 55–56, 66, 70,
    244, 247, 250, 256, 258
Arabic literature on Egypt, 85–87, 109, 155–
    59, 161–62, 175, 177
Arabic studies, 22, 106, 107, 108, 152–53. See
    also Kircher, Athanasius: as Oriental
    philologist; Kircher, Athanasius: study
    of Arabic
Aramaic, 6, 22, 88, 98, 108, 116, 127, 145, 162.
    See also Chaldean language; Estrangelo;
    Syriac language
Arcangelo da Borgonovo, 169
Aristotelianism, 4, 26, 28, 49, 50, 60, 133,
    182, 242
Aristotle, 139, 214
Armenian language, 22, 53, 94, 106, 118, 126
Armenians in Rome, 109, 111
Arnobius, 145
Art of Magnesia (Kircher), 15
Asclepius, 42n12, 133–34, 219. See also Hermetic Corpus
astral medicine, 37, 134
astrology, 26n64, 37, 46, 66, 132, 149, 156,
    167, 196, 255, 256
astronomy, 4, 24, 66, 131, 223
Aufrère, Sydney, 83
Augustine, Saint, 25, 28, 133, 140n32, 154, 219
August of Braunschweig-Lüneburg (duke), 13
Augustus (emperor), 42
Averroes, 177, 185
Avicenna, 177
Azubi, Salomon (rabbi), 117, 195–96

Bacon, Francis, 24, 242, 243
Bandini, Angelo Maria, 233
Bangius, 127
Barachias Nephi (rabbi). *See* Abenephius
Barberini circle, 45n22, 59, 61, 63, 112
Barberini, Francesco (cardinal), 77, 79, 81, 93,
    102, 112–14, 117, 126, 145, 152. *See also*
    Barberini circle
Barberini, Maffeo (Pope Urban VIII), 2, 12,
    107, 112

Bartholin, 127
Basa, Domenico, 158
Baudier, Michel, 77
Bayle, Pierre, 243, 248
Bellori, Giovanni Pietro, 126
Bembine Table, 46, 47, 48, 59, 60, 63, 67, 79,
    81, 95, 96, 140, 143–47, *144*, 234, 240
Bembo, Pietro (cardinal), 143
ben Yochai, Simon, 146
Bernini, Gian Lorenzo, 121, *122*
biblical scholarship, 20–21, 22, 63–66, 67–69,
    118–19, 243–44, 251–52, 255–56, 258.
    See also sacred history
Biondo, Flavio, 45
Blavatsky, Madame, 252
Bloemaert, Cornelis, *125*, 147
Bonfrère, Jacques, 184
*Book of the Garden of Marvels of the World
    and its Regions*, 157–59
Botrel, Moses, 164
Bouchard, Jean-Jacques, 91
Bouvet, Joachim, *234–35*, 236, 250
Boxhorn, 127
Boyle, Robert, 168, 169
Brahe, Tycho, 4
Brucker, Jacob, 247–48, 250, 253
Bruno, Giordano, 4, 50, 131n4
Buxtorf, Johannes (the younger), 101

Cabeo, Niccolò, 253
Cain, 135–36, 137, 160
Cairo, 75, 80, 84, 88, 116, 259
Cambyses, 139
Canaan, 136, 154
Canini, Giovanni Angelo, 147
Capuchin missionaries involved in Oriental
    studies, 72, 118. *See also* de Losches,
    Gilles; de Nantes, Cassien; de Vendôme,
    Agathe
Caraffa, Vincenzo (general of the Society of
    Jesus), 123
Cartari, Vincenzo, 60, 67
Casali, Giovanni Battista, 126, 229
Casaubon, Isaac, 29, 41, 48, 56, 63, 215–19,
    225, 227, 238, 241, 248, 249, 253, 255
Cassian, John, 160, 161
Catholic Church, Catholicism, 45, 49–50,
    107, 118–19, 142–43, 180, 195, 220
Caussin, Nicolas, 44, 140, 199
censorship, 101, 120, 158–59, 167, 180–81,
    192–97; Jesuit censorship of Kircher's

hieroglyphic studies, 168–69, 180–92, *191. See also* Index of Prohibited Books
Chaldean language, 53, 94, 98, 106, 126
Chaldean Oracles, 26, 28, 63, 130, 132–33, 135, 136n23, 139, 145–46, 182, 183, 209, 240n33
Champollion, Jean-François, 38, 224, 226, 228, 233, 234, 255
Chigi, Fabio (Pope Alexander VII), 3
China, 50, 68, 95, 236, 238
China Illustrated (Kircher), 15, 143n40, 227n3
Christina of Sweden (queen), 13
church fathers, 25, 57, 134, 138, 219, 220, 224. *See also specific names*
Cipriani, Giovanni, 49
Clavius, Christoph, 12
Clement VIII (pope), 184, 196
Clement of Alexandria, 46, 140, 145, 205, 219
Colberg, Daniel, 256–57
College of Maronites, 107–8, 126
College of Neophytes, 107–8, 126, 170
College of Revisers, 181–82, 192. *See also* censorship: Jesuit censorship of Kircher's hieroglyphic studies
Collegio Romano, 3, 12, 13, 32, 51, 106, 108, 109, 126, 190
Collegio Urbano, 107
comparison, comparative method, 37, 53, 55, 66–67, 233, 238. *See also* Kircher, Athanasius: combinatory method
Congregation for the Propagation of the Faith, 90, 92, 106, 107, 118. *See also* Polyglot Press of the Propaganda Fide
Congregation of the Holy Office of the Inquisition, 12, 32, 180, 181, 183, 188, 193. *See also* Index of Prohibited Books
Constantinople, 37, 53
Coptic-Arabic lexicon and grammar (manuscript belonging to Della Valle), 88–94, 101, 104, 109, 111, 115–16. *See also Egyptian Language Restored*
Coptic Forerunner (Kircher), 3, 94–98, 96, 97, 100, 106, 109, *110*, 113, 119, 208, 229
Coptic studies, 88–94, 90, 117–19. *See also under* Kircher, Athansius
Copts, 109, 111, 118
Cordovero, Moses, 169–75, *171, 172, 173*
Corpus Hermeticum (*Pimander*), 29, 41–42, 133–34, 146, 150, 204, 205, 209, 215, 219, 246. *See also* Casaubon, Isaac; Hermetica; Hermetic Corpus; Kircher, Athanasius: defense of *Corpus Hermeticum* and pseudo-Aristotle's *Theology*
Court de Gébelin, Antoine, 234, 249
critical philology. *See* textual criticism (critical philology)
Cudworth, Ralph, 206, 238–40, 250
Curran, Brian, 43n14
Cusanus, Nicolaus, influence on Kircher, 51

d'Alembert, Jean le Rond, 18–20, 248, 251
dal Pozzo, Cassiano, 59, 60, 63, 77, 79, 81, 112, 114, 126
d'Aquin, Philippe, 116
d'Arcos, Thomas, 74, 80
de Brosses, Charles, 249
Dee, John, 239
de la Mare, Philibert, 54–55
Della Valle, Pietro, 77, 79, 88–92, 114, 117, 119, 120, 126. *See also* Coptic-Arabic lexicon and grammar (manuscript belonging to Della Valle)
de Losches, Gilles, 117, 152
Democritus, 139
demons, demonism, 65–66, 133, *191*, 194–95. *See also* Devil, the
de Nantes, Cassien, 80
Descartes, René, 2–3, 4, 16, 21, 24, 31, 35
de Thou, François-Auguste, 91
de Vendôme, Agathe, 80, 93
Devil, the, 64. *See also* demons, demonism
Diodorus, 41, 254
Diogenes Laertius, 145
Dionysius. *See* pseudo-Dionysius the Areopagite
doctors of the church. *See* church fathers
Domitian (emperor), 200, *201*
Drummond, William, 249
Du Choul, Guillaume, 59
Dupuis, Charles, 249
Dupuy, Jacques, 91, 118

Ecchellensis, Abraham (Ibrahim al-Haqilani), 101, 108–11, *110*, 115–16, 126–27, 153, 156
ecclesiastical history, 21, 50n35, 142–43, 175, 215
Ecstatic Journey (Kircher), 4
Egypt: 23, 111, 116, 120, 126, 238; Egyptian antiquities, 6, *8, 9, 10*, 13, 36–37, 42–43, 44–46, 47, 51, 53, 68, 76, 106, 213; Egyp-

Egypt (continued)
  tology after Kircher, 226, 229–34, 257–58;
  history, culture, and religion, 65, 67, 85,
  95, 132–37, 139–41, 145–46, 243–49, 251;
  Renaissance Egyptology, 41–48, 55, 59,
  67. *See also* Arabic literature on Egypt;
  Bembine Table; Coptic studies; Hermes
  Trismegistus; hieroglyphs and hieroglyphic writing
Egyptian Language Restored (Kircher), 94–96,
  115–18, 127, 228–29
Egyptian Oedipus (Kircher), *passim*; circumstances of publication, 120–21, 123–24;
  conceived as compendium of Oriental
  texts, 101, 193–94; date of publication,
  4n11; described and summarized, 4–16,
  29, 36–38, 67–68, 131–37, 225; Jesuit censorship of, 168–69, 180–92; manuscript
  (BNCR Ms. Ges. 1235), 32, 181, 189–90,
  191; reception, 177, 229–53; relation
  to *Pamphilian Obelisk*, 123. *See also*
  Kircher, Athanasius: theory and translations of hieroglyphs; *Pamphilian Obelisk*
  (Kircher)
Elichmann, Johannes, 101
emblems, emblem tradition, 43–44, 57–58,
  60, 132, 209–10, 213, 242–43, 245. *See
  also* hieroglyphs and hieroglyphic writing; *imprese*; symbols and symbolism
Emerald Tablet, 177n67, 222–23
empiricism, 6, 20–21, 29–31, 34, 46–48, 69,
  175, 255, 257; esotericism and, 6, 29–31,
  257
encyclopedism, 103, 142, 250–53. *See also*
  Kircher, Athanasius: as polymath
Enlightenment: attitude toward erudition, 16–
  19, 250–51; and esotericism, 30, 243, 249,
  257; theories of language and civilization,
  248–49. *See also* Egypt: Egyptology after
  Kircher
Enoch, 85, 87, 135–36, 152, 154–56, 161, 236
Epeeis, 204
erudition, 6, 18–21, 37, 55–70, 103, 111, 141,
  152, 175–76, 178–79, 189, 221–22, 252–
  53; occult philosophy and, 29–31, 63,
  147, 150, 152, 178, 255–56. *See also* antiquarianism; empiricism; Enlightenment:
  attitude toward erudition; humanism;
  Oriental studies; textual criticism
  (critical philology)
esoteric antiquarianism, 51–67, 138, 150, 197.

  See also Kircher, Athanasius: as esoteric
  antiquary
esoteric wisdom, 6, 25–26, 29, 49, 55, 65, 181,
  246, 248
Estrangelo, 106, *110*
Ethiopian language and literature, 6, 22, 94,
  106, 117, 126, 152–53
Ethiopians in Rome, 109–11
etymology, 65, 68, 85, 205
Eudoxus, 134, 139
Euhemerism, 63–66, 131, 160
Euripides, 134
Eusebius, 25, 66, 140, 146, 204–5
Evans, R. J. W., 49

Fabri, Honoré, 183
Fakhr al-Din (emir), 108
fathers of the church. *See* church fathers
Faunus, 160
Ferdinand III (Holy Roman Emperor), 114–15,
  120–21, 124, 126, 192
Ferdinando II de' Medici of Tuscany (grand
  duke), 116, 120, 126
Ferrucci, Girolamo, 58
Festus, 145
Ficino, Marsilio, 24–25, 41–42, 49, 55, 67,
  127, 133–34, 137–38, 151, 154n10, 167,
  175, 178, 207
Figurism, 234, 236
Flood (biblical), 37, 45, 67, 136, 141, 154–56,
  159–62, 244
Fontenelle, Bernard le Bovier de, 243, 248
Foucquet, Jean-François, 236
Fountain of Four Rivers (Bernini), 121–23, *122*
Frederick of Hesse-Darmstadt (landgrave), 1–2
Fulvio, Andrea, 58

Gale, Theophilus, 240–41, 250
Galileo Affair, 12, 111, 119
Galileo Galilei, 12, 24, 35, 59, 223
Gaon, Saadia (rabbi), 164
Garden of Marvels, 157–59
Gassendi, Pierre, 74
genies, 39, 51, 87, 134, 200, 203, 206
Georgian language, 106
Gibbes, James Alban, 129
Gibbon, Edward, 19
Giehlow, Karl, 213
Giggeius, 75
Gilbert, William, 253
Gnostic amulets, 51–53, *52*. *See also* amulets

Golius, Jacob, 109
Gombrich, E. H., 58
Gorp, Jan van (Goropius), 44
Gottifreddi, Alessandro (general of the Society of Jesus), 190
Grafton, Anthony, 69–70
Granjon, Robert, 120, 158
Great Art of Knowing (Kircher), 15
Great Art of Light and Shadow (Kircher), 15, 120, 227n3
Greaves, John, 46n26, 127
Greek language and literature, 6, 11, 19, 23, 37, 41–42, 53, 89, 126, 132, 152, 175, 209, 217, 233, 257; relation to Coptic, 94, 96, 116
Gregory XIII (pope), 107
Grienberger, Christoph, 12
Gruterus, Johannes, 68
Gualdo, Francesco, 126

Habsburgs, 13, 49, 115, 120. *See also* Ferdinand III (Holy Roman Emperor)
Ham, 37, 65, 85, 136–37, 139, 154, 159–61
Hamilton, Alastair, 229
Hanegraaff, Wouter, 256–57
Harpocrates, *148*
Hebrew language and literature: 6, 22, 88–89, 98, 106–8, 116, 126–27, 135, 183, 209. *See also* Kabbalah; Kircher Athanasius: as Oriental philologist; Kircher, Athanasius: study of Hebrew and the Kabbalah; Oriental studies
Heinsius, 127
Hemphta, 200, 204
Hendreich, Christoph, 251
Hermes (god), 147, *148*
Hermes, Arabic traditions about, 155–57, 159, 161–62; identified with Idris and Enoch, 85, 87, 136, 154–56. *See also* Arabic literature on Egypt
Hermes Trismegistus: 24–25, 29, 37, 42, 46, 48, 49, 63, 130, 131, 133–34, 135–41, 151, 160–62, 175, 183, 217, 219, 222–23, 238, 239, 241; as inventor of obelisks, 71, 85, 87, 137, 154, 161–62
Hermetica, 42n12, 177. See also *Corpus Hermeticum (Pimander)*; Emerald Tablet; Hermetic Corpus
Hermetic Corpus, 26, 28–29, 42, 50n35, 56, 130, 133–34, 215–17, 255. See also *Corpus Hermeticum (Pimander)*; Hermetica
Hermeticism, Hermetic tradition, 6, 25, 29, 48–51, 66, 150, 255. *See also* Kircher, Athansius: as Hermetic philosopher
Hermetism, 25, 48, 240
Herodotus, 41, 254
Herwart von Hohenburg, Johann Georg, 11n18, 46, 48, 77, *78*, 99n78, 120, 140, 143, 145, 199n3
hieroglyphic arithmetic, 27, *191*
hieroglyphic astrology, 187, 196
hieroglyphic doctrine, 6, 36–37, 55, 126, 129–35, 137–41, 143, 147, 211
hieroglyphic medicine, 132, 134
hieroglyphic theology, 132–33
hieroglyphs and hieroglyphic writing: according to modern Egyptology, 38–39; relation to emblems, 43–44; theories of their nature and function, 41, 43–44, 46–48, 129, 198–99, 207, 209–13, 230, 233, 240, 241, 243, 245–47; theories of their origin, 38, 46, 137, 154, 245–47. *See also* Egypt: Egyptology after Kircher; Egypt: Renaissance Egyptology; Kircher, Athanasius: theory and translations of hieroglyphic writing
hierogram, 202, 203, 205
historia litteraria, 141–43
historia philosophica, 29, 141–43, 241, 247–48
historical scholarship, 29–30, 37, 67–69, 224, 253, 255–57. *See also* antiquarianism; erudition
Hobbes, Thomas, 244
Holstenius, Lucas, 3, 60–63, 62, 67, 109, 112, 126
Holy Office. *See* Congregation of the Holy Office of the Inquisition
Homer, 58, 60–63, 66, 134, 243n41
Horapollo, 41–44, 76, 83, 84–85, 133, 146, 199, 204, 207, 246
Horn, Georg, 141
Hottinger, Johann Heinrich, 251
Huet, Pierre Daniel, 220, 238, 250
humanism, 4, 19–21, 41, 69–70, 152. *See also* antiquarianism; erudition; textual criticism (critical philology)
Hume, David, 249
Huygens, Constantijn, 4

Iamblichus, 24, 41, 46, 134, 140, 145–46, 208, 217
Ibn Wahshiyya, 85, 87, 140, 145–46, 157, 159, 177

I Ching, 236
iconography. *See* antiquarianism: study of ancient imagery
ideal reading, 202, 203, 206–9
idolatry, 6, 65–67, 85, 117, 135, 137, 138, 157, 160. *See also* magic; paganism; superstition
Idris, identified with Hermes and Enoch, 85, 87, 136, 154–56
imprese, 43, 132, 210–11, 213. *See also* emblems, emblem tradition; symbols and symbolism
Index of Prohibited Books, 28, 168, 180–81, 184–85
India, 68, 95
Innocent X (pope), 13, 121–24, 192, 201
Inquisition. *See* Congregation of the Holy Office of the Inquisition
Iona, Giovanni Battista, 108–9, 127
Irenaeus, Saint, 140, 183
Isachor Beer (rabbi), 164
Isis, temple and cult of, 43, 145, 200. *See also* Bembine Table
Isis and Osiris, myth of, 37, 38. *See also* Bembine Table; *On Isis and Osiris* (Plutarch); Osiris
Istanbul, 37, 53
Iversen, Erik, 49–50

Jansson van Waesberghe, Jan, 15
Japan, 37, 68, 95, 238
Jaucourt, Louis de, 248
Jewish scholars, 107, 109; employed by Kircher as research assistants, 108, 127, 168. *See also* Azubi, Salomon (rabbi); Iona, Giovanni Battista; Kircher, Athanasius: study of Hebrew and the Kabbalah
Joseph, 155
Josephus, 154, 161
Justin Martyr, Saint, 140, 145, 219

Kabbalah: Christian attitudes toward, 163, 184, 187–88; kabbalistic amulets, 188, 197; kabbalistic astrology, 167; occult philosophy and, 26, 28, 130, 135, 146, 175, 209, 236; practical Kabbalah, 166–67, 187; relation to Christian doctrines, 26. *See also* Kircher, Athanasius: study of Hebrew and the Kabbalah
Kircher, Athanasius. *See also titles of individual books by Kircher*
—ambiguity about birth year, 11n15
—citation methods, 161, 164–69, *165*, 175–76, 178
—combinatory method, 37, 51, 98, 213
—defense of *Corpus Hermeticum* and pseudo-Aristotle's *Theology*, 215–19, 221–25
—desire to visit Orient, 22–23, 120
—as esoteric antiquary, 51–56, 138, 149–50, 197
—as Hermetic philosopher, 6, 48–51, 54–56, 66–67
—lack of critical approach to sources, 29, 81, 153, 174–77, 218–25
—life until 1633, 11–16
—on Malta, 1–3, 23, 34, 101, 113–14, 157
—manuscript of Barachias Nephi. *See* Abenephius
—missionary aspirations, 22–23
—modern scholarship on, 31, 48–51, 66, 68, 69, 127
—Musaeum Kircherianum, 13–15, *14*, 106–7
—occult philosophy as framework of his hieroglyphic studies, 6, 51, 55, 129–35, 138–42, 143–46, 150, 155, 178–79, 209–10, 250
—as Oriental philologist, 6, 7, 11, 22–23, 34, 53–56, 67, 71, 74, 88, 100–102, 116, 126–28, 151–79, 194, 252–53
—patronage, 13, 81, 93, 102, 104, 112–15, 120–24. *See also* Barberini, Francesco (cardinal); dal Pozzo, Cassiano; Della Valle, Pietro; Ferdinand III (Holy Roman Emperor); Innocent X (pope); Peiresc Nicolas-Claude Fabri de
—as polymath, 4, 11, 101–2
—portraits and depiction in frontispieces, *5*, *14*, *17*, *125*, *148*
—quest for fame, 56, 99, 225, 250
—reception of his work, 16–18, 100–101, 177, 226–53, 227–29
—relation to Peiresc. *See under* Peiresc, Nicolas-Claude Fabri de
—relation to seventeenth-century scholarship, 3–4, 6, 24, 29–30, 35, 41, 48, 56–70, 74, 127–28, 152–53, 199, 225, 234–45, 249–50, 252–53, 255–56
—and Republic of Letters, 13–15, 32, 54, 99, 116, 151, 179, 194, 223
—self-image and intellectual persona, 5, 16, 33, 51, 56, 146–49, 162–63, 179
—study of Arabic, 22, 33, 79, 86, 88, 101–2,

116, 127, 153, 156–62, 175, 176–77, 178–79, 183–84, 185, 187, 190, *191*, 251, 252. See also Abenephius (Barachias Nephi)
—study of Coptic, 33, 91–96, 100–101, 104, 105, 106, 109, 111, 115–19, 153, 227–29. See also *Coptic Forerunner* (Kircher); Coptic studies; *Egyptian Language Restored* (Kircher)
—study of Hebrew and the Kabbalah, 22, 53, 79, 98, 102, 108, 132, 135, 146, 149, 153, 162–74, *165*, *171*, *172*, *173*, *174*, 178–79, 183–90, 195–97, 237. See also Hebrew language and literature
—theory and translations of hieroglyphic writing, 37–39, 76–77, *82*, *86*, 95, 129–30, 137, 143–46, 198–213, *202*, *212*, 225
Kollàr, Adam František, 1
Kozi, Iuhana, 111, 115

La Croze, Marthurin Veyssière de, 87n43, 228
La Peyrère, Isaac, 244
Lactantius, 25, 133, 140n32, 154, 219
Lafitau, Joseph-François, 237–38, *239*, 250
Lambeck, Peter, 198, 214
Lead Tablets of Granada, 116
Leibniz, Gottfried, 21, 236
Leinkauf, Thomas, 50–51
libraries: Oriental, 23, 80, 116, 119, 152–53, 259; Roman, 88, 107–8, 112, 119, 124–26; Vatican library, 105–8, 112–13
Liceti, Fortunio, 63, *64*, 68
Ligorio, Pirro, 45, 88
Linus, 134, 139
Livy, 56, 153
Lodestone, or On the Magnetic Art (Kircher), 115, 227n3
Lucian, 145
Lullism, 4, 51, 102

Macrobius, 41, 60, *61*, 85, 140n32,
magic: 51, 54–56, 102, 134, 157, 181, 187, 194; illicit, 80, 81, 101, 180–81, 183, 187–94; natural, 26, 149, 175; origins and diffusion of, 37–38, 46, 85, 135–36, 137, 139, 160, 161; relation to occult philosophy, 26, 41, 129–31, 134, 138; theurgy, 26, 133, 146, 195, 197. See also amulets; idolatry; superstition
Maimonides, Moses, 85, 139–40, 157, 164, 177, 251
Manuel, Frank, 6

Marci, Johannes Marcus von, 115, 120
Maronites, 107, 109, 117. See also College of Maronites; Ecchellensis, Abraham (Ibrahim al-Haqilani)
Martinic, Bernard (grand burgrave), 115
Mascardi, Vitale, 36
Maurice, Thomas, 234
Mazarin, Jules (cardinal), 153
Medicean Press, 106, 120, 158
Medici, Cosimo de', 42
Meir Abulafia (rabbi), 166
Melissus, 134
Mencken, J. B., 16
Mensa Isiaca. See Bembine Table
Mentorella, site of shrine to Saint Eustace, 16
Mercati, Michele, 45–46
Mesramuthisis, 137
metaphysical bombast, 247. See also scholastic mumbo jumbo
Methuselah, 136
Mexico, 37, 245
Milesi, Marzio, 126
Miller, Peter, 67
Misraim, 85, 136–37, 155, 159–61
missions and missionaries, 3, 50, 106–8, 236–38. See also Capuchin missionaries involved in Oriental studies; Kircher, Athanasius: missionary aspirations
Mithras, 58
Momigliano, Arnaldo, 20n42, 21, 68–70
monotheism among ancient pagans, 37, 60, 65–67, 112, 156, 249
Montfaucon, Dom Bernard de, 230, 233–34
moral demonstration, 214–15, 218, 220–22
More, Henry, 238
Morin, Jean, 72, 89, 120
Moses, 25, 28, 65, 138, 155, 238, 241
Mount Horeb inscription, 96–98, *97*, 100
Murtadi, 87
Musaeum Kircherianum, 13, *14*, 15, 106–7
Musaeus, 134
museums. See antiquities, collections and museums of
myth, mythology, 37, 58–60, 85, 133, 149, 160, 238, 243, 246, 254–55. See also allegory and allegorism; Euhemerism; mythography
mythography, 44, 57, 63, 249

Nabatean Agriculture. See Ibn Wahshiyya
Naudé, Gabriel, 126

Near Eastern languages. *See* Oriental studies; *see also* specific languages
Neoplatonism, 24–29, 49–50, 132, 134, 137, 138, 170, 255, 256; cosmology and metaphysics, 26, 27, 131, 134, 145, 205; criticism of 59, 242–43, 247–48; hermeneutics and theory of hieroglyphic symbolism, 41, 43, 57–58, 60–63, 129–30, 145–46, 199, 207–9, 242–43, 256. *See also* Agrippa, Cornelius; Ficino, Marsilio; Iamblichus; occult philosophy; Patrizi, Francesco; Plotinus; Porphyry; Proclus
Newton, Isaac, 1, 21, 31, 229, 248
Nihus, Barthold, 244
Noah, 30, 37, 135, 141, 162n36, 236, 241
nondiscursive communication, 26, 199, 207–9, 214, 224
Nozze Aldobrandini, 60

obelisks, *14, 18,* 39, 53, 77, 83, 84, 85, *86, 87,* 137, 154, 158, 161–62, 209, 230–33; Flaminian obelisk, 229; Lateran obelisk, 77, 78; obelisk of the Villa Celimontana, 39, 40; Pamphilian obelisk (Piazza Navona), 121, 122, 123, 131, 147, *148,* 198, 200–206, *201, 202,* 211, *212,* 214; of Rome, 42, 45, 76, 79, 81, 106, 112; Solarium obelisk, 230, 232; Vatican obelisk, 42–43, 45. *See also* Egypt: Egyptian antiquities; Hermes Trismegistus: as inventor of obelisks; Kircher, Athanasius: theory and translations of hieroglyphic writing
Obicini, Tomasso, 89–91, *90,* 96, 106
occult philosophy, 6, 24–29, 34, 42, 50, 127, 135, 151, 154, 179, 194, 225, 255–56; Aristotelianism and, 26, 28, 50; Christianity and, 26–28, 48–50, 138; erudition and, 29–31, 63, 129–32, 138, 141, 143–50, 152, 178, 255–56; decline of, 30, 243–50, 255–58; persistence in seventeenth century, 29–31, 50, 141–42, 234–42, 249–50, 255–56. *See also* esoteric antiquarianism; Kircher, Athanasius: occult philosophy as framework of his hieroglyphic studies; Neoplatonism; *prisca theologia*
occult sciences, 26, 255–56
Odysseus, 61
On Isis and Osiris (Plutarch), 41, 83n31. *See also* Plutarch
Orientalism, 21n50

Oriental studies, 6, 21–24, 67, 72, 74, 89, 152–53, 177, 251–53, 258–59; in Rome, 104–11, 118–20; decline in eighteenth century, 24, 233, 258. *See also* Arabic studies; Coptic studies; Hebrew language and literature; Kircher: Athanasius: as Oriental philologist
Origen, 140, 219
Orpheus, 24, 63, 134, 139, 183, 240n33
Orphic hymns, 26, 37, 133, 139, 146, 183, 240, 246
Osiris, 37, 39, 85, 200, 211

paganism: as field of early modern scholarship 29, 37, 44, 57, 59–60, 63–69, 112, 141; relationship to Christianity and the Bible, 25–28, 37, 42, 50, 63–66, 131, 141, 216–18, 238, 240–41, 243–44, 250, 252. *See also* antiquarianism: study of ancient imagery; idolatry; myth, mythology; mythography
Pamphilian Obelisk (Kircher), *passim*; circumstances of publication, 121–24; described and summarized, 123, 131, 136–37, 141, 225; Jesuit censorship of, 180–92; relation to *Egyptian Oedipus*, 123. *See also Egyptian Oedipus* (Kircher); Kircher, Athanasius: theory and translations of hieroglyphic writing
Pardes Rimmonim (Moses Cordovero), 126, 169–75, *171, 172, 173*
Parker, Samuel, 244–45, 247, 250
Parmenides, 134
Paschal Chronicle, 160
Pastine, Dino, 49
Patrizi, Francesco, 28, 49, 50, 55, 67, 127, 132–33, 154, 178, 196, 216, 218, 253
Peiresc, Nicolas-Claude Fabri de, 12, 68–69, 71–85, 73, 89–94, 96, 101, 109, 116, 117, 119, 152, 177; his doubts about Kircher's scholarship, 72, 79, 81, 83, 98–100, 105, 177; as Kircher's mentor and patron, 12, 72, 75, 77–83, 101–3, 113–14, 126, 176
perennial philosophy, 29–30, 63, 69, 142, 240–41, 246–47. *See also* Adamic wisdom; antediluvian knowledge; *prisca theologia*
Persian, 106, 120, 149
Petau, Denis, 101, 117, 127
Petit, Samuel, 72, 74, 90–92
Pherecydes, 204

Phidias, 217
Philo of Alexandria, 139
philology. *See* erudition; humanism; Oriental studies; textual criticism (critical philology)
physico-mathematics, 3, 13
Pico della Mirandola, Giovanni, 25, 55, 133, 135, 147
Picques, Louis, 228
Picus Jupiter, 160
Pignoria, Lorenzo, 46, 47, 48, 59–60, 65–67, 140, 143, 145, 199n3, 230
Pimander. See *Corpus Hermeticum (Pimander)*
Pius VI (pope), 230
plagiarism thesis, 65
Plantavit de la Pause, Jean, 116
Plato, 24, 25, 28, 134, 139, 204, 218, 238, 241
Platonism. *See* Neoplatonism
Pletho, Gemisthus, 132
Pliny, 41, 43, 140n32
plomos (lead tablets) of Granada, 116
Plotinus, 24, 41, 140, 146, 207, 216
Plutarch, 41, 83n31, 85, 147, 149, 205
Polybius, 20
Polyglot Press of the Propaganda Fide, 90, 92, 94, 106, 115, 117. *See also* Congregation for the Propagation of the Faith
Porphyry, 24, 41, 61, 63, 66, 140, 146
Postel, Guillaume, 106, 107n8, 152n3
Prémare, Joseph Henri Marie de, 236
prisca theologia, 24–26, 28, 30–31, 34, 42, 49, 55, 63, 65, 130, 132, 137–39, 142–43, 149–50, 152, 154, 162, 175, 197, 205n15, 209, 236–41, 243, 246–50, 256, 259. *See also* occult philosophy
Proclus, 24, 28, 41, 63, 85, 133–34, 146
Propaganda Fide. *See* Congregation for the Propagation of the Faith
Protheories (Kircher), 74, 76–77
Psellus, Michael, 146, 183
pseudo-Aristotle's *Theology*, 146, 204–5, 211, 212, 215–16, 218, 222, 240
pseudo-Dionysius the Areopagite, 26, 28–29, 58, 140
Ptolemy III Euergetes (pharaoh), 200
pyramids, 87, 136–37, 154–56, 161–62
Pythagoras, 66, 134, 139, 155, 183, 238
Pythagoreans, 63, 126, 149, 188, 219
Pythagorean verses, 37, 133, 183, 246

Quatremère, Étienne Marc, 228–29
Quintillian, 221

Rabelais, 243
Raimondi, Giovanni Battista, 116, 158
Ramesses II (pharaoh), 39, 200
Ranke, Leopold von, 178
Raphael, 45
real character, 242–43, 245
Reitzenstein, Richard, 177
Republic of Letters, 3, 15, 32, 54, 72, 94–95, 99, 151, 179, 194, 223
Reuchlin, Johannes, 135, 168, 184
rhetoric, 58, 221, 242
Rhode, Johann, 64, 126
Richelieu, Cardinal, 153
Ricius, Paulus, 166–69, 175, 195
Ripa, Cesare, 58
Rittangel, Johannes Stephan, 164–66
Rivosecchi, Valeriano, 49
Roman College of the Society of Jesus. *See* Collegio Romano
Romano, Giovanni Battista, 126
Rome: as cosmopolis, 105, 107–11, 117–18, 127; Egyptian antiquities, 42–46, 230; intellectual and scientific culture, 12–13, 60–63, 103, 104–13, 117–20, 124–28, 180, 183, 192–94. *See also under* libraries; obelisks; Oriental studies
Rosetta Stone, 38, 257
Rowland, Ingrid, 131
Ruska, Julius, 177

Sabians, 157, 177
sacred history, 37, 63–65, 141–42, 234–36, 243–44, 247–50, 256–59
sacred philology, 21, 66, 258
Salamas, 140, 156, 157–59
Samaritan language, 6, 7, 22, 89, 98, 106, 118
Sanchuniathon, 140, 206
Santi, Leone, 127
Sapienza University, 107–8
Saumaise, Claude, 72, 80, 89–94, 105, 117–19, 127
Saurid, 156
Scaliger, Joseph Juste, 66, 89, 127, 168–69, 253
Schatta, Michael, 111, 115
scholasticism, 4, 184n14, 220–21. *See also* Aristotelianism; Thomism

scholastic mumbo jumbo, 208. *See also* metaphysical bombast
Schott, Kaspar, 51–55, 124, 146–47
Scialac, Vittorio, 117
Scott, Walter, 177
Selden, John, 24n59, 65–67, 127, 253
Seth, 66, 135, 154, 160–61, 162n36, 236
Sibylline Oracles, 50n35
Sino-Syriac (Nestorian) monument, 95
Sionita, Gabriel, 117
Sixtus V (pope), 45
Society of Jesus: intellectual and scientific culture, 4, 11–15, 50, 56, 107, 131, 236–38; internal censorship system, 168–69, 181–92, 194–97; overseas missions, 3, 50, 107, 236–38. *See also* Collegio Romano
Spencer, John, 251
Sphinx mystagoga (Kircher), 33n77
Spinoza, Benedict, 220, 244, 247
Spitzel, Gottlieb, 251
Stanley, Thomas, 141
Stefanoni, Pietro, 126
Steuco, Agostino, 25
Stoicism, 58
Strabo, 41
Stuckrad, Kocku von, 149
Stukeley, William, 229, *231*
Suarès, Joseph Maria, 112
superstition, 37–38, 45, 54, 65, 67, 81–83, 101, 135–38, 141, 143, 158, 160, 181, 183, 185–90, 192–96, 257
symbols and symbolism, 25–26, 37, 43–44, 57–63, 66, 68–69, 131–32, 140, 145–46, 199, 207–14; soft theory of, 58, 60; strong theory of, 57–58, 63, 198, 204, 242–43; symbolic wisdom, 6, 58, 63, 132, 229, 233, 238, 242. *See also* allegory and allegorism; emblems, emblem tradition; hieroglyphs and hieroglyphic writing; Neoplatonism: hermeneutics and theory of hieroglyphic symbolism; occult philosophy
Synesius, 24
Syriac language, 7, 22, 53, 89, 94, 98, 106, 108–9, *110*, 116, 118, 126, 157, 160

Tacitus, 20
talismans. *See* amulets
Tenison, Thomas, 229
Tertullian, 60, 140, 219
textual criticism (critical philology), 19–20, 21, 29–30, 34, 56, 103, 153, 176–77, 218–19, 221–23, 224, 238, 246, 255. *See also* erudition; Kircher, Athanasius: defense of *Corpus Hermeticum* and pseudo-Aristotle's *Theology*; Kircher, Athanasius: lack of critical approach to his sources
Thales, 139
Thomas Aquinas, Saint, 28, 182. *See also* Thomism
Thomism, 50, 149, 220. *See also* Aristotelianism
Toland, John, 247–48
Tomasini, Giacomo Filippo, 63
Tower of Babel (Kircher), 15
Trinity, doctrine of: in Kircher's interpretation of hieroglyphs, 37, *38*, 68, 203–6, 227, 229, 247; known to ancient Jews, 28, 164, 166, 167, 170; know to Egyptians and other pagans, 28, 87, 131, 133, 182, 186, 216, 229, 231, 236–38, 252
Typographia Medicea, 106, 120, 158

Underground World (Kircher), 15–16, 121, 227n3
Universal Musurgia (Kircher), 120, 227n3
University of Rome, 107–8
Urban VIII (pope), 2, 12, 107, 112

Valeriano, Pierio, 43–44, 59, 67, 140, 199
Valla, Lorenzo, 19, 29, 221
Varro, 20
Vatican Library. *See under* libraries
Ventimiglia, Carlo di, 126
Vickers, Brian, 256
Vico, Enea, 143
Vico, Giambattista, 246–50
Vita Admodum Reverendi P. Athanasii Kircheri (Kircher), 11n14, 15–16, 76, 98
Vitelleschi, Hippolito, 126
Vitelleschi, Muzio (general of the Society of Jesus), 75
Viva, Jacques, 127
Vossius, G. J., 66–67, 127, 141, 253

Warburton, William (bishop), 56, 226, 241, 243, 245–50
Western esotericism, 26n65, 175, 256–57
Wilkins, David, 228
Wilkins, John, 243, 245
Winckelmann, Johann Joachim, 230

Wits, Hermann, 227, 250–51
Wolf, J. C., 252
Wysing, Nicolaus, 186, 190
Wyttenbach, Daniel, 234

Yates, Frances, 48–49, 255
Young, Thomas, 233–34, 258
Yverdon Encyclopedia, 16, 227

Zoega, Georg, 229–33, 258
Zohar, 146, 169
Zorn, Peter, 252
Zoroaster, 26, 63, 85, 132–33, 136, 140–41, 151, 154, 161, 183, 206n18, 240n33, 246

Printed and bound by CPI Group (UK) Ltd, Croydon, CR0 4YY
09/06/2025
14685704-0003